Patterns and Processes of Speciation in Ancient Lakes

Developments in Hydrobiology 205

Series editor
K. Martens

Patterns and Processes of Speciation in Ancient Lakes

Proceedings of the Fourth Symposium on Speciation in Ancient Lakes, Berlin, Germany, September 4–8, 2006

Editors

Thomas Wilke[1], Risto Väinölä[2] & Frank Riedel[3]

[1]*Justus Liebig University Giessen, Department of Animal Ecology and Systematics, Heinrich-Buff-Ring 26-32 (IFZ), D-35392 Giessen, Germany*

[2]*Finnish Museum of Natural History, PO Box 26, FI-00014 University of Helsinki, Finland*

[3]*Free University of Berlin, Department of Earth Sciences, Malteserstr. 74-100, House D, D-12249 Berlin, Germany*

Reprinted from Hydrobiologia, Volume 615 (2008)

Springer

Library of Congress Cataloging-in-Publication Data

A C.I.P. Catalogue record for this book is available from the Library of Congress.

ISBN: 978-1-4020-9581-8

Published by Springer,
P.O. Box 17, 3300 AA Dordrecht, The Netherlands

Cite this publication as Hydrobiologia vol. 615 (2008).

Cover illustration: Eastern shore of Lake Ohrid near Veli Dab – one of the lake's major hotspots of endemic biodiversity. Photograph: Thomas Wilke, 2004.

Printed on acid-free paper

TABLE OF CONTENTS

Hydrobiologia (2008) 615:1–3
DOI 10.1007/s10750-008-9559-x

Preface

Thomas Wilke · Risto Väinölä · Frank Riedel

Ancient lakes, that is, extant lakes that typically have continuously existed since before the Pleistocene, for more than two million years— such as lakes Baikal, Tanganyika, Malawi, Biwa, and Ohrid, have long been recognized as centres of biodiversity. Moreover, during the past years, speciation in ancient lakes has emerged as an important and fruitful topic in studies of evolutionary biology. The current understanding of a number of key evolutionary concepts, such as sympatric speciation, sexual selection, adaptive radiation, hybridization, and punctuated equilibrium, is in part based on insights from speciation studies in these lakes, and this is evident in the increasing number of citations of articles pertaining to speciation in ancient lakes. Thomson Reuter's Web of Science®, for example, lists under this topic, only one cited paper in 1996, but 190 for the year 2007.

Guest editors: T. Wilke, R. Väinölä & F. Riedel
Patterns and Processes of Speciation in Ancient Lakes: Proceedings of the Fourth Symposium on Speciation in Ancient Lakes, Berlin, Germany, September 4–8, 2006

T. Wilke (✉)
Department of Animal Ecology and Systematics, Justus Liebig University Giessen, Heinrich-Buff-Ring 26-32 (IFZ), 35392 Giessen, Germany
e-mail: Tom.Wilke@allzool.bio.uni-giessen.de

R. Väinölä
Finnish Museum of Natural History, University of Helsinki, PO Box 17, 00014 Helsinki, Finland

F. Riedel
Department of Earth Sciences, Free University of Berlin, Malteserstr. 74-100, House D, 12249 Berlin, Germany

This volume represents the outcome of a fourth meeting in a series of conferences on speciation in ancient lakes (SIAL). The series was started by a meeting that took place in 1993 in Mont-Rigi, Belgium, and was brought together to review the state of art in the field for the first time since the classical seminal review of Brooks (1950). The proceedings of that meeting were published as a special volume of *Archiv für Hydrobiologie* (Martens et al., 1994); apart from summarizing both theoretical concepts of evolution in ancient lakes and practical activities in these systems, it also initiated a wealth of renewed interest in ancient lake issues.

The second SIAL meeting entitled "International Conference on Ancient Lakes: their Biological and Cultural Diversities" was held in 1997 near Lake Biwa in Japan. It focused on the cultural diversity of ancient lakes (Kawanabe et al., 1999) as well as on their biodiversity, ecology, and evolution (Rossiter & Kawanabe, 2000).

The third conference, SIAL-3, took place in Irkutsk. Naturally, the focus of that conference was the oldest and arguably most famous of the ancient lakes Lake Baikal. The conference proceedings were published in two special issues (Ivanov et al., 2003; Coulter et al., 2006), adding to the reputation of Baikal as one of the best-studied lakes in the world.

Finally, the latest meeting SIAL-4 was held on September 4–8, 2006, at the Freie Universität Berlin, Germany, organized by Frank Riedel (Germany) together with George Coulter (New Zealand), Matthias Glaubrecht (Germany), Doug Haffner (Canada), Hiroya Kawanabe (Japan), Koen Martens (Belgium), Oleg Timoshkin (Russia), Risto Väinölä (Finland), and Thomas Wilke (Germany); 67 scientists from 16 different countries contributed to the scientific program. While the abstracts of SIAL-4 were published in a special issue of *Berliner Paläobiologische Abhandlungen* (Kossler et al., 2006), selected full papers are subject of the present special issue "Patterns and processes of Speciation in Ancient Lakes." The contributions in this issue deal with three ancient lakes systems. One system is the famous East African Great Lakes, which have been studied extensively over the past decades, resulting in significant new findings about evolutionary processes in general. On the other hand, we include a set of papers on two other prominent lake systems that have so far received much less attention, the European ancient sister lakes Ohrid and Prespa, which have long been recognized as hotspots of biodiversity but were largely neglected in the international scientific literature until recently, and the two central lake systems on the Indonesian island of Sulawesi, the rich biodiversity and interesting evolutionary patterns of which only very recently have started to draw the attention of evolutionary biologists.

East African Great Lakes

The African Great Lakes—a series of lakes in or near the Great Rift Valley, for example Lake Victoria, Lake Tanganyika, and Lake Malawi—are major hotspots of freshwater biodiversity, with splendid examples of large radiations of cichlid fishes. While not all of the African Great Lakes qualify as ancient (e.g., the relatively young Lake Victoria), these lakes have long been a "playground" for evolutionary biologists.

In the current issue, Koblmüller et al. give an account of the higher taxonomy and phylogeny of the Lake Tanganyika cichlid species assemblage, its relationship to the African cichlid fauna, and key factors leading to the astonishing diversity of cichlids in the lake. Several other workers report on the emerging field of sexual selection and reproductive strategies of cichlid fish species in Lake Tanganyika (Sefc, Egger et al. and Sturmbauer et al.) and Lake Malawi (Anseeuw et al.). An endemic radiation of freshwater crabs in Lake Tanganyika is the subject of the study of Marijnissen et al., who discuss data on interspecific spatial and ecological segregation in the framework of adaptive radiation theory. Finally, Sturmbauer reviews and discusses conservation strategies for aquatic biota and their biological impact on species flocks in the East African Great Lakes.

The Balkan lakes Ohrid and Prespa

One of the most outstanding European ancient lakes is the oligotrophic and karstic Lake Ohrid situated in the central Balkans. Together with its sister lake, Lake Prespa, it represents one of the most significant hotspots of endemic biodiversity in the world. Although the recognition of Lake Ohrid as a peculiar lake with an outstanding biodiversity began in the late 19th century, and basic taxonomical and evolutionary studies were carried out prior to and after World War II, the political instabilities in the Balkans in the 1980s and 1990s led to a sharp decline of projects carried out at these lakes. Only the last decade has seen a revival of scientific activity in lakes Ohrid and Prespa.

Here, Albrecht and Wilke summarize the current knowledge of the limnological and faunal history of Lake Ohrid, update the existing data on its biodiversity and endemism, and review patterns and processes of speciation in the lake. In primary research studies, several groups of endemic molluscs are investigated with regard to the sister-lake relationship of lakes Ohrid and Prespa (Albrecht et al.), patterns of endemic biodiversity (Hauswald et al.), and evolutionary patterns in inconspicuous (cryptic) taxa (Schultheiß et al.).

Sulawesi lake systems

While evolutionary studies in ancient lakes such as Lake Baikal and Lake Tanganyika have been ongoing for many decades, patterns of endemic diversity in the two central lake systems on the Indonesian island of Sulawesi, that is, Lake Poso and the lakes of the Malili lake system, have been largely neglected for a century. Recently, however, there has been a renewed interest in these lakes and in their endemic species flocks.

In the present issue, Glaubrecht and Rintelen review and discuss the evolutionary and taxonomic implications of the *Tylomelania* gastropod species flocks and demonstrate that they provide instructive model cases for the study of speciation mechanisms, adaptive radiation, and niche exploitation. Schwarzer et al. utilized the adaptive radiation of sailfin silversides in Lake Matano (Malili lake system) to study the effect of gene flow on speciation processes. The data presented suggest that genetic exchange between lacustrine and riverine taxa takes place, potentially allowing introgression into Lake Matano from lower lakes and streams. The rich endemic diatom flora of the Malili lake system is the subject of the study of Bramburger et al. The workers reviewed the mechanisms contributing to the development and maintenance of community composition and relative abundance of taxa. Finally, Sabo et al. studied the phytoplankton and zooplankton communities of Lake Matano and substantiated the assumption that resource limitation plays an important role in shaping the unique endemic assemblages currently observed in the food web of the lake.

We hope that the present special issue can deepen the awareness of ancient lakes in the scientific community in general, and of their role as natural laboratories and prime hotspots of biodiversity in particular. It is also hoped that this issue stimulates further research that may lead to a better understanding of key evolutionary processes in these lakes and that ultimately may help in mitigating the decline of biodiversity seen today in many ancient lakes. These topics will certainly continue to be focal points during the fifth conference, SIAL-5, which is scheduled to take place at Lake Ohrid in September 2009.

We would like to thank the other members of the organizing committee for their hard work. We also would like to thank Koen Martens, who kindly gave us the opportunity to publish this issue and who assisted us during preparation. Eugene Coan is gratefully acknowledged for improving a number of manuscripts as a native English speaker and biologist, and Claudia Wilke helped with proofreading.

The Guest Editors

References

Brooks, J. L., 1950. Speciation in ancient lakes. Quarterly Review of Biology 25(30–60): 131–176.

Coulter, G., O. Timoshkin, L. Timoshkina & K. Martens (eds), 2006. Species and speciation in ancient lakes. Hydrobiologia 568(S): 287 pp.

Ivanov, A., G. Coulter, O. Timoshkin & F. Riedel (eds), 2003. Proceedings of the international symposium speciation in ancient lakes, SIAL III. Berliner Paläobiologische Abhandlungen 4: 200 pp.

Kawanabe, H., G. W. Coulter & A. C. Roosevelt (eds), 1999. Ancient lakes: their cultural and biological diversity. Kenobi Productions, Ghent, Belgium.

Kossler, A., R. Kohring & F. Riedel (eds), 2006. Abstracts of the international symposium speciation in ancient lakes, SIAL IV. Berliner Paläobiologische Abhandlungen 9: 70 pp.

Martens K., B. Goddeeris & G. Coulter (eds), 1994. Speciation in ancient lakes. Advances in Limnology 44: 508 pp.

Rossiter, A. & H. Kawanabe (eds), 2000. Ancient lakes: biodiversity, ecology and evolution. Advances in Ecological Research 31: 624 pp.

Hydrobiologia (2008) 615:5–20
DOI 10.1007/s10750-008-9552-4

SPECIATION IN ANCIENT LAKES

The Lake Tanganyika cichlid species assemblage: recent advances in molecular phylogenetics

Stephan Koblmüller · Kristina M. Sefc · Christian Sturmbauer

Published online: 30 September 2008

Abstract Lake Tanganyika is not the most species-rich of the Great East African Lakes, but comprises the greatest diversity of cichlid fishes in terms of morphology, ecology, and breeding styles. The lake contains a polyphyletic assemblage of cichlid lineages, which evolved from several ancient species that colonized the emerging lake some 9–12 million years ago. Based on morphological characteristics, the Tanganyikan cichlids have been classified into 12, or, more recently, 16 tribes, which are largely supported by molecular data. The radiations of East African cichlids are believed to be driven by complex interactions between extrinsic factors, such as climatic changes and geological processes, and intrinsic biological characteristics of the involved organisms. Diversification within different lineages occurred simultaneously in response to drastic habitat changes such as the establishment of lacustrine deep-water conditions 5–6 MYA and subsequent major lake-level fluctuations. This seems particularly true for the mouthbrooding lineages whereas the substrate breeders underwent a more gradual process of diversification. This review presents an account of the taxonomy and phylogeny of the Lake Tanganyika cichlid species assemblage, its relationship to the African cichlid fauna, key factors leading to the astonishing diversity and discusses recently proposed alternative age estimates for the Lake Tanganyika cichlid species assemblage.

Keywords Adaptive radiation · Cichlidae · Diversification · Phylogeny · Speciation

Introduction

With about 3,000 species, distributed from Central and South America, across Africa to Madagascar and southern India, the family Cichlidae represents the most species-rich family of vertebrates. Cichlids are outstanding in terms of variation of body shapes, color patterns and behavior, and their enormous diversity of trophic and ecological specializations (Fryer & Iles, 1972; Meyer, 1993; Kornfield & Smith, 2000; Kocher, 2004; Salzburger & Meyer, 2004). Throughout their range of distribution cichlids have repeatedly demonstrated their capacity for rapid speciation with niche partitioning, a process termed adaptive radiation (reviewed in Turner, 2007), but their greatest diversity

Guest editors: T. Wilke, R. Väinölä & F. Riedel
Patterns and Processes of Speciation in Ancient Lakes: Proceedings of the Fourth Symposium on Speciation in Ancient Lakes, Berlin, Germany, September 4–8, 2006

S. Koblmüller (✉) · K. M. Sefc · C. Sturmbauer
Department of Zoology, Karl-Franzens University of Graz, Universitätsplatz 2, 8010 Graz, Austria
e-mail: stephan.koblmueller@uni-graz.at

S. Koblmüller
Department of Evolutionary Biology, Evolutionary Biology Centre (EBC), Uppsala University, Norbyvägen 18D, Uppsala 752 36, Sweden

counting almost 2,000 species arose within the three Great East African Lakes, Tanganyika, Malawi, and Victoria (Turner et al., 2001; Kocher, 2004).

Adaptive radiation is believed to be induced by extrinsic environmental factors such as geologic and climatic events, which create novel and vacant niches, and intrinsic traits of the organism, providing a key to the success over other organisms, termed key innovations. After its initiation, radiation is modulated by extrinsic factors interacting with biological characteristics of the involved species such as ecological specialization, site fidelity, territoriality, homing and mating behavior, and social organization (Fryer & Iles, 1972; McKaye & Gray, 1984; Rossiter, 1995; Sturmbauer, 1998). Extrinsic factors provide the opportunity and intrinsic factors the potential for radiation and both have to coincide to trigger a radiation (Martens et al., 1994; Sturmbauer, 1998).

Two key innovations are thought to be responsible for the evolutionary success of cichlids. The particular anatomy of the pharyngeal apophysis (Liem, 1973; Greenwood, 1973, 1978; Rice & Lobel, 2003) provides a second set of jaws decoupled from the oral jaws. This derived morphological trait is shared with the families Embiotocidae, Pomacentridae, and Labridae and allows for efficient processing of prey (Yamaoka, 1978; Liem & Greenwood, 1981; Liem, 1986; Liem & Sanderson, 1986; Stiassny & Jensen, 1987; Drucker & Jensen, 1991; Galis & Drucker, 1996; Koblmüller et al., 2003; Wainwright, 2005). The elements of the pharyngeal jaw apparatus were used as the primary taxonomic characters uniting the Cichlidae, Embiotocidae, Pomacentridae, and Labridae into the suborder Labroidei (Kaufman & Liem, 1982; Stiassny & Jensen, 1987). Molecular data, on the other hand, indicate that the Labroidei do not constitute a monophyletic assemblage, suggesting a repeated independent evolution of a specialized pharyngeal jaw apparatus (Streelman & Karl, 1997; Mabuchi et al., 2007). Minor structural modifications allow the utilization of novel food resources, so that new trophic niches can be rapidly occupied (Stiassny, 1991; Sturmbauer, 1998; Albertson et al., 1999). Moreover, the pharyngeal jaws are involved in sound production and are thus important for communication (Lobel, 2001; Kaatz, 2002; Rice & Lobel, 2002, 2003).

The second key factor believed to be associated with the diversification of cichlids is their highly specialized reproductive behavior (Crapon de Crapona, 1986), in particular, the various modes of brood-care behavior found in the East African cichlids (Fryer & Iles, 1972; Goodwin et al., 1998; Barlow, 2000; Kornfield & Smith, 2000; Klett & Meyer, 2002). Especially in polygamous mouthbrooding lineages, sexual selection via female choice of nuptial male coloration seems to play an important role in species diversification (McKaye et al., 1993; Turner & Burrows, 1995; Deutsch, 1997; Seehausen et al., 1997, 1998, 1999; Knight et al., 1998; Seehausen & van Alphen, 1999; Allender et al., 2003; Knight & Turner, 1999, 2004). However, a recent simulation study (Sefc, 2008) demonstrated that a mating system such as that of the sexually monomorphic, sequential monogamous, mouthbrooding cichlid genus *Tropheus* can also provide a considerable opportunity for sexual selection.

Due to their enormous species number and the complex behavioral and morphological characteristics, enabling the utilization of specific niches within a species-rich assemblage, the family Cichlidae has become a popular and well-known model system in various disciplines of biological sciences (Fryer & Iles, 1972; Sturmbauer, 1998; Kornfield & Smith, 2000; Kocher, 2004). Currently, 200 cichlid species are described for Lake Tanganyika, with several more awaiting scientific description. By recent estimates, the total number is expected to amount to 250 endemic cichlid species in Lake Tanganyika, and represents the morphologically, behaviorally, and ecologically most diverse species assemblage (Snoeks, 2000; Turner et al., 2001), even though the number of endemic cichlids is much higher for Lake Malawi (500–1,000) and Lake Victoria (>500; Turner et al., 2001). In contrast to the mono- or di-phyletic origin of the haplochromine cichlids in the species flock of Lake Malawi and the "superflock" of Lake Victoria, which includes the cichlid fauna of neighboring rivers and lakes such as Lake Albert, Edward, George, Kyoga, and Kivu (Meyer et al., 1990; Moran et al., 1994; Nagl et al., 2000; Seehausen, 2002; Verheyen et al., 2003; Takahashi et al., 2001a; Terai et al., 2004), the cichlids of Lake Tanganyika are of polyphyletic origin and comprise both substrate breeding and mouthbrooding lineages (Salzburger et al., 2002a; Koblmüller et al., 2005). While the cichlid species radiation in Lake Tanganyika enjoys a longstanding reputation among biologists, the lake's small flocks of other organisms—catfish, spiny eels, Nile perches, as well as groups of gastropods, crabs, and shrimps (Coulter, 1991)—have been less intensively studied. Only recently, studies explicitly focused on some of these less species-rich flocks of Lake Tanganyika: the squeaker catfishes of the genus *Synodontis* (Day & Wilkinson, 2006; Koblmüller et al., 2006), thiarid and thalassoid gastropods (e.g., West & Michel, 2000; Wilson et al., 2004), platythelphusid crabs (Marijnissen et al., 2006), and atyid prawns (Fryer, 2006). To date, however, the most comprehensive data exist for Lake Tanganyika's cichlids, and this article presents a synthesis of recent molecular phylogenetic work on the lake's cichlid species assemblage.

Classification of Lake Tanganyika cichlids into tribes and their phylogenetic relationships

Based on morphological characteristics, the Tanganyikan cichlids have been classified into 12 (Poll, 1986) or, more recently, 16 tribes (Takahashi, 2003), which are largely supported by molecular data (see Table 1). However, some disagreement exists. Poll (1986) assigned the two species *Oreochromis tanganicae* and *Boulengerochromis microlepis* to the tribe Tilapiini. In a mitochondrial phylogeny of the Tilapiini, *B. microlepis* clusters with *Tilapia sparrmanii*, a species widely distributed in southern African rivers (Klett & Meyer, 2002), but bootstrap support was very low. Takahashi (2003) erected a new tribe—Boulengerochromini—for the monotypic genus *Boulengerochromis*. The erection of a new tribe is indeed supported by nuclear genetic data, which place *B. microlepis* within the Lake Tanganyika cichlid species flock, albeit at different positions (Nishida, 1997; Terai et al., 2003; Clabaut et al., 2005; Sugawara et al., 2005). These studies, however, included only a few if any tilapiine species, and the systematic placement of *B. microlepis* in the context of other nonmouthbrooding tilapiine cichlids will require further phylogenetic analyses with representative taxon sampling of both tilapiine and Lake Tanganyika lineages.

Table 1 Classification of Lake Tanganyika cichlids into tribes according to Poll (1986), Takahashi (2003), and as suggested by molecular data

Poll, 1986	Takahashi, 2003	Molecular classification[a]	
Tylochromini	Tylochromini	Tylochromini	
Tilapiini	Tilapiini	Tilapiini	
	Boulengerochromini	Boulengerochromini	
Bathybatini	Bathybatini	Bathybatini	Bathybatini
			Hemibatini
Trematocarini			Trematocarini
Eretmodini	Eretmodini	Eretmodini	
Lamprologini	Lamprologini	Lamprologini	
Ectodini	Ectodini	Ectodini	
Cyprichromini	Cyprichromini	Cyprichromini	
Perissodini	Perissodini	Perissodini	
Limnochromini	Limnochromini	Limnochromini	
	Greenwoodochromini		
	Benthochromini	Benthochromini	
Haplochromini	Haplochromini	Haplochromini	
	New tribe[b]	?	
Tropheini	Tropheini	Tropheini form monophyletic sub-group within the Haplochromini	
	Cyphotilapiini	Cyphotilapiini	

[a] References for molecular data: Nishida, 1997 (note that in this study based on allozyme data *Gnathochromis permaxillaris* was not resolved within the Limnochromini); Salzburger et al., 2002a, 2005; Clabaut et al., 2005; Koblmüller et al., 2005

[b] Takahashi (2003) erected a new tribe for '*Ctenochromis*' *benthicola*. Since its genus name is unlikely to remain valid in future, Takahashi (2003) refrained from assigning a name to this new tribe

According to Poll's (1986) classification, the Bathybatini include the genera *Bathybates* and *Hemibates*, and the Trematocarini comprise *Trematocara* and *Telotrematocara*, whereas Takahashi (2003) joins all four genera into a single tribe, the Bathybatini. In contrast, mitochondrial data (Koblmüller et al., 2005) indicate that Poll's Trematocarini and the genera *Bathybates* and *Hemibates* constitute three equally divergent lineages. Hence, the molecular data support Takahashi's comprehensive Bathybatini, or, alternatively, suggest a split of Poll's (1986) Bathybatini into the Bathybatini and a new tribe, Hemibatini.

Among the remaining tribes, the Eretmodini, Lamprologini, Ectodini, Cyprichromini, and Perissodini are identical in both Poll's and Takahashi's classification. Their respective monophyly is also well supported by molecular data (Sturmbauer & Meyer, 1993; Sturmbauer et al., 1994; Takahashi et al., 1998; Salzburger et al., 2002a; Koblmüller et al., 2004, 2007b; Brandstätter et al., 2005; Duftner et al., 2005). In Takahashi's classification, Poll's Limnochromini were split into the tribes Limnochromini, Greenwoodochromini (comprising *Greenwoodochromis bellcrossi* and *G. christyi*), and Benthochromini (including *Benthochromis melanoides* and *B. tricoti*), of which the Greenwoodochromini are not supported by molecular data. While, based on mitochondrial DNA sequences, the genus *Benthochromis* constitutes a distinct lineage, the genus *Greenwoodochromis* is clearly nested within the Limnochromini (Duftner et al., 2005).

The Haplochromini, which comprise the major cichlid species flocks in the remaining East African rift lakes and the vast majority of the African riverine cichlid fauna (except for western Africa), are the most species rich cichlid tribe in Africa. Nevertheless, only a few haplochromine cichlids occur in the Lake Tanganyika basin, and even these are usually found in river mouths of inflowing rivers rather than in the lake itself. Only two truly lacustrine species were originally assigned to the Haplochromini by Poll (1986), *Ctenochromis benthicola* and *C. horei*. Takahashi (2003) erected a new tribe for *C. benthicola*, but as the genus name will not remain valid, he did not suggest a name for the new tribe. Molecular work on the phylogenetic placement of *C. benthicola* is still needed. The second species, *Ctenochromis horei*, is currently assigned to the Tropheini (Takahashi, 2003), which is also confirmed by both mitochondrial and nuclear molecular data (Sturmbauer et al., 2003; Koblmüller, Egger & Sefc, unpublished data). Takahashi split the Tropheini into two tribes, Tropheini and Cyphotilapiini, a hypothesis strongly supported by lepidological (Lippitsch, 1998) and molecular data (Nishida, 1997; Salzburger et al., 2002a; Clabaut et al., 2005; Duftner et al., 2005). Furthermore, a recent molecular study showed that the Tropheini constitute a monophyletic lineage within the Haplochromini (Salzburger et al., 2005).

Hypotheses regarding the molecular phylogenetic relationships between the distinct tribes differ slightly depending on whether nuclear or mitochondrial data are used. Nuclear data support, with the exception of *Tylochromis polylepis* and *Oreochromis tanganicae*, a monophyletic group including the whole Lake Tanganyika cichlid species flock (including the Orthochromini from the Malagarazi system and the Haplochromini, which are distributed from northern to southern Africa, but are supposed to have their origin in Lake Tanganyika). One should note, however, that all nuclear phylogenies available to date are based on a rather incomplete taxon sampling concerning riverine species that might potentially cluster within the Lake Tanganyika cichlid species flock. Thus, the available nuclear data allow only limited conclusions concerning the phylogenetic inter-relationships among the major ancestral Tanganyikan lineages and riverine tilapiine lineages. Both mitochondrial (Klett & Meyer, 2002; Salzburger et al., 2002a; Koblmüller et al., 2005) and nuclear data (Takahashi & Okada, 2002; Terai et al., 2003) suggest that the Trematocarini, Bathybatini, and *Hemibates* are ancestral to the so-called MVhL-clade (Takahashi et al., 2001b), which includes the species flocks of Lake Malawi and Victoria, the H-lineage (comprising the Eretmodini, Limnochromini, Benthochromini, Ectodini, Perissodini, Tropheini, Cyphotilapiini, and Haplochromini; Nishida, 1991), and the Lamprologini. The H-lineage was defined by Nishida (1991) based on allozyme data. However, mitochondrial data rejected the H-lineage in its original composition, since the Eretmodini were placed either as most ancestral lineage within the MVhL-clade or as sister group to the Lamprologini (Salzburger et al., 2002a, 2005; Clabaut et al., 2005; Koblmüller et al., 2005). This led Clabaut et al. (2005) to define the C-lineage, which is equivalent to the H-lineage except for the exclusion of the

Eretmodini (Fig. 1). All phylogenetic hypotheses suggest, regardless of the markers used, a period of extremely rapid cladogenesis at the base of the radiation of the C-lineage, leading to the formation of several tribes, whose inter-tribal phylogenetic relationships could not be resolved with confidence so far (Sturmbauer & Meyer, 1993; Streelman et al., 1998; Takahashi et al., 2001b; Salzburger et al., 2002a,

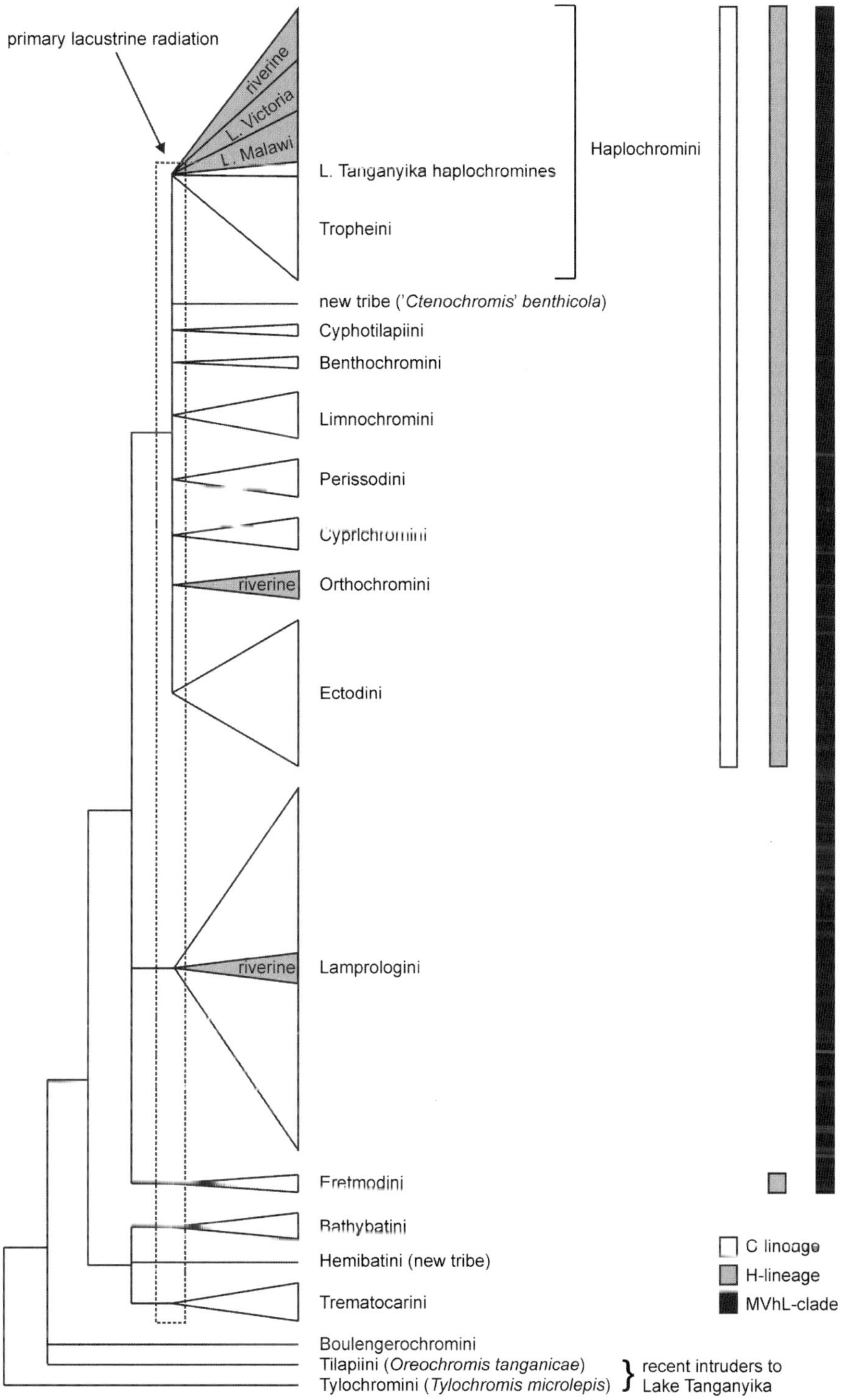

Fig. 1 Schematic molecular phylogeny of the Lake Tanganyika cichlid species assemblage based on combined evidence from several studies. No molecular data exist for '*Ctenochromis*' *benthicola*, and this species was placed as a distinct tribe within the Lake Tanganyika cichlid species assemblage according to Takahashi's (2003) morphological classification. Lineages that underwent radiation are indicated by triangles, whose size corresponds to the species number within the lineage (except for the very species-rich nonTanganyika haplochromines). Gray triangles indicate lineages that do not occur in Lake Tanganyika. The "primary lacustrine radiation" is assumed to have coincided with the establishment of a true lacustrine habitat with deep-water conditions about 5–6 MYA. Bars to the right indicate taxonomic groups that have been proposed based on combined mitochondrial and nuclear DNA sequence data (C-lineage; Clabaut et al., 2005), allozyme data (H-lineage; Nishida, 1991), and SINEs (MVhL-lineage; Takahashi et al., 2001b). Note that the phylogenetic history of most tribes is much better resolved than illustrated and that branch lengths and proportional times of radiation are not to be taken as accurate

2005; Takahashi & Okada, 2002; Terai et al., 2003; Duftner et al., 2005; Sugawara et al., 2005). However, there is strong evidence that the Tropheini form a monophyletic group within the Haplochromini and that the members of the genus *Orthochromis* from the Malagarazi River system, which were originally assigned to the Haplochromini, actually represent a distinct tribe within the C-lineage (Salzburger et al., 2005).

The temporal context of the Lake Tanganyika cichlid diversification

Lake Tanganyika is the oldest of the African rift lakes, with a highly complex geological history (Scholz & Rosendahl, 1988; Gasse et al., 1989; Tiercelin & Mondeguer, 1991; Cohen et al., 1993, 1997; Lezzar et al., 1996; Scholz et al., 2003). It is now well established that several lineages of cichlids independently colonized the emerging lake and seeded the so-called "primary lacustrine radiation," a synchronous diversification within several lineages (Salzburger et al., 2002a; Fig. 1), which may have coincided with the establishment of deep-water conditions in a clear lacustrine habitat about 5–6 MYA (Tiercelin & Mondeguer, 1991; Cohen et al., 1993, 1997). With the growth of phylogenetic information on the Lake Tanganyika species assemblage in recent years, seven seeding lineages were identified: the ancestor of the substrate-breeding *Boulengerochromis microlepis*, the ancestor of the genus *Hemibates* (Koblmüller et al., 2005), and the ancestors of the Trematocarini, Bathybatini, Eretmodini, Lamprologini, and the C-lineage (Clabaut et al., 2005). Molecular studies demonstrated that *Oreochromis tanganicae* and *Tylochromis polylepis* were not yet present in the lake at the time of the primary radiation, but colonized Lake Tanganyika only recently (Klett & Meyer, 2002; Koch et al., 2007) and filled ecological niches that were occupied by only a few other species.

In the course of the "primary lacustrine radiation," the substrate breeding Lamprologini and the mouthbrooding C-lineage diversified rapidly into several lineages (Salzburger et al., 2002a; Takahashi & Okada, 2002; Terai et al., 2003; Clabaut et al., 2005; Sugawara et al., 2005). Interestingly, the *Orthochromis* species, which are presently distributed in the Malagarazi River system, also originated within the Lake Tanganyika radiation in the course of the "primary lacustrine radiation" (Salzburger et al., 2002a, 2005). Furthermore, the most ancient diversification within the tribe Bathybatini between the small *Bathybates minor* and the ancestor of the remaining large *Bathybates* species is also part of the primary radiation (Koblmüller et al., 2005).

Considerable differences in the patterns of diversification have been observed among the Lake Tanganyika cichlid tribes. After rapid formation of the main lineages of the Lamprologini, further diversification in this substrate breeding tribe proceeded gradually (Day et al., 2007; Koblmüller et al., 2007a). In contrast, distinct periods of rapid cladogenesis were detected in the mouthbrooding lineages (Sturmbauer et al., 2003; Koblmüller et al., 2004, 2005, 2007b; Brandstätter et al., 2005; Duftner et al., 2005). Most conspicuously, a "secondary radiation," which took place almost simultaneously in most mouth-brooding tribes, becomes evident at about half the age of the primary radiation (Koblmüller et al., 2004, 2005, 2007a; Duftner et al., 2005; Seehausen, 2006). Assuming an age of about 5–6 MY for the primary radiation, the secondary radiation was dated to roughly 2.5–3 MYA (for a discussion on age estimates derived from alternative calibrations see below). It should be noted that also the radiation of platythelphusid crabs in Lake Tanganyika is likely to have occurred at the same time (Marijnissen et al., 2006). One of the mouth-brooding tribes, the Perissodini, diversified slightly later (1.5–2 MYA), which may be related to intrinsic characteristics specific to this tribe (Koblmüller et al., 2007b; for discussion see below).

Modes of speciation

For decades, evolutionary biologists have been pondering about the causes of the species richness of African cichlids. Since many species are extremely young, locally distributed and morphologically similar, inferring species status in cichlid fishes is often problematic (Sturmbauer, 1998; Genner et al., 2004). Also, the mechanisms leading to reproductive isolation and enabling the rapid formation of species flocks in the East African Great Lakes have been under discussion for a long time (Mayr, 1984;

Ribbink, 1994). Allopatric diversification is certainly a major pathway of speciation in many of Lake Tanganyika's cichlid lineages (Sturmbauer, 1998). The importance of habitat fragmentation for the diversification of the Tanganyikan cichlids was shown by a series of studies on littoral species (Sturmbauer & Meyer, 1992; Snoeks et al., 1994; Kohda et al., 1996; Sturmbauer et al., 1997). Habitat barriers prevent gene flow between populations to a degree that depends on species-specific traits such as dispersal ability, stenotopy, and innate migration behavior. Differences in the patterns of population structure between sympatric and even syntopic species (Duftner et al., 2006; Koblmüller et al., 2007c; Sefc et al., 2007) suggest that species-specific responses to habitat structure play a role in determining the potential for allopatric speciation (Mayr, 1984) of different lineages.

Furthermore, several studies inferred an impact of lake-level fluctuations on the diversification of Lake Tanganyika cichlids (Sturmbauer & Meyer, 1992; Rossiter, 1995; Verheyen et al., 1996; Sturmbauer et al., 1997, 2001, 2003, 2005; Rüber et al., 1998; Baric et al., 2003; Duftner et al., 2007). Lake level changes in Lake Tanganyika have been related to regional and even global climatic events, and/or tectonic activities (Tiercelin & Mondeguer, 1991; Cohen et al., 1993, 1997; Lezzar et al., 1996; Scholz & Rosendahl, 1988; Nicholson, 1999; Scholz et al., 2003). A large body of geological data for the last 1.1 MY consistently suggests repeated water-level fluctuations of several hundreds of meters, although the proposed magnitude of the drops varies between studies. At least two of the reconstructed lake-level fluctuations in the more recent past ($\sim$ 1.1 MYA, 18–23 KYA) had a magnitude of >600 m (Cohen et al., 1997), resulting in a separation of the three sub basins of the lake. Hardly any data are available for the time prior to the last million years, although it is very likely that lake-level fluctuations occurred in the earlier history of Lake Tanganyika as well. It is well established that the climate in eastern Africa shifted toward more arid conditions roughly 2.5–3 MYA (DeMenocal, 1995; Bobe & Behrensmeyer, 2004; Sepulchre et al., 2006). Since Lake Tanganyika's water level strongly depends on the rainfall in its catchment area (e.g., Nicholson, 1999), aridification in eastern Africa would lead to a severe drop of the lake level. Drastic lowstands certainly promoted allopatric speciation in the isolated basins. Less severe lake-level fluctuations have the strongest effect on populations of littoral species, which experience cycles of population admixture and fragmentation as their habitat shrinks and rises with the changing shoreline. When the succession of population differentiation and secondary admixis results in speciation, lake-level fluctuations acquire the function of a "species pump" (Rossiter, 1995). Indeed, diversification of the littoral cichlid tribes Eretmodini (Rüber et al., 1998) and Tropheini (Sturmbauer et al., 2003, 2005; Egger et al., 2007) was probably severely affected by lake-level fluctuations, coupled with parallel evolution of equivalent trophic types (Rüber et al., 1999; Sturmbauer et al., 2003). It has been assumed that species living above sandy bottom or in the pelagic zone were less affected by lake-level fluctuations than those inhabiting rocky littoral habitats (Coulter, 1991; Rossiter, 1995). This is probably true for minor fluctuations. However, recent studies suggested that the diversification of the pelagic deep-water cichlids of the tribe Bathybatini (Koblmüller et al., 2005), the benthic deep-water cichlids of the tribe Limnochromini (Duftner et al., 2005) and the pelagic, but shore-associated cichlids of the tribe Cyprichromini (Brandstätter et al., 2005) coincided with a period of aridification 2.5–3 MYA, during which the lake may have been split into its three basins.

Sympatric speciation, driven by sexual selection of male nuptial coloration or by natural selection creating adaptation to different niches, has been suggested to have contributed to the cichlid diversity in the species flocks of Lakes Malawi and Victoria as well as of West African and Central American crater lakes, although convincing evidence comes from a few studies only (Schliewen et al., 1994, 2001; Schliewen & Klee, 2004; Barluenga et al., 2006). So far, there is no evidence for sympatric speciation in Lake Tanganyika, although it has been suggested to have played a role in the evolution of its cichlid species flock (Sturmbauer, 1998).

The supposition that the primary Tanganyika radiation, which gave rise to the extant tribes (Fig. 1), was driven by natural selection derives support from the observation that most tribes are limited to particular ecological niches (Table 2): all Bathybatini are pelagic piscivorous deep-water species; all Trematocarini are pelagic zooplanktivorous

Table 2 Lake Tanganyika cichlid tribes as suggested by molecular data, with species numbers and biological characteristics

Tribe	Species no.[a]	Biological characteristics
Tylochromini	1[c]	Invertebrate-feeder; found in river mouths and in the shallow sandy habitat; maternal mouthbrooder
Tilapiini	1[c]	Found in vegetated areas; feeds on plants; maternal mouthbrooder
Boulengerochromini	1	The largest species; piscivorous; found on sandy bottom; substrate breeder
Bathybatini	7	Pelagic piscivorous deep-water species; maternal mouthbrooders
Hemibatini[b]	1	Pelagic deep-water species; feeds on shrimps and fish; maternal mouthbrooder
Trematocarini	9	Pelagic zooplanktivorous deep-water species with several species migrating into shallow waters during night; maternal mouthbrooders
Eretmodini	4	Restricted to very shallow rocky habitat; reduced swim-bladder; aufwuchs- and/or invertebrate-feeders; maternal and biparental mouthbrooders
Lamprologini	81[c]	Found in shallow to deep waters on sandy and rocky substrate and shell-beds; substrate breeders
Ectodini	34	Found in various types of habitat, although the ancestor of this tribe most likely inhabited sandy bottom in shallow waters; feed on aufwuchs, plankton and shrimps; maternal and biparental mouthbrooders
Cyprichromini	7	Semipelagic to pelagic, but shore-associated planktivorous species; maternal mouthbrooders
Perissodini	9	Scale-eaters (except for the two zooplanktivorous shallow water species of the genus *Haplotaxodon*), with the origin of scale-eating in the deep-water; biparental mouthbrooders
Limnochromini	11	Benthic (sandy/muddy bottom) deep-water species; feed on invertebrates and small fish; biparental mouthbrooders
Benthochromini	3	Semipelagic planktivorous species, associated with sediment-rich bottom; maternal mouthbrooders
Haplochromini	4[c]	Zooplanktivorous species, found in river mouths and vegetated areas; maternal mouthbrooders
'*Ctenochromis*' *benthicola*[d]	1	Little known; lives in the deep-rocky habitat
Tropheini	24	Species of the shallow (mostly rocky) habitat; feed on aufwuchs or invertebrates; maternal mouthbrooders
Cyphotilapiini	2	Piscivorous species of the deep-rocky habitat; maternal mouthbrooders
Total	200	

[a] Only valid species are considered

[b] Based on molecular data (Koblmüller et al., 2005), the erection of the tribe Hemibatini seems justified. Alternatively, the Bathybatini, Trematocarini, and the genus *Hemibates* might be included in a single tribe, the Bathybatini, as suggested by Takahashi (2003)

[c] Some tribes are not exclusively distributed in Lake Tanganyika: Tylochromini, western to central Africa; Tilapiini, whole Africa; Lamprologini, Congo system (eight species) and Malagarazi River (one species); Haplochromini, whole Africa except for the western part of the continent

[d] Tribal status for '*C.*' *benthicola* is based on morphological data only (Takahashi, 2003); no molecular data available

deep-water species with several species migrating into shallow waters at night; all Eretmodini are restricted to very shallow rocky habitat; all Limnochromini are benthic (sandy/muddy bottom) deep-water species feeding on invertebrates or small fish; the three species of Benthochromini are semipelagic planktivorous species, associated with sediment-rich bottom; all Cyprichromini are semipelagic to pelagic, but shore-associated planktivorous species; the Perissodini are scale-eaters except for the two shallow water species of the genus *Haplotaxodon*; all Tropheini are species of the shallow (mostly rocky) habitat; the Cyphotilapiini are piscivorous species of the deep-rocky habitat. Among the mouthbrooding lineages, only the Ectodini are found in various types of habitat, although the ancestor of this tribe most likely inhabited the sandy shallow benthal (Koblmüller et al., 2004). A wide range of habitat types is also utilized by the Lamprologini, which are the most species-rich of the Lake Tanganyika cichlid tribes, comprise all substrate breeding cichlids of Lake Tanganyika (except for

Boulengerochromis microlepis) and are equally diverse in terms of morphology, ecology and behavior as the whole C-lineage.

Later on in the history of Lake Tanganyika's cichlid species assemblage, specialization to particular foraging areas and prey types may have driven the radiations within the Limnochromini (Duftner et al., 2005), Bathybatini (Koblmüller et al., 2005) and the Perissodini (Koblmüller et al., 2007b; Takahashi et al., 2007), and specialization to different breeding sites (shell-beds versus rocks) may underlie the more recent diversification of the lamprologine cichlid genus *Telmatochromis* (Takahashi, 2004). As mentioned above, the secondary radiation within the tribe Perissodini occurred slightly after the secondary radiation within other mouth-brooding tribes. It has been suggested that speciation within this predominantly scale-eating tribe might have been initiated by trophic diversification following (both causally and chronologically) the radiations in other cichlid tribes. There is phylogenetic evidence that scale-eating originated in the deep-pelagic habitat, and the proliferation of other deep-water cichlids might have induced the evolution and diversification of the scale-eaters (Koblmüller et al., 2007b).

Recently, evidence has been accumulating that introgressive hybridization is more common in East African cichlid fishes than previously thought and plays an important role in the evolution of these fishes (Seehausen, 2004). Homogenizing effects of hybridization have been documented in some haplochromine cichlids of Lake Victoria, where species barriers based on visual discrimination break down with increasing water turbidity due to pollution and erosion (Seehausen et al., 1997), and in Lake Malawi, where introgressive hybridization followed faunal translocation (Stauffer et al., 1996; Streelman et al., 2004). In Lake Tanganyika, introgression appears to be most common in the substrate-breeding lamprologines, where discrepancies between morphological similarities and the placement of taxa in mitochondrial and nuclear phylogenies indicate hybridization even among distantly related taxa (Day et al., 2007; Koblmüller et al., 2007a, unpublished data). Molecular phylogenies of mouthbrooding lineages so far revealed only one convincing case of introgression (Eretmodini; Rüber et al., 2001). Under favorable circumstances, transgressive segregation can equip hybrid populations with novel traits allowing the exploitation of extreme niches and even result in new species, as has been demonstrated in tilapiine cichlids of a West African crater lake (Schliewen & Klee, 2004). Evidence for a role of hybridization in speciation and even hybrid speciation resulting in morphologically well-defined species has also been found in the Lamprologini of Lake Tanganyika (Salzburger et al., 2002b; Schelly et al., 2006; Koblmüller et al., 2007a). Especially gastropod-shell-breeding lamprologines were shown to be particularly prone to—presumably accidental—hybridization: while lake-level fluctuations have been proposed to induce hybridization by merging populations and by providing novel niches for arising hybrid taxa (Rüber et al., 2001; Salzburger et al., 2002b), which theoretically applies to any littoral lineage, the breeding behavior of the shell-breeding lamprologines bears additional potential for cross-species fertilization. Many shell-breeding species live and breed in closest vicinity within the same shell bed, and often, large males release sperm above the opening of the shells in which their small-bodied female mates spawn. Water currents and wave action may direct some of the sperm into another than the intended shell, resulting in the occasional fertilization of another species' eggs. Phylogenetic analyses demonstrated that the phylogeny of the shell-breeding clade was to a considerable degree shaped by the high rate of introgressive hybridization and that in some instances, the evolution of novel taxa was launched by hybridization (Koblmüller et al., 2007a).

New dates for cichlid radiations and their implications for Lake Tanganyika

Dating of cladogenetic events in most existing phylogenetic studies of Lake Tanganyika cichlids was based on the assumption that the simultaneous origin of the extant tribes—the primary radiation (Salzburger et al., 2002a)—coincided with the shift from the shallow and swampy wetland conditions to lacustrine deep-water habitat at 5–6 MYA (Tiercelin & Mondeguer, 1991, Cohen et al., 1993, 1997), consistent with the observation that cichlid radiations are usually associated with truly lacustrine conditions (reviewed, e.g., in Seehausen, 2006). All these studies were based on mitochondrial DNA—or amino acid (Duftner et al., 2005)—sequences, assuming a strictly clock-like model of molecular evolution. In most cases, the linearized tree approach as described in Takezaki et al. (1995) was applied. It should be emphasized that the assumption that the primary Tanganyika radiation coincided with the formation of a truly lacustrine habitat around 5–6 MYA is not based on independent evidence. Applying a strict molecular clock and setting the age of the primary Tanganyika radiation to 5–6 MYA yields time estimates for cladogenesis events in African cichlids, e.g., radiations in Lake Malawi, paleo-Lake Makgadikgadi (Joyce et al., 2005) that are very much in accordance with the geological record. Strict molecular-clock based models suffer from the problem that, in any dataset, there is usually some degree of heterogeneity in substitution rates. Usually, datasets with more closely related taxa show less heterogeneity in substitution rates than datasets with distantly related taxa. In order to overcome this problem, taxa that show significantly slower or faster substitution rates have to be removed from the analysis. However, tests for rate-constancy (e.g., Takezaki et al., 1995; Robinson-Rechavi & Huchon, 2000) might fail to detect small gradual changes of substitution rates and thus, the resulting age estimates might be biased. Recently, more and more sophisticated methods for deriving age estimates based on molecular data have been developed. Especially, those methods that relax the assumption of a molecular clock (e.g., Thorne et al., 1998; Kishino et al., 2001; Sanderson, 2002; Britton, 2005) and combine molecular data with time constraints (upper and lower bounds), usually derived from fossils or geological data, have become increasingly popular. It will be interesting to see whether the previously published dates still hold true when these new methods are applied. It should be noted that empirical studies that compared approaches that account for rate variation with models that assume a strict molecular clock have not identified consistent differences in their outcome—in some cases no differences were observed (e.g., Wiegman et al., 2003), whereas other studies demonstrated that results differ with respect to the method used (e.g., Douzery et al., 2004). When applying relaxed molecular-clock models it is important to constrain both recent and more ancient calibration points to obtain proper age estimates (Renner & Zhang, 2004).

Recently, new dates for African cichlid radiations were proposed on the basis of combined nuclear and mitochondrial DNA sequences, applying a 'relaxed molecular clock' calibrated by the Gondwanan fragmentation, which separated Madagascan and Asian cichlids from their African ancestors (Genner et al., 2007). Dating comparatively recent cladogenesis events like the Tanganyika radiation might be problematic with the Gondwana calibration, since, as mentioned above, both deep and shallow splits should be constrained to obtain less biased age estimates and avoid a distorted chronology of diversification events (e.g., Renner & Zhang, 2004). Indeed, these new age estimates would have huge implications for the reconstruction of the evolutionary history of Lake Tanganyika's species assemblage. The primary radiation would be moved to 20 to >35 MYA, which is considerably older than the proposed age of the truly lacustrine habitat in Lake Tanganyika (5–6 MYA) and even much older than the emergence of small and shallow lakes in the Lake Tanganyika area due to increased tectonic activity 9–12 MYA (Tiercelin & Mondeguer, 1991; Cohen et al., 1993, 1997). This poses the question what triggered the observed synchronous radiations and where they should have taken place. So far, all studied cichlid radiations have been associated with truly lacustrine habitats, and it has been proposed that rivers are not a suitable habitat to host a radiation (see, e.g., Joyce et al., 2005). Prior 20 MYA the swampy Proto-Malagarazi River was suggested to have been present in this area (Tiercelin & Mondeguer, 1991). Thus, if the ancient age estimates hold true, the primary Tanganyika radiation would be the first example for a rapid radiation in rivers and/or small water bodies. Based on biogeographic patterns of riverine cichlid species distributions, the existence of a large paleo-lake in the Congo region at some point in the past has been postulated (Beadle, 1981; Coulter, 1991), but there are no geological data supporting this hypothesis.

The observed resource partitioning between most cichlid tribes in Lake Tanganyika presents another problem for any scenario placing the secondary radiation in conditions other than in a single deep-water lake, for which there is no evidence in the area before 5–6 MYA. If both the primary radiation into the extant tribes and the secondary radiation within some tribes predate the beginning of lake formation in the Lake Tanganyika area (as implied by the Gondwanan calibration), most of the diversification must have taken place in different surrounding water bodies, and each tribe would be expected to span several ecological niches rather than being clearly specialized for one type of habitat (Table 2).

Biogeographic patterns of cichlid species distributions are hard to reconcile with the novel age estimate. Previous studies, assuming the recent age of 5–6 MY for the primary radiation, have already had difficulties explaining the absence of ancient lineages related to the seeders of the Tanganyika radiation in the surrounding rivers. With the novel, more ancient dates of Genner et al. (2007), the extant cichlid tribes would have originated and radiated in rivers, and then colonized the emerging lake without leaving any relatives in their native areas. Serious extinction events have been quite common in southern Africa (Stewart, 2001), but the new model would require the complete elimination of a species assemblage near the size of the extant Tanganyika cichlid fauna from their native habitat.

Another puzzle is presented by the occurrence of natural hybrids between distantly related lamprologine cichlids in Lake Tanganyika (Koblmüller et al., 2007a). With the new time estimates, hybridization yielding viable offspring would involve species that have diverged 15–20 MYA. However, it should be noted that fertilization success can be very high between species over very long periods of separation: Bolnick & Near (2005) reported a fertilization success of ~60% between centrarchid species that diverged 20 MYA. This makes the Centrarchidae outstanding in terms of hybrid viability among animals, and does not necessarily imply that the same phenomenon holds true for cichlid fishes, in particular, the small lamprologine cichlids with their—even compared to the Centrarchidae—short generation time. Additionally, unlike the cichlid fishes, which are considered one of the prime examples for rapid diversification and speciation, the Centrarchidae experience a comparatively slow diversification rate, even relative to other vertebrates with longer generation times (Bolnick & Near, 2005).

In the same study, Genner et al. (2007) also explore an alternative dating based on fossil calibration, which they reject in favor of the Gondwanan calibration because of a worse match with—fossil based—substitution rates in other fishes. Dates based on the fossil calibration are better reconcilable with the observed resource partitioning among tribes. However, although several fossils were used, they are comparatively old and only distantly related to the Tanganyika radiation. Unfortunately, the fossil record is very poor for cichlid fishes. Additionally, there is often a great amount of uncertainty in fossil dating and in their phylogenetic placement, which, obviously, might have a great impact on proposed age estimates, when applying them as calibration point. The fossil calibration dates the primary Tanganyika radiation within the period when a series of small and shallow lakes existed in the Lake Tanganyika area 9–12 MYA, and the secondary radiation of the mouthbrooding tribes to the time of the onset of a lacustrine deep-water habitat 5–6 MYA. Prior to the secondary radiation, competition between tribes would have resulted in the ecological divergence of the different tribes. Only those tribes, which are not restricted to relatively narrow trophic niches (i.e., the Ectodini and Lamprologini), diversified prior to the formation of the deep-water lacustrine habitat. However, the absence of ancient lamprologines and ectodines outside Lake Tanganyika remains unexplained. Although a number of lamprologine species are found in the Congo system and one species in the Malagarazi River, the split of these species from their sister lineages in Lake Tanganyika is, regardless of the calibration used, too recent for them to represent the remnants of an originally riverine species assemblage (Koblmüller et al., unpublished data), suggesting that lamprologines secondarily colonized the river systems from Lake Tanganyika (see also Sturmbauer et al., 1994). The fossil calibration also rejects the hypothesis that the haplochromine cichlids evolved in Lake Tanganyika and later on colonized surrounding rivers and lakes (Salzburger et al., 2002a, 2005), but instead is consistent with a hypothesis that the Haplochromini originated and diversified prior to the formation of the deep-water habitat, with only one lineage (Tropheini) colonizing the emerging Lake Tanganyika.

It is remarkable that the two most commonly used methods of molecular-clock calibration—by geological and by fossil evidence—are both at odds with each other and the biology and the phylogeny of the Lake Tanganyika species assemblage. Whether similar problems exist for the younger species flock in Lake Malawi, whose origin would be dated to 4 MYA by the Gondwanan and to 2 MYA by the fossil calibration, depends on the exact geological history of the lake. If, as some sources imply, Lake Malawi desiccated at about 1 MYA (Delvaux, 1995), refugia for the species flock must be found, whereas a calibration setting the primary Tanganyika radiation to 5–6 MYA dates the Lake Malawi flock to about 1 MYA in accordance with the geologic record (Koblmüller et al., 2008). Genner et al.'s (2007) study is of crucial importance in revealing the inconsistencies between the established age estimates and alternative calibration methods, and identifies the need for a resumption of work on the dating of cichlid radiations.

Acknowledgments We thank P. Ngalande, H. Phiri, and the team at the Mpulungu station of the Ministry of Agriculture and Cooperatives, Republic of Zambia, for their longstanding cooperation during fieldwork. We are further grateful to B. Egger and two anonymous reviewers for valuable comments on earlier versions of the manuscript. Financial support was provided by the provincial government of Styria to KMS and the Austrian Science Foundation (grants P17380 to KMS and P17680 to CS).

References

Albertson, R. C., J. A. Markert, P. D. Danley & T. D. Kocher, 1999. Phylogeny of a rapidly evolving clade: the cichlid fishes of Lake Malawi, East Africa. Proceedings of the National Academy of Sciences USA 96: 5107–5110.

Allender, C. J., O. Seehausen, M. E. Knight, G. F. Turner & N. Maclean, 2003. Divergent selection during speciation of Lake Malawi cichlid fishes inferred from parallel radiations in nuptial coloration. Proceedings of the National Academy of Sciences USA 100: 14074–14079.

Baric, S., W. Salzburger & C. Sturmbauer, 2003. Phylogeography and evolution of the Tanganyikan cichlid genus *Tropheus* based upon mitochondrial DNA sequences. Journal of Molecular Evolution 56: 54–68.

Barlow, G. W., 2000. The Cichlid Fishes: Nature's Grand Experiment in Evolution. Perseus, Cambridge, MA.

Barluenga, M., K. N. Stölting, W. Salzburger, M. Muschick & A. Meyer, 2006. Sympatric speciation in Nicaraguan crater lake cichlid fish. Nature 439: 719–723.

Beadle, L. C., 1981. The Inland Waters of Tropical Africa. Longman, London.

Bobe, R. & A. K. Behrensmeyer, 2004. The expansion of grassland ecosystems in Africa in relation to mammalian evolution and the origin of the genus *Homo*. Palaeogeography, Palaeoclimatology, Palaeoecology 207: 399–420.

Bolnick, D. I. & T. J. Near, 2005. Tempo of hybrid inviability in centrarchid fishes (Teleostei: Centrarchidae). Evolution 59: 1754–1767.

Brandstätter, A., W. Salzburger & C. Sturmbauer, 2005. Mitochondrial phylogeny of the Cyprichromini, a lineage of open-water cichlid fishes endemic to Lake Tanganyika. Molecular Phylogenetics and Evolution 34: 382–391.

Britton, T., 2005. Estimating divergence times in phylogenetic trees without a molecular clock. Systematic Biology 54: 500–507.

Clabaut, C., W. Salzburger & A. Meyer, 2005. Comparative phylogenetic analyses of the adaptive radiation of Lake Tanganyikan cichlid fish: nuclear sequences are less homoplasious but also less informative than mitochondrial DNA. Journal of Molecular Evolution 61: 666–681.

Cohen, A. S., M. J. Soreghan & C. A. Scholz, 1993. Estimating the age of ancient lakes: an example from Lake Tanganyika, East African rift system. Geology 21: 511–514.

Cohen, A. S., K.-E. Lezzar, J.-J. Tiercelin & M. J. Soreghan, 1997. New palaeogeographic and lake-level reconstructions of Lake Tanganyika: implications for tectonic, climatic and biological evolution in rift lakes. Basin Research 9: 107–132.

Coulter, G. W., 1991. Lake Tanganyika and its Life. Oxford University Press, Oxford.

Crapon de Crapona, M.-D., 1986. Are 'preferences' and 'tolerances' in cichlid mate choice important for speciation? Journal of Fish Biology 29(Supplement A): 151–158.

Day, J. J. & M. Wilkinson, 2006. On the origin of the *Synodontis* catfish species flock from Lake Tanganyika. Biology Letters 22: 548–552.

Day, J. J., S. Santini & J. Garcia-Moreno, 2007. Phylogenetic relationships of the Lake Tanganyika cichlid tribe Lamprologini: the story from mitochondrial DNA. Molecular Phylogenetics and Evolution 45: 629–642.

Delvaux, D., 1995. Age of Lake Malawi (Nyasa) and water level fluctuations. Annual Rapports of the Royal Museum of Central Africa, Department of Geology and Mineralogy 1993 & 1994: 99–108.

DeMenocal, P. B., 1995. Plio-Pleistocene African climate. Science 270: 53–59.

Deutsch, J. C., 1997. Colour diversification in Malawi cichlids: evidence for adaptation, reinforcement or sexual selection? Biological Journal of the Linnean Society 62: 1–14.

Douzery, E. J. P., E. A. Snell, E. Bapteste, F. Delsuc & H. Philippe, 2004. The timing of eukaryotic evolution: does a relaxed molecular clock reconcile proteins and fossils? Proceedings of the National Academy of Sciences USA 101: 15386–15391.

Drucker, E. G. & J. S. Jensen, 1991. Functional analysis of a specialized prey processing behavior: winnowing by surfperches (Teleostei: Embiotocidae). Journal of Morphology 210: 267–287.

Duftner, N., S. Koblmüller & C. Sturmbauer, 2005. Evolutionary relationships of the Limnochromini, a tribe of benthic deep water cichlid fishes endemic to Lake Tanganyika, East Africa. Journal of Molecular Evolution 60: 277–289.

Duftner, N., K. M. Sefc, S. Koblmüller, B. Nevado, E. Verheyen, H. Phiri & C. Sturmbauer, 2006. Distinct population structure in a phenotypically homogeneous rock-dwelling cichlid fish from Lake Tanganyika. Molecular Ecology 15: 2381–2396.

Duftner, N., K. M. Sefc, S. Koblmüller, W. Salzburger, M. Taborsky & C. Sturmbauer, 2007. Parallel evolution of facial stripe patterns in the *Neolamprologus brichardi/pulcher* species complex endemic to Lake Tanganyika. Molecular Phylogenetics and Evolution 45: 706–715.

Egger, B., S. Koblmüller, C. Sturmbauer & K. M. Sefc, 2007. Complementary AFLP and mtDNA sequence data distinguish evolutionary processes in the Lake Tanganyika cichlid genus *Tropheus*. BMC Evolutionary Biology 7: 137.

Fryer, G., 2006. Evolution in ancient lakes: radiation of Tanganyikan atyid prawns and speciation of pelagic cichlid fishes in Lake Malawi. Hydrobiologia 568: 131–142.

Fryer, G. & T. D. Iles, 1972. The Cichlid Fishes of the Great Lakes of Africa. T.H.F., Neptune, NJ.

Galis, F. & E. G. Drucker, 1996. Pharyngeal biting mechanics in centrarchid and cichlid fishes: insights into a key evolutionary innovation. Journal of Evolutionary Biology 9: 641–670.

Gasse, F., V. Ledee, M. Massault & J.-C. Fontes, 1989. Water-level fluctuations of Lake Tanganyika in phase with oceanic changes during the last glaciation and deglaciation. Nature 342: 57–59.

Genner, M. J., O. Seehausen, D. F. R. Cleary, M. E. Knight, E. Michel & G. F. Turner, 2004. How does the taxonomic status of allopatric populations influence species richness within African cichlid species assemblages. Journal of Biogeography 31: 93–102.

Genner, M. J., O. Seehausen, D. H. Lunt, D. A. Joyce, P. W. Shaw, G. R. Carvalho & G. F. Turner, 2007. Age of cichlids: new dates for ancient lake fish radiations. Molecular Biology and Evolution 24: 1269–1282.

Goodwin, N. B., S. Balshine-Earn & J. D. Reynolds, 1998. Evolutionary transitions in parental care in cichlid fish. Proceedings of the Royal Society London, Series B 265: 2265–2272.

Greenwood, P. H., 1973. Morphology, endemism and speciation in African cichlid fishes. Verhandlungen der Deutschen Zoologischen Gesellschaft 66: 115–124.

Greenwood, P. H., 1978. A review of the pharyngeal apophysis and its significance in the classification of cichlid fishes. Bulletin of the British Museum of Natural History (Zoology) 33: 297–325.

Joyce, D. A., D. H. Lunt, R. Bills, G. F. Turner, C. Katongo, N. Duftner, C. Sturmbauer & O. Seehausen, 2005. Riverine cichlid fish diversity of southern Africa arose during an adaptive radiation in a lost lake. Nature 435: 90–95.

Kaatz, I. M., 2002. Multiple sound-producing mechanisms in teleost fishes and hypotheses regarding their behavioural significance. Bioacoustics 12: 230–233.

Kaufman, L. S. & K. F. Liem, 1982. Fishes of the suborder Labroidei (Pisces: Perciformes): phylogeny, ecology and evolutionary significance. Breviora 472: 1–19.

Kishino, H., J. L. Thorne & W. J. Bruno, 2001. Performance of a divergence time estimation method under a probabilistic model of rate evolution. Molecular Biology and Evolution 18: 352–361.

Klett, V. & A. Meyer, 2002. What, if anything, is a Tilapia?—Mitochondrial ND2 phylogeny of Tilapiines and the evolution of parental care systems in the African cichlid fishes. Molecular Biology and Evolution 19: 865–883.

Knight, M. E. & G. F. Turner, 1999. Reproductive isolation among closely related Lake Malawi cichlids: can males recognize conspecific females by visual cues? Animal Behaviour 58: 761–768.

Knight, M. E. & G. F. Turner, 2004. Laboratory mating trials indicate incipient speciation by sexual selection among populations of the cichlid fish *Pseudotropheus zebra* from Lake Malawi. Proceedings of the Royal Society London, Series B 271: 675–680.

Knight, M. E., G. F. Turner, C. Rico, J. J. H. van Oppen & G. M. Hewitt, 1998. Microsatellite paternity analysis on captive Lake Malawi cichlids supports reproductive isolation by direct mate choice. Molecular Ecology 7: 1605–1610.

Koblmüller, S., N. Duftner, C. Sturmbauer, H. Sammer, N. Gantner, R. Kopp, S. Voigt, B. Stadlbauer, A. Brandstätter & R. Hanel, 2003. Vergleichende Betrachtung von Ernährungsmorphologie und Nahrungsspezifität ausgewählter mediterraner Lippfisch-Arten (Perciformes, Labridae). Berichte des naturwissenschaftlich-medizinischen Vereins Innsbruck 90: 219–230.

Koblmüller, S., W. Salzburger & C. Sturmbauer, 2004. Evolutionary relationships in the sand-dwelling cichlid lineage of Lake Tanganyika suggest multiple colonization of rocky habitats and convergent origin of biparental mouthbrooding. Journal of Molecular Evolution 58: 79–96.

Koblmüller, S., N. Duftner, C. Katongo, H. Phiri & C. Sturmbauer, 2005. Ancient divergence in bathypelagic Lake Tanganyika deepwater cichlids: mitochondrial phylogeny of the tribe Bathybatini. Journal of Molecular Evolution 60: 297–314.

Koblmüller, S., C. Sturmbauer, E. Verheyen, A. Meyer & W. Salzburger, 2006. Mitochondrial phylogeny and phylogeography of East African squeaker catfishes (Siluriformes: *Synodontis*). BMC Evolutionary Biology 6: 49.

Koblmüller, S., N. Duftner, K. M. Sefc, M. Aibara, M. Stipacek, M. Blanc, B. Egger & C. Sturmbauer, 2007a. Reticulate phylogeny of gastropod-shell-breeding cichlids from Lake Tanganyika—the result of repeated introgressive hybridization. BMC Evolutionary Biology 7: 7.

Koblmüller, S., B. Egger, C. Sturmbauer & K. M. Sefc, 2007b. Evolutionary history of Lake Tanganyika's scale-eating cichlid fishes. Molecular Phylogenetics and Evolution 44: 1295–1305.

Koblmüller, S., K. M. Sefc, N. Duftner, M. Warum & C. Sturmbauer, 2007c. Genetic population structure as indirect measure of dispersal ability in a Lake Tanganyika cichlid. Genetica 130: 121–131.

Koblmüller, S., U. K. Schliewen, N. Duftner, K. M. Sefc, C. Katongo & C. Sturmbauer, 2008. Age and spread of the haplochromine cichlid fishes in Africa. Molecular Phylogenetics and Evolution. doi:10.1016/j.ympev.2008.05.045.

Koch, M., S. Koblmüller, K. M. Sefc, N. Duftner, C. Katongo & C. Sturmbauer, 2007. Evolutionary history of the endemic Lake Tanganyika cichlid fish *Tylochromis polylepis*: a recent intruder to a mature adaptive radiation. Journal of Zoological Systematics and Evolutionary Research 45: 64–71.

Kocher, T. D., 2004. Adaptive evolution and explosive speciation: the cichlid fish model. Nature Reviews Genetics 5: 288–298.

Kohda, M., Y. Yanagisawa, T. Sato, K. Nakaya, Y. Rimura, K. Matsumoto & H. Ochi, 1996. Geographical color variation in cichlid fishes at the southern end of Lake Tanganyika. Environmental Biology of Fishes 45: 237–248.

Kornfield, I. & P. F. Smith, 2000. African cichlid fishes: model systems for evolutionary biology. Annual Reviews in Ecology and Systematics 31: 163–196.

Lezzar, K. E., J.-J. Tiercelin, M. de Batist, A. S. Cohen, T. Bandora, P. van Rensbergen, C. le Turdu, W. Mifundu & J. Klerkx, 1996. New seismic stratigraphy and Late Tertiary history of the north Tanganyika basin, East African rift system, deduced from multichannel and high-resolution reflection seismic data and piston core evidence. Basin Research 8: 1–28.

Liem, K. F., 1973. Evolutionary strategies and morphological innovations: cichlid pharyngeal jaws. Systematic Zoology 22: 425–441.

Liem, K. F., 1986. The pharyngeal jaw apparatus of the Embiotocidae (Teleostei): a functional and evolutionary perspective. Copeia 1986: 311–323.

Liem, K. F. & P. H. Greenwood, 1981. A functional approach to the phylogeny of pharyngognath teleosts. American Zoologist 21: 83–101.

Liem, K. F. & S. L. Sanderson, 1986. The pharyngeal jaw apparatus of labrid fishes: a functional morphological perspective. Journal of Morphology 187: 143–158.

Lippitsch, E., 1998. Phylogenetic study of cichlid fishes in Lake Tanganyika: a lepidological approach. Journal of Fish Biology 53: 752–766.

Lobel, P. S., 2001. Acoustic behavior of cichlid fishes. Journal of Aquariculture and Aquatic Sciences 9: 167–186.

Mabuchi, K., M. Miya, Y. Azuma & M. Nishida, 2007. Independent evolution of the specialized pharyngeal jaw apparatus in cichlid and labrid fishes. BMC Evolutionary Biology 7: 10.

Marijnissen, S. A., E. Michel, S. R. Daniels, D. Erpenbeck, S. B. Menken & F. R. Schram, 2006. Molecular evidence for recent divergence of Lake Tanganyika endemic crabs (Decapoda: Platythelphusidae). Molecular Phylogenetics and Evolution 40: 628–634.

Martens, K., G. Coulter & B. Goddeeris, 1994. Speciation in ancient lakes—40 years after Brooks. In Martens, K., B. Goddeeris & G. Coulter (eds), Speciation in Ancient Lakes. Archiv für Hydrobiologie, Vol. 44: 75–96.

Mayr, E., 1984. Evolution of fish species flocks: a commentary. In Echelle, A. A. & I. Kornfield (eds), Evolution of Fish Species Flocks. University of Maine at Orono Press, Orono: 3–11.

McKaye, K. R. & W. N. Gray, 1984. Extrinsic barriers to gene flow in rock-dwelling cichlids of Lake Malawi: macrohabitat heterogeneity and reef colonization. In Echelle, A. A. & I. Kornfield (eds), Evolution of Fish Species Flocks. University of Maine at Orono Press, Orono: 169–183.

McKaye, K. R., J. H. Howard, J. R. Stauffer Jr., R. P. Morgan & F. Shonhiwa, 1993. Sexual selection and genetic relationships of a sibling species complex of bower building cichlids in Lake Malawi, Africa. Japanese Journal of Ichthyology 40: 15–21.

Meyer, A., 1993. Phylogenetic relationships and evolutionary processes in East African cichlids. Trends in Ecology and Evolution 8: 279–284.

Meyer, A., T. D. Kocher, P. Basasibwaki & A. C. Wilson, 1990. Monophyletic origin of Lake Victoria cichlid fishes suggested by mitochondrial DNA sequences. Nature 347: 550–553.

Moran, P., I. Kornfield & P. N. Reinthal, 1994. Molecular systematics and radiation of the haplochromine cichlids (Teleostei, Perciformes) of Lake Malawi. Copeia 1994: 274–288.

Nagl, S., H. Tichy, W. E. Mayer, N. Takezaki, N. Takahata & J. Klein, 2000. The origin and age of haplochromine fishes in Lake Victoria, East Africa. Proceedings of the Royal Society London, Series B 267: 1059–1061.

Nicholson, S. E., 1999. Historical and modern fluctuations of Lakes Tanganyika and Rukwa and their relationship to rainfall variability. Climatic Change 41: 53–71.

Nishida, M., 1991. Lake Tanganyika as an evolutionary reservoir of old lineages of East African cichlid fishes: inferences from allozyme data. Experientia 47: 974–979.

Nishida, M., 1997. Phylogenetic relationships and evolution of Tanganyikan cichlids: a molecular perspective. In Kawanabe, H., M. Hori & M. Nagoshi (eds), Fish Communities in Lake Tanganyika. Kyoto University Press, Kyoto, Japan: 1–23.

Poll, M., 1986. Classification des Cichlidae du lac Tanganika: tribus, genres et espèces. Memoires de la classe des sciences. T. XLV. Fascicule 2 edn. Académie Royale de Belgique, Brussels.

Renner, S. S. & L.-B. Zhang, 2004. Biogeography of the *Pistia* clade (Araceae): based on chloroplast and mitochondrial DNA sequences and Bayesian divergence time inference. Systematic Biology 53: 422–432.

Ribbink, A. J., 1994. Alternative perspectives on some controversial aspects of cichlid fish speciation. Advances in Limnology 44: 101–125.

Rice, A. N. & P. S. Lobel, 2002. Enzyme activities of pharyngeal jaw musculature in the cichlid *Tramitichromis intermedius*: implications for sound production in cichlid fishes. Journal of Experimental Biology 205: 3519–3523.

Rice, A. N. & P. S. Lobel, 2003. The pharyngeal jaw apparatus of the Cichlidae and Pomacentridae: function in feeding and sound production. Reviews in Fish Biology and Fisheries 13: 422–444.

Robinson-Rechavi, M. & D. Huchon, 2000. RRTree: Relative-Rate Test between groups of sequences on a phylogenetic tree. Bioinformatics 16: 296–297.

Rossiter, A., 1995. The cichlid fish assemblages of Lake Tanganyika: ecology, behaviour and evolution of its species flock. Advances in Ecological Research 26: 187–252.

Rüber, L., E. Verheyen, C. Sturmbauer & A. Meyer, 1998. Lake level fluctuations and speciation in rock-dwelling cichlid fish in Lake Tanganyika, East Africa. In Grant, P. (ed.), Evolution on Islands. Oxford University Press, Oxford: 225–240.

Rüber, L., E. Verheyen & A. Meyer, 1999. Replicated evolution of trophic specializations in an endemic cichlid fish lineage from Lake Tanganyika. Proceedings of the National Academy of Sciences USA 96: 10230–10235.

Rüber, L., A. Meyer, C. Sturmbauer & E. Verheyen, 2001. Population structure in two sympatric species of the Lake Tanganyika cichlid tribe Eretmodini: evidence for introgression. Molecular Ecology 10: 1207–1225.

Salzburger, W. & A. Meyer, 2004. The species flocks of East African cichlid fishes: recent advances in molecular phylogenetics and population genetics. Naturwissenschaften 91: 277–290.

Salzburger, W., A. Meyer, S. Baric, E. Verheyen & C. Sturmbauer, 2002a. Phylogeny of the Lake Tanganyika cichlid species flock and its relationships to Central and East African haplochromine cichlid fish faunas. Systematic Biology 51: 113–135.

Salzburger, W., S. Baric & C. Sturmbauer, 2002b. Speciation via introgressive hybridization in East African cichlids? Molecular Ecology 11: 619–625.

Salzburger, W., T. Mack, E. Verheyen & A. Meyer, 2005. Out of Taganyika: genesis, explosive speciation, key-innovations and phylogeography of the haplochromine cichlid fishes. BMC Evolutionary Biology 5: 17.

Sanderson, M. J., 2002. Estimating absolute rates of molecular evolution and divergence times: a penalized likelihood approach. Molecular Biology and Evolution 19: 101–109.

Schelly, R., W. Salzburger, S. Koblmüller, N. Duftner & C. Sturmbauer, 2006. Phylogenetic relationships of the lamprologine cichlid genus *Lepidiolamprologus* (Teleostei: Perciformes) based on mitochondrial and nuclear sequences, suggesting introgressive hybridization. Molecular Phylogenetics and Evolution 38: 426–438.

Schliewen, U. K. & B. Klee, 2004. Reticulate speciation in Cameroonian crater lake cichlids. Frontiers in Zoology 1: 5

Schliewen, U. K., D. Tautz & S. Pääbo, 1994. Sympatric speciation suggested by monophyly of crater lake cichlids. Nature 368: 629–632.

Schliewen, U. K., K. Rassmann, M. Markmann, J. Markert, T. D. Kocher & D. Tautz, 2001. Genetic and ecological divergence of a monophyletic cichlid species pair under fully sympatric conditions in Lake Ejagham, Cameroon. Molecular Ecology 10: 1471–1488.

Scholz, C. A. & B. R. Rosendahl, 1988. Low lake stands in Lakes Malawi and Tanganyika, East Africa, delineated with multifold seismic data. Science 240: 1645–1648.

Scholz, C. A., J. W. King, G. S. Ellis, P. K. Swart, J. C. Stager & S. M. Colman, 2003. Palaeolimnology of Lake Tanganyika, East Africa, over the past 100 k yr. Journal of Palaeolimnology 30: 139–150.

Seehausen, O., 2002. Patterns in fish radiation are compatible with Pleistocene desiccation of Lake Victoria and 14, 600-year history for its species flock. Proceedings of the Royal Society London, Series B 269: 491–497.

Seehausen, O., 2004. Hybridization and adaptive radiation. Trends in Ecology and Evolution 19: 198–207

Seehausen, O., 2006. African cichlid fish: a model system in adaptive radiation research. Proceedings of the Royal Society, Series B 273: 1987–1998.

Seehausen, O. & J. J. M. van Alphen, 1999. Can sympatric speciation by disruptive sexual selection explain rapid evolution of cichlids diversity in Lake Victoria? Ecology Letters 2: 262–271.

Seehausen, O., J. J. M. van Alphen & F. Witte, 1997. Cichlid fish diversity threatened by eutrophication that curbs sexual selection. Science 277: 1808–1811.

Seehausen, O., F. Witte, J. J. M. van Alphen & N. Bouton, 1998. Direct mate choice maintains diversity among sympatric cichlids in Lake Victoria. Journal of Fish Biology 53(Supplement A): 37–55.

Seehausen, O., J. J. M. van Alphen & R. Lande, 1999. Color polymorphism and sex ratio distortion in a cichlid fish as an incipient stage in sympatric speciation by sexual selection. Ecology Letters 2: 367–378.

Sefc, K. M., 2008. Variance in reproductive success and the opportunity for selection in a serially monogamous species: simulations of the mating system of *Tropheus* (Teleostei: Cichlidae). Hydrobiologia (this volume). doi: 10.1007/s10750-008-9563-1

Sefc, K. M., S. Baric, W. Salzburger & C. Sturmbauer, 2007. Species-specific population structure in rock-specialized sympatric cichlid species in Lake Tanganyika, East Africa. Journal of Molecular Evolution 64: 33–49.

Sepulchre, P., G. Ramstein, F. Fluteau, M. Schuster, J.-J. Tiercelin & M. Brunet, 2006. Tectonic uplift and eastern African aridification. Science 313: 1419–1423.

Snoeks, J., 2000. How well known is the ichthyodiversity of the large East African lakes? Advances in Ecological Research 31: 17–38.

Snoeks, J., L. Rüber & E. Verheyen, 1994. The Tanganyika problem: comments on the taxonomy and distribution patterns of its cichlid fauna. In Martens, K., G. Goddeeris & G. W. Coulter (eds), Speciation in Ancient Lakes. Schweizerbart'sche Verlagsdruckerei, Stuttgart: 355–372.

Stauffer, J. R., N. J. Bowers, T. D. Kocher & K. R. McKaye, 1996. Evidence of hybridization between *Cynotilapia afra* and *Pseudotropheus zebra* (Teleostei: Cichlidae) following an intralacustrine translocation in Lake Malawi. Copeia 1996: 203–208.

Stewart, K. M., 2001. The freshwater fish of Neogene Africa (Miocene-Pleistocene): systematics and biogeography. Fish and Fisheries 2: 177–230.

Stiassny, M. L. J., 1991. Phylogenetic interrelationships of the family Cichlidae: an overview. In Keenleyside, M. H. A. (ed.), Cichlid Fishes, Behaviour, Ecology and Evolution. Chapman & Hall, London: 1–35.

Stiassny, M. L. J. & J. S. Jensen, 1987. Labroid interrelationships revisited: morphological complexity, key innovations, and the study of comparative diversity. Bulletin of the Museum of Comparative Zoology 151: 269–319.

Streelman, J. T. & S. A. Karl, 1997. Reconstructing labroid evolution with single-copy nuclear DNA. Proceedings of the Royal Society London, Series B 264: 1011–1020.

Streelman, J. T., R. Zardoya, A. Meyer & S. A. Karl, 1998. Multilocus phylogeny of cichlid fishes (Pisces: Perciformes): evolutionary comparison of microsatellite and single-copy nuclear loci. Molecular Biology and Evolution 15: 798–808.

Streelman, J. T., S. L. Gmyrnek, M. R. Kidd, C. Kidd, R. L. Robinson, E. Hert, A. J. Ambali & T. D. Kocher, 2004. Hybridization and contemporary evolution in an introduced cichlid fish from Lake Malawi National Park. Molecular Ecology 12: 2471–2479.

Sturmbauer, C., 1998. Explosive speciation in cichlid fishes of the African Great Lakes: a dynamic model of adaptive radiation. Journal of Fish Biology 53(Supplement A): 18–36.

Sturmbauer, C. & A. Meyer, 1992. Genetic divergence, speciation and morphological stasis in a lineage of African cichlid fishes. Nature 358: 578–581.

Sturmbauer, C. & A. Meyer, 1993. Mitochondrial phylogeny of the endemic mouthbrooding lineages of cichlid fishes of Lake Tanganyika, East Africa. Molecular Biology and Evolution 10: 751–768.

Sturmbauer, C., E. Verheyen & A. Meyer, 1994. Mitochondrial phylogeny of the Lamprologini, the major substrate spawning lineage of cichlid fishes from Lake Tanganyika in Eastern Africa. Molecular Biology and Evolution 11: 691–703.

Sturmbauer, C., E. Verheyen, L. Rüber & A. Meyer, 1997. Phylogeographic patterns in populations of cichlid fishes from rock habitats in Lake Tanganyika. In Kocher, T. D. & C. Stepien (eds), Molecular Phylogeny of Fishes. Academic Press, San Diego: 97–111.

Sturmbauer, C., S. Baric, W. Salzburger, L. Rüber & E. Verheyen, 2001. Lake level fluctuations synchronize genetic divergence of cichlid fishes in African lakes. Molecular Biology and Evolution 18: 144–154.

Sturmbauer, C., U. Hainz, S. Baric, E. Verheyen & W. Salzburger, 2003. Evolution of the tribe Tropheini from Lake Tanganyika: synchronized explosive speciation producing multiple evolutionary parallelism. Hydrobiologia 500: 51–64.

Sturmbauer, C., S. Koblmüller, K. M. Sefc & N. Duftner, 2005. Phylogeographic history of the genus *Tropheus*, a lineage of rock-dwelling cichlid fishes endemic to Lake Tanganyika. Hydrobiologia 542: 335–366.

Sugawara, T., Y. Terai, H. Imai, G. F. Turner, S. Koblmüller, C. Sturmbauer, Y. Shichida & N. Okada, 2005. Parallelism of amino acid changes at the RH1 affecting spectral sensitivity among deep-water cichlids from Lakes Tanganyika and Malawi. Proceedings of the National Academy of Sciences USA 102: 5448–5453.

Takahashi, T., 2003. Systematics of Tanganyikan cichlid fishes (Teleostei: Perciformes). Ichthyological Research 50: 367–382.

Takahashi, T., 2004. Morphological and genetic distinctness of rock and shell-bed dwelling Telmatochromis (Teleostei, Cichlidae) in the south of Lake Tanganyika suggest the existence of two species. Journal of Fish Biology 65: 419–435.

Takahashi, K. & N. Okada, 2002. Mosaic structure and retropositional dynamics during evolution of subfamilies of short interspersed elements in African cichlids. Molecular Biology and Evolution 19: 1303–1312.

Takahashi, K., Y. Terai, M. Nishida & N. Okada, 1998. A novel family of short interspersed elements (SINEs) from cichlids: the patterns of insertion of SINEs at orthologous loci support the proposed monophyly of four major groups of cichlid fishes in Lake Tanganyika. Molecular Biology and Evolution 15: 391–407.

Takahashi, K., M. Nishida, M. Yuma & N. Okada, 2001a. Retroposition of the AFC family of SINEs (Short Interspersed Repetitive Elements) before and during the adaptive radiation of cichlid fishes in Lake Malawi and related inferences about phylogeny. Journal of Molecular Evolution 53: 496–507.

Takahashi, K., Y. Terai, M. Nishida & N. Okada, 2001b. Phylogenetic relationships and ancient incomplete lineage sorting among cichlid fishes in Lake Tanganyika as revealed by analysis of the insertion of retroposons. Molecular Biology and Evolution 18: 2057–2066.

Takahashi, R., K. Watanabe, M. Nishida & M. Hori, 2007. Evolution of feeding specialization in Tanganyikan scale-eating cichlids: a molecular phylogenetic approach. BMC Evolutionary Biology 7: 195.

Takezaki, N., A. Rzhetsky & M. Nei, 1995. Phylogenetic test of the molecular clock and linearized tree. Molecular Biology and Evolution 12: 823–833.

Terai, Y., K. Takahashi, M. Nishida, T. Sato & N. Okada, 2003. Using SINEs to probe ancient explosive speciation: "hidden" radiation of African cichlids? Molecular Biology and Evolution 20: 924–930.

Terai, Y., N. Takezaki, W. E. Mayer, H. Tichy, N. Takahata, J. Klein & N. Okada, 2004. Phylogenetic relationships among East African haplochromine fish as revealed by short interspersed elements (SINEs). Journal of Molecular Evolution 58: 64–78.

Thorne, J. L., H. Kishino & I. S. Painter, 1998. Estimating the rate of evolution of the rate of molecular evolution. Molecular Biology and Evolution 15: 1647–1657.

Tiercelin, J.-J. & A. Mondeguer, 1991. The geology of the Tanganyika trough. In Coulter, G. W. (ed.), Lake Tanganyika and its Life. Oxford University Press, London: 7–48.

Turner, G. F., 2007. Adaptive radiation of cichlid fish. Current Biology 17: 827–831.

Turner, G. F. & M. T. Burrows, 1995. A model of sympatric speciation by sexual selection. Proceedings of the Royal Society London, Series B 260: 287–292.

Turner, G. F., O. Seehausen, M. E. Knight, C. J. Allender & R. L. Robinson, 2001. How many species of cichlid fishes are there in African lakes? Molecular Ecology 10: 793–806.

Verheyen, E., L. Rüber, J. Snoeks & A. Meyer, 1996. Mitochondrial phylogeography of rock dwelling cichlid fishes reveals evolutionary influence of historic lake level fluctuations in Lake Tanganyika, Africa. Philosophical Transactions of the Royal Society London, Series B 351: 797–805.

Verheyen, E., W. Salzburger, J. Snoeks & A. Meyer, 2003. Origin of the superflock of cichlid fishes from Lake Victoria, East Africa. Science 300: 325–329.

Wainwright, P. C., 2005. Functional morphology of the pharyngeal jaw apparatus. In Shadwick, R. & G. V. Lauder (eds), Biomechanics of Fishes. Elsevier, Chicago: 77–101.

West, K. & E. Michel, 2000. The dynamics of endemic diversification: molecular phylogeny suggests an explosive origin of the thiarid gastropods of Lake Tanganyika. Advances in Ecological Research 31: 331–354.

Wiegman, B. M., D. K. Yeates, J. L. Thorne & H. Kishino, 2003. Time flies, a new molecular time-scale for brachyceran fly evolution without a clock. Systematic Biology 52: 745–756.

Wilson, A. B., M. Glaubrecht & A. Meyer, 2004. Ancient lakes as evolutionary reservoirs: evidence from the thalassoid gastropods of Lake Tanganyika. Proceedings of the Royal Society London, Series B 271: 529–536.

Yamaoka, K., 1978. Pharyngeal jaw structure in labrid fishes. Publications of the Seto Marine Biological Laboratory 24: 409–426.

Hydrobiologia (2008) 615:21–35
DOI 10.1007/s10750-008-9563-1

SPECIATION IN ANCIENT LAKES

Variance in reproductive success and the opportunity for selection in a serially monogamous species: simulations of the mating system of *Tropheus* (Teleostei: Cichlidae)

Kristina M. Sefc

Published online: 10 September 2008

Abstract Sexual selection is believed to play a major role in speciation processes of cichlid fishes in the East African Great Lakes, in particular Lakes Malawi and Victoria, by driving diversification in male nuptial coloration. In Lake Tanganyika, a high rate of color pattern diversification is found in the genus *Tropheus*, where more than 100, mostly allopatric, color morphs have been described. Whether color pattern differentiation in *Tropheus* could follow from sexual selection as well, has not yet been conclusively shown. Unlike typical sexually selected species, *Tropheus* are sexually monomorphic, establish temporary pair bonds and spawn monogamously. Variance in mating success among individuals would still allow for sexual selection, but due to their high population census sizes, parentage data from natural populations are difficult to obtain. Simulations designed according to existing data on mating behavior in *Tropheus* suggested that variance in male reproductive success can be substantial and lead to levels of opportunity for selection, which are similar to values estimated, for example, from natural populations of sexually dimorphic songbirds. Variance in male success was mostly affected by the way females perceived and reacted to differences between males, and to a lesser extent by the duration of components of the reproductive cycles of males and females. These results indicate that future work on the importance of sexual selection in *Tropheus* should concentrate on mate choice cues and female mate choice behavior, but does not depend critically on the acquisition of more detailed life history data.

Keywords Sexual selection · Mating success · Standardized variance in reproductive success · Monogamy · Speciation · Mate choice

Guest editors: T. Wilke, R. Väinölä & F. Riedel
Patterns and Processes of Speciation in Ancient Lakes: Proceedings of the Fourth Symposium on Speciation in Ancient Lakes, Berlin, Germany, September 4–8, 2006

K. M. Sefc (✉)
Department of Zoology, Karl-Franzens University of Graz, Universitätsplatz 2, 8010 Graz, Austria
e-mail: kristina.sefc@uni-graz.at

Introduction

The cichlid fishes of the Great Lakes in East Africa feature an unequaled rate of speciation and are renowned model systems for studies of evolution (e.g., Meyer, 1993; Kornfield & Smith, 2000; Turner et al., 2001; Kocher, 2004; Seehausen, 2006). The oldest of the Great Lakes, Lake Tanganyika, possesses ~250 endemic cichlid species with a high level of morphological and genetic divergence (e.g., Salzburger et al., 2002; Koblmüller et al., 2008). Many of the cichlid species in Lake Tanganyika display geographic variation in coloration (Konings, 1998), most strikingly realized in the over 100, mostly allopatric, color variants within species of the genus *Tropheus*

(Schupke, 2003). *Tropheus* is a member of the tribe Tropheini and closely related to the haplochromine cichlids (Sturmbauer et al., 2003; Salzburger et al., 2005), which make up the species-rich flocks of Lakes Malawi and Victoria. Intra- and inter-specific diversification of male nuptial coloration in the lacustrine Haplochromini has been attributed to sexual selection (e.g., Seehausen & van Alphen, 1999; Knight & Turner, 2004), and it has been suggested that sexual selection is involved in the evolution and maintenance of color morphs in *Tropheus* (Meyer et al., 1994; Salzburger et al., 2006).

However, *Tropheus* do not display all of the traits that are generally associated with sexual selection. First, *Tropheus* are sexually monomorphic, and both genders exhibit concurrent repertoires of context-dependent color patterns signaling motivation and social status employed, for example in territory defense (Wickler, 1969; Nelissen, 1976; Sturmbauer & Dallinger, 1995). Second, *Tropheus* are maternal mouthbrooders and female investment in reproduction is high (Schürch & Taborsky, 2005), but males also invest by allowing their mates to feed from their territory for a period of stable pair-bonding prior to spawning (Yanagisawa & Nishida, 1991). After spawning, females abandon their mates to mouthbrood and eventually establish or regain their own feeding territories. Genetic analyses of mouthbrooding *Tropheus* mothers and their offspring demonstrated that clutches are sired by a single male each (Egger et al., 2006), and variance in male reproductive success is therefore not affected by extra-pair spawning (e.g., Jones et al., 2001; Singer et al., 2006). Third, although positive color-assortative mating between highly distinct *Tropheus* morphs has recently been demonstrated in a population consisting of artificially admixed morphs, reproductive isolation was weak between less distinct morphs (Salzburger et al., 2006). Similar results were obtained in laboratory mate choice experiments, where morphs differed in the strength of assortative preferences, and discrimination between homotypic and heterotypic individuals decreased with decreasing color differentiation between morphs (Egger et al., 2008, and in prep.). However, field observations indicated that a male's mating success also depends on the size and quality of his territory (Yanagisawa & Nishida, 1991). This is particularly plausible, because females apparently need the resources obtained by heavy feeding from their mates' territories to mature their eggs and become ready for spawning (Yanagisawa & Nishida, 1991). It is conceivable that competition for territories favors conspicuous dominant coloration in *Tropheus*, in particular, because body color intensity is strongly enhanced during aggressive interactions (Sturmbauer & Dallinger, 1995). Hence, if, for example, female mate preferences are influenced by male territory quality, and successful defense of a valuable territory is correlated with differences in color traits among individuals, sexual selection on coloration may act through social selection for territorial success.

Irrespective of the exact cues employed in mate choice, selection must be associated with variance in reproductive success among the individuals of the selected sex, and the opportunity for selection (*I*), setting the upper limit for the intensity of selection, can be expressed as the standardized variance in mating and reproductive success (Wade & Arnold, 1980; Fairbairn & Wilby, 2001). Monogamous mating entails evenly distributed mating success among individuals, but in a serially monogamous system with change of partners between successive spawnings, variance may be introduced by repeated failure or success to secure a mate. The variance in reproductive success in natural populations is estimated via molecular genetic parentage analyses, which depend crucially on exhaustive sampling of juveniles and adults in the population. This difficulty may be one of the reasons why the majority of published studies are concerned with bird and mammal species, where offspring and candidate parents can often be sampled from the nest or from a small herd (e.g. references in Table 4). Although *Tropheus* populations display high genetic structure on a small geographic scale (Sefc et al., 2007), their high census size, with more than 100 adult *Tropheus* counted in a 20 × 20 m square transect (Sturmbauer et al., 2008), precludes the exhaustive sampling required for paternity analyses. However, extensive field work provided detailed information on mating and breeding behavior of *Tropheus* (Yanagisawa & Sato, 1990; Yanagisawa & Nishida, 1991; Sturmbauer & Dallinger, 1995), which can be employed to simulate mating in a breeding population. Modeling and computer simulations have been already proved as valuable tools to examine the effects of behavioral and environmental variables on the opportunity for selection and on operational sex ratios (OSR) (Webster et al., 1995; Jones et al., 2001; Kvarnemo &

Merilaita, 2006). In the present study, reproductive success was simulated under the assumption that individuals differ in a trait, which determines the duration of their reproductive cycle, clutch size, and for males, their attractiveness to potential female mating partners. This trait can, for example be interpreted as one that is associated with success in the acquisition of valuable territories containing rich food resources. Males with high-quality territories would be more attractive to females, but in addition, the duration of the burst feeding period needed for the maturation of the female would be reduced. Hence, males with good trait values would also benefit from investing less time in each reproductive effort. For females, the ability to secure their own high-quality feeding territories would improve their nutritional status, reduce the time needed for recovery from mouthbrooding and enhance their fertility.

The simulations assess the level of opportunity for selection that can be reached in the monogamous mating system, and investigate the importance of variables for which only limited field-based knowledge exists. Thus, the results also provide guidance for further fieldwork to address those factors, which are identified as the most crucial determinants of the variance in mating and reproductive success.

Methods

Tropheus life history data

Reproductive cycles of male and female *Tropheus* have been documented in detailed field studies. For a period of a few days, a female inspects several males before choosing her mate. Males usually court visiting females. Once she has made her choice, she moves into the male's territory and feeds heavily under his protection, which provides her with the energy required to mature her eggs (Yanagisawa & Nishida, 1991). In the field, the duration of stable pairing varied considerably among pairs, ranging from 4 to 21 days (Yanagisawa & Nishida, 1991). Brood sizes range from 2 to 14 eggs. Genetic data indicated one sire per brood, such that each female probably spawns exclusively with her social mate (Egger et al., 2006). Upon spawning, the female leaves the male and mouthbroods her eggs without feeding, whereby she suffers substantial enfeeblement. The brooding period from spawning to independence of fry averaged 33.2 days in a sample of seven *Tropheus moorii* females, with little variance among females (SD = 1.1; Yanagisawa & Sato, 1990). After release of the fry, the female either resumes the defense of her previous feeding territory or establishes a new territory. Data from a field study by Yanagisawa & Sato (1990) suggest long spawning intervals for females: ~50% of the females spawned only once in a survey period of 4–5 month, one female spawned twice within that period, and the other females didn't spawn at all during that period.

Simulation procedure

Reproductive success of males and females was simulated under parameter combinations representing different operational sex rations (OSR) and female mate choice behaviors. Mating populations comprised 200 males and 200 females, and were simulated for 1000 days. The 1:1 adult sex ratio corresponds to observations in the field (Yanagisawa & Nishida, 1991). Table 1 lists parameters and parameter values used in the simulations. The simulations were programmed in perl v5.8.2 (included in the cygwin package, and run under Windows NT).

Design of male and female reproductive cycles

Reproductive cycles of males and females were designed according to the field data cited above. The reproductive cycle of males was divided into the mating phase, during which the male was available for pairing, and a time-out phase (Kvarnemo & Merilaita, 2006), which comprised the time paired with a female (paired-time), and an enforced solitary period after break-up of the previous pair bond, allowing males to court visiting females. Time-out in the female cycle consisted of paired-time, a period of mouthbrooding, and a solitary recovery period. The recovery period also included the time the female takes to visit and inspect males, such that, at the end of her recovery period, she can choose among those males that have been solitary for the enforced male solitary period or longer.

Individual trait values

Males and females were assigned values of a 'quality trait', which determined the duration of components

Table 1 Parameters entered in the simulations

Parameter	Explanation	Settings
Underlying mating trait		
Trait value distribution[a]	Distribution, from which values representing individual quality trait scores are drawn	χ^2, normal (assumed; represents two distinct alternatives, as the distributions underlying fertility and mating success are not known; see text)
Reproductive cycles		
Paired-time	Duration of the pre-spawning pair bond between males and females	Min. 4-max. 21 days (based on field data, Yanagisawa & Nishida, 1991)
Enforced male solitary period	Period after pair-bonding, during which males are required to remain solitary to receive and court visiting females	5 days (educated guess based on observation that females inspect potential mates for a few days, Yanagisawa & Nishida, 1991)
Mouthbrooding period	Time during which the female mouthbroods her eggs and fry	33 days (based on field data, Yanagisawa & Sato, 1990)
Female recovery period[a]	Time needed by the female to recover from mouthbrooding, until she is ready to mate again	Max. 60 days, and max. 150 days, for simulations of short and long recovery, respectively (based on field data, Yanagisawa & Sato, 1990)
Mate choice and reproduction		
Mate sample size[a]	Number of males visited and assessed for their mate quality by females before their mating decision	5, 20 (assumed; no field data available)
Female discrimination ability[a]	Factor determining the ability of females to discriminate between small differences in male quality trait values and to choose the best available mate; zero discrimination ability corresponds to random mate choice	0, 1, 2 (assumed; exponent for male trait value, see text)
Clutch size	Number of offspring produced by each spawning	Min. 2-max. 14 fry (based on field data, Egger et al., 2006)

Minimum or maximum values are given for variables that were influenced by trait values

[a] Settings that varied between simulation runs

See *Methods* for details on rationales and simulation procedures

of their reproductive cycles, clutch size, and for males, their attractiveness to females. This trait could for example be associated with the individuals' ability to acquire and defend high-quality territories. Mapping of *Tropheus* territories revealed considerable variance of territory size between individuals, but sample sizes were too low to infer the underlying distribution (Yanagisawa & Nishida, 1991; Sturmbauer & Dallinger, 1995; Sturmbauer et al., 2008). Male body size, a trait which might be expected to be normally distributed, was not correlated with territory size and mating success in the field (Yanagisawa & Nishida, 1991). Body coloration is an important component of communication between *Tropheus* individuals and is a very likely determinant of territorial and reproductive success, but there are no data on the color pattern variation within populations. Because of the lack of information, upon which the choice of the trait value distribution for the simulations could be based, two alternative distributions were tested. Trait values were drawn randomly from a normal distribution (mean = 20, SD = 5), and from a χ^2 distribution with 3 degrees of freedom to obtain a skewed distribution with high trait values in only a few individuals (Fig. 1). Individual trait values remained unchanged during the duration of the simulated period. This corresponds to observations of Yanagisawa & Nishida (1991) that males remained in their territory even if they never received visits from females. To make different simulations comparable, trait values were drawn only once, and then used for replicate runs and different combinations of

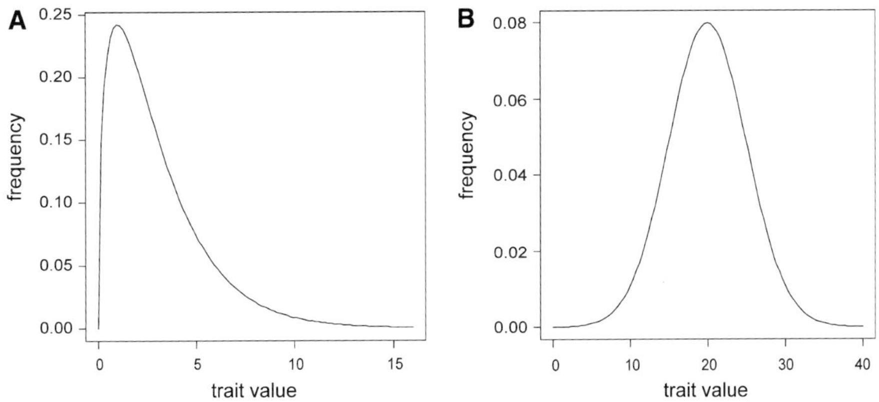

Fig. 1 χ^2 (**A**) and normal distributions (**B**), from which individual trait values were drawn

the other parameters. The assigned trait values had a coefficient of variance of CV = 1.21 in the normal distribution, and CV = 1.77 in the χ^2 distribution.

Trait values determined a male's chance to be chosen as mate, and influenced the duration of paired-time and female recovery time in different individuals, and the number of eggs per spawning. The following assumptions were made: a male with a high-quality territory offers a rich food resource to his mate, such that she can feed efficiently and her eggs mature more quickly than if she had paired with an inferior male. Similarly, females with high trait values will succeed in securing superior territories during their recovery phase, and will sooner be ready to mate than inferior females. Additionally, superior females were supposed to produce more viable offspring per spawning. Therefore, high trait values reduce paired-time and recovery time and increase the number of offspring.

To modify an individual's reproductive cycle and fertility according to its trait value, each individual's value was first expressed as a proportion of the maximum trait value in the breeding population (i.e., as relative trait value, 0–1). For each male, the defined maximum paired-time (21 days) was then reduced by the product of (0.8 × maximum paired time × relative trait value of this male), and the result was rounded to the nearest integer. Consequently, individual paired-time ranged from 4 to 21 days and was negatively correlated with individual trait values. Similarly, individual female recovery time was obtained by reducing the maximum recovery time (60 or 150 days) by the product of (0.5 × maximum recovery time × relative trait value of this female), and ranged from 30–60 and 75–150 days in simulations of fast and slow recovery, respectively. Individual female clutch sizes ranged from 2 to 14 and were derived from a defined minimum clutch size of 2 by adding the product of (6 × minimum clutch size × relative female trait value) and rounding off to the nearest integer.

Start conditions of the simulation

Mating in *Tropheus* is not synchronized, and mouthbrooding females, paired individuals and solitary territory defenders can be found at the same time. Hence, the simulated individuals had to be assigned different stages in their reproductive cycle at the beginning of the simulation. To create starting conditions, all males were ready for mating, and each female was randomly assigned a position in her reproductive cycle. If she was either within the maximum paired-time, had left a previous mate <5 days ago (the male's enforced solitary period), or was ready to mate, a male was chosen to be (or have been) her mate by the below described procedure, and was assigned the matching position within his paired-time or enforced solitary period. Mating success during this procedure was not scored for analyses.

Simulation of mating decisions

When a female had completed her recovery time, she inspected a number of males (defined by the parameter 'mate sample size') and chose a mate from this sample. The mate sample was drawn randomly from all males in mating phase (i.e., neither pairing nor in enforced solitary status). Yanagisawa & Nishida (1991) report that females visited several males before formation of a stable pair, but numbers of mate sample sizes in the field are not available and may be difficult to assess, because a female may not need to conspicuously inspect a male's territory to evaluate his quality. To assess the influence of female knowledge on male mating success, two different mate sample sizes (5 and 20) were analyzed.

Within a mate sample, each male's share in the summed trait values of the sample equaled his probability to be chosen by the female. Thus, female mate evaluation was not assumed to be infallible, and males with similar trait values had similar chances to be selected. Different degrees of female discrimination ability were simulated by raising the male trait values to the power of a defined 'discrimination' value (0, 1, or 2). Zero discrimination simulated a situation where all males are equally likely to be chosen for mating, whereas the discrimination value of 2 increased the differences between male trait values, and hence increased the mating probability for males with higher trait values relative to inferior males. Operational sex ratios were calculated as the proportion of males among all individuals ready to mate (Kvarnemo & Ahnesjö, 1996) on each day when at least one female was ready to mate, and averaged across the simulated period. Upon mate choice, the pair-bonding duration for the pair was determined as the average of the individual paired-time of the male and the female. At the end of their paired-time period, the female produced the number of offspring defined by her individual clutch size value.

The spatial structure of territories was neglected in the simulations. Yanagisawa & Nishida (1991) report that females visit males' beyond their immediate neighborhood and that females do not always return to their previous territory after break-up of the current pair bond. Hence, although the spatial structure of male territories may persist for long periods of time, individual neighborhoods change through the movements of females, justifying random composition of the mate samples.

Statistical analysis

Number of mates and number of offspring were recorded for males and females to summarize individual mating success and reproductive success, respectively, over the simulated period. The opportunity for selection I was calculated as the variance in reproductive success, i.e., number of offspring produced, divided by the square of the mean. Additionally, the variance in mating success, i.e., number of mates, was standardized by the squared mean to obtain an estimate for the opportunity for sexual selection I_S (e.g., Wade & Arnold, 1980; Whittingham & Dunn, 2005). The correlations of I with the four tested factors—trait value distribution, female recovery time, mate sample size, and discrimination ability—were analyzed in SPSS by calculating variance components and the proportion of total variance explained by each factor and their interactions.

Results

Operational sex ratios

Although males and females form stable pair bonds prior to spawning, the mating and brood care behavior of *Tropheus* involves longer time-out phases of females than of males. In the simulations, the time-out periods of males ranged between 9 and 26 days and consisted of the duration of the stable pair bond (4 to 21 days) and an additional 5-day solitary period reserved for court-visiting females. For females, the simulated time-out period included the duration of the pair bond (4 to 21 days), a mouthbrooding period of 33 days, and a recovery period. Spawning intervals vary considerably among females and last up to several months (Yanagisawa & Sato, 1990). The simulations compared the effect of fast recovery (30–60 days) and slow recovery (75–150 days) on operational sex ratios and the opportunity for selection. Time-out of females lasted from 67 to 114 days assuming fast recovery, and 112 to 204 days assuming slow recovery. The resulting OSR was strongly male-biased in all simulations. Averaged across days and replicates, 98.47 to 99.08% of the individuals ready to mate were males, assuming fast and slow female recovery, respectively (Table 2). In addition to the strong impact of female recovery time on OSR, there were significant effects of the trait value distribution and the interaction between recovery time and trait value distribution (Fig. 2A).

Opportunity for selection on males in the simulated mating system of *Tropheus*

The opportunity for selection, I was calculated as the variance of the total number of offspring per male standardized by the squared mean number of

Table 2 Operational sex ratio (OSR) in simulations of the *Tropheus* mating system in comparison to values from simulations of a polygynous system

Mating system	Recovery	Trait value distribution	
		χ^2	Normal
Tropheus (pair-bonding)	Fast	98.61	98.47
	Slow	99.08	99.07
Polygynous	Fast	98.73	98.60
	Slow	99.14	99.07

Recovery time and trait value distribution were the only parameters affecting OSR values, and values were averaged across replicates of simulations with different mate sample size and discrimination settings

offspring across males. In the simulations, variance in reproductive success among males resulted from both direct female mate choice, and from variance in territorial success among males and females, as the possession of a high-quality territory was associated with higher reproductive output in both genders. Consequently, the trait value distribution was the most important determinant of I, followed by female discrimination ability, female recovery time, and mate sample size, with significant contributions by interactions between the factors (Fig. 2B). In the simulations assuming normal distributed trait values, discrimination ability and recovery time had a much stronger effect on I than mate sample size, which did not contribute to the total variation (Fig. 2C). In contrast, with χ^2 distributed traits, both female recovery time and mate sample size had a smaller impact on I than discrimination ability (Fig. 2D).

Values of I ranged from 0.09 to 0.29 in the simulations with normal distributed trait values, and from 0.30 to 1.31, when trait values where χ^2 distributed (Fig. 3). Scaled variance in mating success (I_S, based on number of mates per male) was similar to I (I_S = 0.09–0.28 for normal distributed

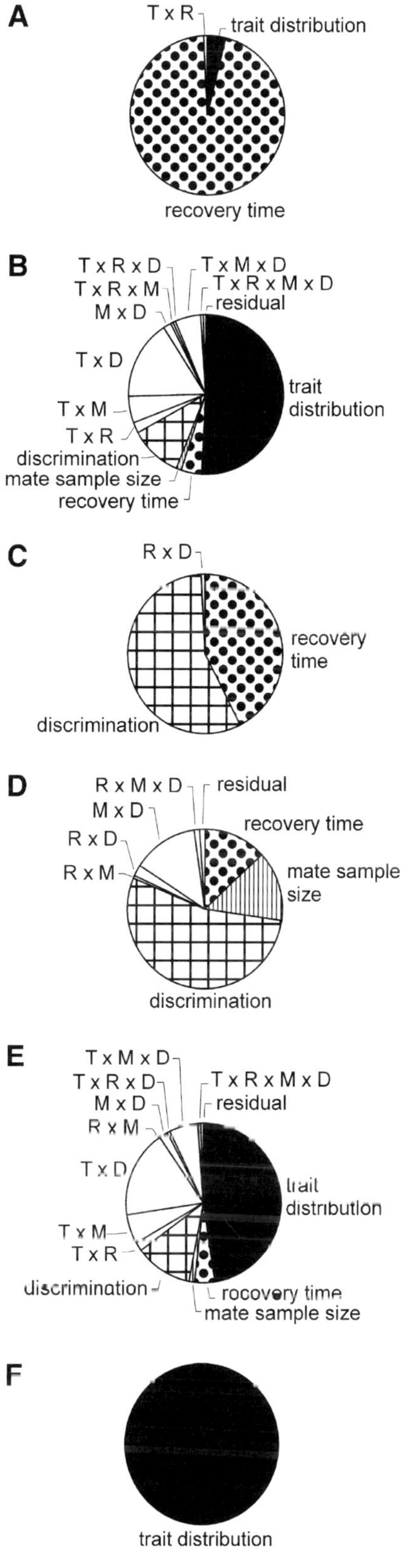

Fig. 2 The effects of trait distribution, female recovery time, mate sample size, and discrimination ability on (**A**) operational sex ratios, (**B**) the opportunity for selection on males based on offspring numbers, (**C**) the opportunity for selection on males based on offspring numbers, when traits are normally distributed, (**D**) the opportunity for selection on males based on offspring numbers, when traits are χ^2 distributed, (**E**) the opportunity for selection on males based on number of mates, and (**F**) opportunity for selection on females based on offspring numbers. The pie charts illustrate the proportion of total variance explained by each factor (shaded fields) and by factor interactions

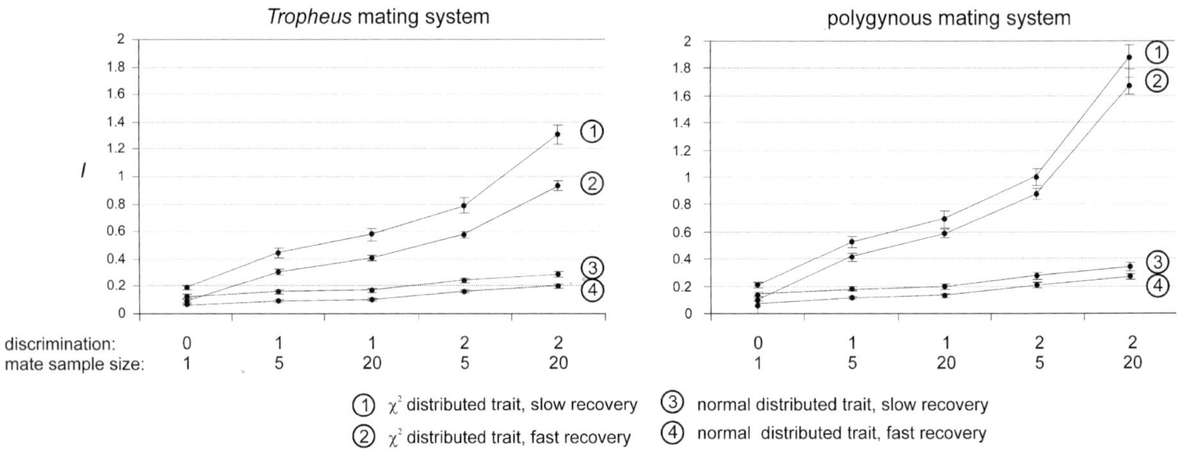

Fig. 3 Opportunity for selection *I* in the *Tropheus* mating system and in a polygynous mating system without pair bonding. Lines labeled 1 to 4 connect values obtained with different combinations of trait distributions and female recovery times for increasingly selective female mate choice. The simulations with a discrimination factor of zero and a mate sample size of 1 correspond to random mate choice. Simulation results are shown as average *I* values with error bars representing standard deviations over 100 replicates

traits, and $I_S = 0.28$–1.26 for χ^2 distributed traits), and determined by the same variables, i.e., trait value distribution had the largest effect, followed by female discrimination ability, female recovery time, and mate sample size, with significant contributions of factor interactions (Fig. 2E). When females performed well in identifying high-quality males (20 males sampled, discrimination factor of 2), the highest quality male was able to secure up to 3.4% of all matings, and more than 20% of all males remained without any mates (Fig. 4).

The above model makes the assumption that male attractiveness and the reproductive cycle and fertility of males and females are determined by the same trait, for example territorial success, which is not necessarily true. An additional set of simulations examined reproductive success, when the trait serving as cue for female mate choice (i.e., male attractiveness) was not correlated with the trait influencing time-out periods and fertility. Decoupling the female mate choice cues from reproductive traits yielded almost the same *I* values as the first model, when male attractiveness was χ^2 distributed, and reduced *I* by 6% when male attractiveness was normal distributed (Table 3). A third set of simulations assumed that males differed only in mate choice cues, whereas all individuals had the same time-out duration and fertility. Here, the reduction of *I* was by 13 and 7% for χ^2 and normal distributed male attractiveness, respectively, compared to the first model (Table 3). Hence, female mate choice alone can produce most of the variance in male reproductive success.

For comparison: opportunity for selection on males in a simulated polygynous mating system

The investment in reproduction, which male *Tropheus* accept by pairing with a female for a period of time prior to spawning, was considered to diminish the variance in male mating and reproductive success, and therefore the strength of sexual selection on males (Egger et al., 2006). Simulations of a polygynous mating system, which was congruent with the above simulations in all respects except that males did not contract pair bonds (time-out = 1 day), were conducted to evaluate the magnitude of the reduction of *I* incurred by male–female pair bonding. OSR values were only slightly higher than in the *Tropheus* mating system (98.60–99.14%, Table 2), and the values of *I* in the polygynous system ranged from 0.12 to 0.34 in the simulations with normal distributed trait values, and from 0.41 to 1.85, when trait values where χ^2 distributed (Fig. 3). Under simulated conditions, pair-bonding never reduced the opportunity for selection by more than 50%.

Opportunity for selection on females

In the simulated model, variance in reproductive success between females was mostly due to

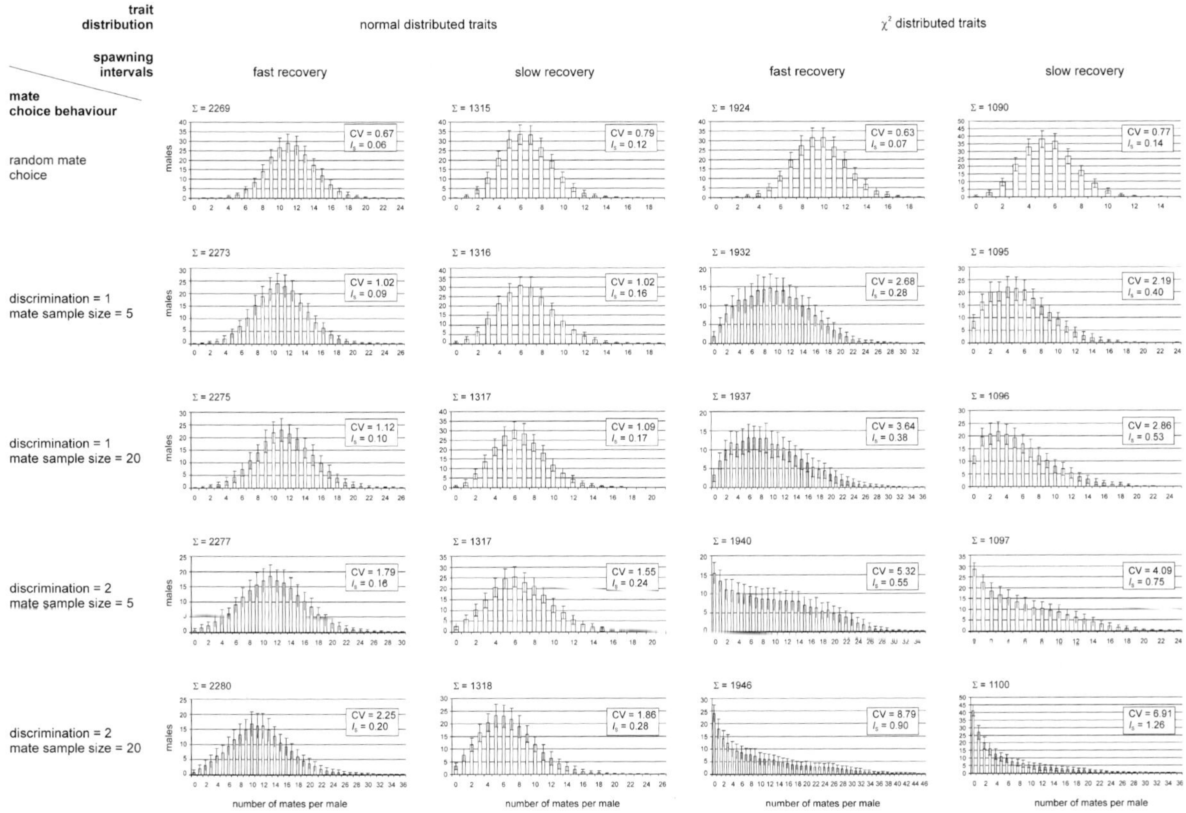

Fig. 4 Frequency distribution of mating success among males obtained in simulations of different components of female mate choice, spawning intervals (determined by female recovery times) and trait distributions. Σ, total number of matings; CV, coefficient of variance of male mating success; I_S, standardized variance in male mating success. Average mating success and standard deviations across 100 replicate simulations are shown

differences in brood size, and consequently, the main determinant of the scaled variance of female reproductive success I was the trait value distribution (Fig. 2F). I values for females were 0.38 and 0.43 with χ^2 distributed trait values, and 0.06 and 0.08 with normal distributed traits, for fast and slow recovery, respectively. In contrast, there was very little variance in mating success ($I_S = 0.008$ and 0.015 with χ^2 distributed trait values, and 0.005 and 0.010 with normal distributed traits, for fast and slow recovery, respectively).

Discussion

Simulations of mating and reproduction according to behavioral observations of *Tropheus* suggest that temporary pair-bonding between males and females does not dispose of the potential for considerable variance in reproductive success between males. Although males are not available for mating during and shortly after their pair-bonding period, the OSR is strongly male biased, and females can choose mates from a large number of males at any time. Consequently, the highest variance was achieved, when male quality had a skewed distribution and females were able to identify and mate with the best males. In this case, the opportunity for sexual selection on males in the simulated *Tropheus* mating system was only slightly over 50% of that achieved in a polygynous system without pair-bonding. In contrast, when simulation conditions did not allow for high variance, i.e., when male quality was normal distributed or when females were unable to recognize the best males, there was little difference in the opportunity for selection between a polygynous system and the *Tropheus* mating system (Fig. 3). A comparison with the opportunity for sexual selection (I) estimated from natural populations (Table 4) shows that the simulated values ($I = 0.09$–1.31) are in the range of those found in different species

Table 3 Opportunity for selection estimated in simulations of the *Tropheus* mating system (TMS) in comparison to values from simulations assuming that mate choice cues are not correlated with fertility and duration of the reproductive cycle (UNCORR), and to values from simulations assuming that males differ only in mate choice cues and individuals have identical fertility and reproductive cycles (MCC only)

Recovery time	Discrimination ability	Mate sample size	Opportunity for selection (I)		
			TMS	UNCORR	MCC only
χ^2 distribution					
Long	Low	5	0.304	0.311	0.257
Long	Low	20	0.404	0.410	0.347
Long	High	5	0.578	0.575	0.511
Long	High	20	0.930	0.896	0.806
Short	Low	5	0.445	0.457	0.385
Short	Low	20	0.578	0.590	0.504
Short	High	5	0.790	0.799	0.709
Short	High	20	1.307	1.269	1.165
Normal distribution					
Long	Low	5	0.094	0.085	0.087
Long	Low	20	0.102	0.097	0.094
Long	High	5	0.161	0.149	0.148
Long	High	20	0.201	0.182	0.182
Short	Low	5	0.162	0.156	0.152
Short	Low	20	0.172	0.168	0.164
Short	High	5	0.241	0.235	0.227
Short	High	20	0.289	0.276	0.270

In the latter simulation (MCC only), the recovery time of all females was set to 53 and 133 days for fast and slow recovery, respectively, for comparison with data under the χ^2 distribution, and to 43 and 107 for the normal distribution. Values were averaged across replicates of simulations

of birds and in the pumpkinseed sunfish, but much lower than values obtained from highly polygynous mammals. The largest number of studies was conducted on songbirds, where I values ranged from 0.10 to 1.02 with a mean of 0.46 in sexually monomorphic species, and were only slightly higher ($I = 0.14$–2.8, mean $= 0.69$) in sexually dimorphic species, although sexual dimorphism is considered to indicate the presence of sexual selection on the ornamented sex (e.g., Barraclough et al., 1995). It must be noted that comparison of I values across different studies are somewhat compromised by methodological differences, such as consideration of non-mated adults and the time-span for which reproductive success is measured. Additionally, variance estimates from the same population can vary considerably within (Lindström, 2001) and between breeding seasons (e.g., Madsen & Shine, 1993; Friedl & Klump, 1999; Coltman et al., 2002; Prosser et al., 2002), and the most relevant measure with respect to evolutionary implications is certainly the variance in life-time reproductive success (e.g., Weatherhead & Boag, 1997; Coltman et al., 1999).

In contrast to the body of data concerning variance among males, only few studies addressed the variance between females. In male-pregnant seahorses (*Hippocampus subelongatus*), sexual selection acts on females, and accordingly, I_S values were 6 times higher in females than in males (Kvarnemo et al., 2007). Similarly, in a water snake, where sexual selection on males follows from competition for females, I was more than 5 times higher in males than in females (Prosser et al., 2002). Higher opportunity for sexual selection on males than on females was also found in experimental breeding populations of the sand goby (*Pomatoschistus minutus*), a species with paternal brood care (Singer et al., 2006). Studies on species with biparental and female-biased

Table 4 Estimates of the opportunity for selection (I, based on number of offspring) in wild populations

Species	Dimorphism	I (I_S) males	I (I_S)[a] females	Comments; references in ()
Fishes				
Pumpkinseed sunfish *Lepomis gibbosus*	Females paler	1.20	0.99	(1)
Western Australian seahorse *Hippocampus subelongatus*	Males with longer tails	0.27[a]	1.59[a]	Male-pregnant; (2)
Birds				
Common yellowthroat *Geothlypis trichas*	Males with black mask	0.48	n.a.	(3)
Hooded warbler *Wilsonia citrina*	Males with black hood and throat	0.46	n.a.	(4)
Yellow warbler *Dendroica petechia*	Males with rusty breast streaks	0.53	n.a.	(5)
Black-throated blue warbler *Dendroica caerulescens*	Males with blue/black/white pattern, females brown	0.68 and 0.74	0.53 and 0.60	Data from two seasons; (6)
Great reed warbler *Acrocephalus arundinaceus*	–	1.02	n.a.	(7)
House wren *Troglodytes aedon*	–	0.22	n.a.	(3)
Red-winged blackbird *Agelaius phoenicus*	Males black with red epaulets, females brown	1.36	1.22	Lifetime success; (8)
		0.39	n.a.	One season; (9)
Bullock's oriole *Icterus galbula bullockii*	Males with orange/black pattern, females brown	0.17	n.a.	(10)
Brown cowbird *Molothrus ater*	Males black with brown head, females dull brown	2.80	1.75	Brood parasitic; (11)
Blue tit *Parus caeruleus*	Dimorphic UV reflectance	0.28	n.a.	(12)
Black-capped chickadee *Poecile atricapillus*	–	0.10	n.a.	(13)
Barn swallow *Hirundo rustica*	–	0.49	n.a.	(14) As cited in (11)
House martin *Delichon urbica*	–	0.31	n.a.	(15)
Purple martin *Progne subis*	Slightly dimorphic	0.33	n.a.	(14) As cited in (11) and (16)
		0.25	n.a.	(17)
Tree swallow *Tachycineta bicolor*	–	0.79	n.a.	(18)
Dark-eyed junco *Junco hyemalis*	Males with black head	0.72	0.51	(19) As cited in (11), (16) and (20)
Red bishop *Euplectes orix*	Males orange-red and black, females brownish	0.59–1.25	n.a.	Data from four seasons; (21)
Bluethroat *Luscinia svecica svecica*	Males with more blue on throat than females	0.37	n.a.	(22)
Collared flycatcher *Ficedula albicollis*	Males with contrast-rich black/white pattern, females duller	0.14	n.a.	(23)
Rosefinch *Carpodacus erythrinus*	Males with red pattern, females brown	0.40	n.a.	(16)
Savannah sparrow *Passerculus sandwichensis*	–	0.48	n.a.	(20)
Mammals				
Greater horseshoe bat *Rhinolophus ferrumequinum*	Female-biased size dimorphism	3.70	n.a.	(24)
Fruit bat *Cynopterus sphinx*	Size dimorphism	4.76	n.a.	Two seasons; (25)
Harbour seal *Phoca vitulina concolour*	Slight male-biased size dimorphism	3.99	n.a.	Calculated from (26)
Pacific harbour seal *Phoca vitulina richardsi*	Slight male-biased size dimorphism	14.29	n.a.	Calculated from (27)

Table 4 continued

Species	Dimorphism	I (I_S) males	I (I_S)[a] females	Comments; references in ()
Soay sheep *Ovis aries*	Male-biased size dimorphism	3.46	n.a.	Lifetime success; (28)
Bighorn sheep *Ovis canadensis*	Male biased size dimorphism, horn dimorphism	4.52	n.a.	Six years; (29)
Red deer *Cervus elaphus*	Male-biased size dimorphism	5.97	n.a.	(30) As cited in (24)
Yellow-pine chipmonk *Tamias amoenus*	Female-biased size dimorphism	1.243	0.928	(31)

[a] Values from one study considering number of mates (I_S)

n.a., no data available

(1) Rios-Cardenas, 2005; (2) Kvarnemo et al., 2007; (3) Whittingham & Dunn, 2005; (4) Strutchbury et al., 1997; (5) Yezerinac et al., 1995; (6) Webster et al., 2001; (7) Hasselquist et al., 1995; (8) Weatherhead & Boag, 1997; (9) Gibbs et al., 1990; (10) Richardson & Burke, 2001; (11) Woolfenden et al., 2002; (12) Kempenaers et al., 1992; (13) Otter et al., 1998; (14) Birkhead & Møller, 1998; (15) Whittingham & Lifjeld, 1995; (16) Albrecht et al., 2007; (17) Wagner et al., 1996; (18) Kempenaers et al., 2001; (19) Ketterson et al., 1997; (20) Freeman-Gallant et al., 2005; (21) Friedl & Klump, 1999; (22) Johnsen et al., 2001; (23) Sheldon & Ellegren, 1999; (24) Rossiter et al., 2006; (25) Storz et al., 2001; (26) Coltman et al., 1998; (27) Hayes et al., 2006; (28) Coltman et al., 1999; (29) Coltman et al., 2002; (30) Marshall, 1998; (31) Schulte-Hostedde et al., 2004

investment in brood care report 10–30% lower I values in females than in males, and I was 40% lower in females than in males of the brood parasitic Brown cowbird (Table 4). In the simulations of the *Tropheus* mating system, all females were assumed to mate after completion of their assigned reproductive cycle, and variance in female reproductive success was mainly due to differences in clutch size and spawning intervals. Consequently, I was substantially lower in females than in males in all simulations except those with the lowest variance in male success. In order to generate more balanced selection pressures on both sexes, the model would have to increase social competition for reproduction-related resources among females, and/or introduce mating success variance through male mate choice. In black swans, an ornamental trait expressed in both males and females was shown to be involved in mutual sexual selection and in social dominance over individuals of the same sex (Kraaijeveld et al., 2004). Similar mechanisms could be responsible for the sexually monomorphic expression and between-population differentiation of conspicuous color patterns in *Tropheus*. Male body size was not correlated with territory size and mating success in the field (Yanagisawa & Nishida, 1991), and alternatively, color pattern intensity may be crucially involved in the determination of individual reproductive success.

In the simulations, variance in male reproductive success was most strongly determined by two components of female mate choice behavior, i.e., the distribution of mate choice cues among males and the ability of the females to discriminate small differences in cue quality. In comparison, female spawning intervals and the number of males assessed as potential mates had a smaller, albeit significant, effect on male variances. Furthermore, quality-dependent variability in paired-time added only little variance to male reproductive success, and variation in female time-out periods and fertility created only low variability in female reproductive success. These results are encouraging for future work addressing the role of social and sexual selection in color diversification in *Tropheus*, since data on spawning intervals, duration of pair-bonding periods, and the number of males assessed by females as potential mates are difficult to obtain, whereas mate choice cues and female discrimination ability can be tested experimentally (e.g., Maan et al., 2004). The variance of male mate quality in the population must be measured in the field after the traits relevant for reproductive success have been identified. Territory size was shown to influence male mating success in *Tropheus* (Yanagisawa & Nishida, 1991), but the association between coloration and territorial success has to be tested and cannot be assumed a priori (see, e.g., Maan et al., 2006). Social selection on color patterns among females and male mate choice (Amundsen & Forsgren, 2001; Werner & Lotem, 2003; Kraaijeveld et al., 2004) will have to be examined to address the evolution of sexually monomorphic color pattern diversification, particularly interesting because sexual monomorphism is a derived trait in *Tropheus*.

Acknowledgments I would like to thank the organisers of SIAL IV (Berlin, 2006) for an outstanding symposium and the editors for compiling this special issue. Many thanks to Stephan Koblmüller, Steven Weiss and Bernd Egger for comments on the analyses and the manuscript, and to two anonymous reviewers for their suggestions regarding data presentation. Financial support for research on *Tropheus* by the Austrian Science Fund (Grant P17380-B06) is thankfully acknowledged.

References

Albrecht, T., J. Schnitzer, J. Kreisinger, A. Exnerová, J. Bryja & P. Munclinger, 2007. Extrapair paternity and the opportunity for sexual selection in long-distant migratory passerines. Behavioural Ecology 18: 477–486.

Amundsen, T. & E. Forsgren, 2001. Male mate choice selects for female coloration in a fish. Proceedings of the National Academy of Sciences USA 98: 13155–13160.

Barraclough, T. G., P. H. Harvey & S. Nee, 1995. Sexual selection and taxonomic diversity in passerine birds. Proceedings of the Royal Society of London Series B 259: 211–215.

Birkhead, T. R. & A. P. Møller, 1998. Sperm competition and sexual selection. Academic Press, San Diego, Calif.

Coltman, D. W., W. D. Bowen & J. M. Wright, 1998. Male mating success in an aquatically mating pinniped, the harbour seal (*Phoca vitulina*), assessed by microsatellite DNA markers. Molecular Ecology 7: 627–638.

Coltman, D. W., M. Festa-Bianchet, J. T. Jorgenson & C. Strobeck, 2002. Age-dependent sexual selection in bighorn rams. Proceedings of the Royal Society of London Series B 269: 165–172.

Coltman, D. W., J. A. Smith, D. R. Bancroft, J. Pilkington, A. D. C. MacColl, T. H. Clutton-Brock & J. M. Pemberton, 1999. Density-dependent variation in lifetime breeding success and natural and sexual selection in Soay rams. The American Naturalist 154: 730–746.

Egger, B., B. Obermüller, E. Eigner, C. Sturmbauer & K. M. Sefc, 2008. Assortative mating preferences between colour morphs of the endemic Lake Tanganyika cichlid genus *Tropheus*. Hydrobiologia (this issue). doi:10.1007/s10750-008-9564-0

Egger, B., B. Obermüller, H. Phiri, C. Sturmbauer & K. M. Sefc, 2006. Monogamy in the maternally mouthbrooding Lake Tanganyika cichlid fish *Tropheus moorii*. Proceedings of the Royal Society of London Series B 273: 1797–1802.

Fairbairn, A. J. & A. E. Wilby, 2001. Inequality of opportunity: measuring the potential for sexual selection. Evolutionary Ecology Research 3: 667–686.

Freeman-Gallant, C. R., N. T. Wheelwright, K. E. Meiklejohn, S. L. States & S. V. Sollecito, 2005. Little effect of extrapair paternity on the opportunity for sexual selection in Savannah sparrows (*Passerculus sandwichensis*). Evolution 59: 422–430.

Friedl, T. W. P. & G. M. Klump, 1999. Determinants of male mating success in the red bishop (*Euplectes orix*). Behavioural Ecology and Sociobiology 46: 387–399.

Gibbs, H. L., P. J. Weatherhead, P. T. Boag, B. N. White, L. M. Tabak & D. J. Hoysak, 1990. Realized reproductive success of polygynous red-winged blackbirds revealed by DNA markers. Science 250: 1394–1397.

Hasselquist, D., S. Bensch & T. von Schantz, 1995. Low frequency of extrapair paternity in the polygynous great reed warbler, *Acrocephalus arundinaceus*. Behavioural Ecology 6: 27–38.

Hayes, S. A., D. E. Pearse, D. P. Costa, J. T. Harvey, B. J. Le Boeuf & J. C. Garza, 2006. Mating system and reproductive success in eastern Pacific harbour seals. Molecular Ecology 15: 3032–3034.

Johnsen, A., J. T. Lifjeld, S. Andersson, J. Örnborg & T. Amundsen, 2001. Male characteristics and fertilization success in bluethroats. Behaviour 138: 1371–1390.

Jones, A. G., D. Walker, C. Kvarnemo, K. Lindström & J. C. Avise, 2001. How cuckoldry can decrease the opportunity for sexual selection: data and theory from a genetic parentage analysis of the sand goby, *Pomatoschistus minutus*. Proceedings of the National Academy of Science USA 98: 9151–9156.

Kempenaers, B., S. Everding, C. Bishop, P. Boag & R. J. Robertson, 2001. Extra-pair paternity and the reproductive role of male floaters in the tree swallow (*Tachycineta bicolor*). Behavioural Ecology and Sociobiology 49: 251–259.

Kempenaers, B., G. R. Verheyen, M. Van den Broeck, T. Burke, C. Van Broeckhoven & A. Dhondt, 1992. Extra-pair paternity results from female preference for high-quality males in the blue tit. Nature 357: 494–496.

Ketterson, E. D., P. G. Parker, S. A. Raouf, V. Nolan Jr., C. Ziegenfus & C. R. Chandler, 1997. The relative impact of extra-pair fertilizations on variation in male and female reproductive success in dark-eyed juncos (*Junco hyemalis*). Ornithological Monographs 49: 81–101.

Knight, M. E. & G. F. Turner, 2004. Laboratory mating trials indicate incipient speciation by sexual selection among populations of the cichlid fish *Pseudotropheus zebra* from Lake Malawi. Proceedings of the Royal Society of London Series B 271: 675–680.

Koblmüller S., K. M. Sefc & C. Sturmbauer, 2008. The Lake Tanganyika cichlid species assemblage: recent advances in molecular phylogenetics. Hydrobiologia (this issue). doi:10.1007/s10750-008-9552-4

Kocher, T. D., 2004. Adaptive evolution and explosive speciation: the cichlid fish model. Nature Genetics Reviews 5: 288–298.

Konings, A., 1998. Tanganyika cichlids in their natural habitat. Cichlid Press, El Paso, Texas.

Kornfield, I. & P. F. Smith, 2000. African cichlid fishes: model systems for evolutionary biology. Annual Review of Ecology and Systematics 31: 163–196.

Kraaijeveld, K., J. Gregurke, C. Hall, J. Komdeur & R. A. Mulder, 2004. Mutual ornamentation, sexual selection, and social dominance in the black swan. Behavioural Ecology 15: 380–389.

Kvarnemo, C. & I. Ahnesjö, 1996. The dynamics of operational sex ratios and competition for mates. Trends in Evolution and Ecology 11: 404–408.

Kvarnemo, C. & S. Merilaita, 2006. Mating distribution and its temporal dynamics affect operational sex ratio: a simulation study. Biological Journal of the Linnean Society 89: 551–559.

Kvarnemo, C., G. I. Moore & A. G. Jones, 2007. Sexually selected females in the monogamous Western Australian seahorse. Proceedings of the Royal Society of London Series B 274: 521–525.

Lindström, K., 2001. Effects of resource distribution on sexual selection and the cost of reproduction in sandgobies. The American Naturalist 158: 64–74.

Maan, M. E., O. Seehausen, L. Söderberg, L. Johnson, E. A. P. Ripmeester, H. D. J. Mrosso, M. I. Taylor, T. J. M. van Dooren & J. J. M. van Alphen, 2004. Intraspecific sexual selection on a speciation trait, male coloration, in the Lake Victoria cichlid *Pundamilia nyererei*. Proceedings of the Royal Society of London Series B 271: 2445–2452.

Maan, M. E., M. van der Spoel, P. Quesada Jimenez, J. J. M. van Alphen & O. Seehausen, 2006. Fitness correlates of male coloration in a Lake Victoria cichlid fish. Behavioural Ecology 17: 691–699.

Madsen, T. & R. Shine, 1993. Temporal variability in sexual selection acting on reproductive tactics and body size in male snakes. American Naturalist 141: 167–171.

Marshall, T. C., 1998. Inbreeding and fitness in wild ungulates. PhD Thesis, University of Edinburgh.

Meyer, A., 1993. Phylogenetic relationships and evolutionary processes in East African cichlid fishes. Trends in Ecology and Evolution 8: 3–8.

Meyer, A., C. Montero & A. Spreinat, 1994. Evolutionary history of the cichlid fish species flocks of the East African great lakes inferred from molecular phylogenetic data. Archiv für Hydrobiologie Beiheft Ergebnisse der Limnologie 44: 407–423.

Nelissen, M., 1976. Contribution to the ethology of *Tropheus moorii* Boulenger (Pisces, Cichlidae) and a discussion of the significance of its colour pattern. Revue de Zoologie Africaine 90: 17–29.

Otter, K., L. Ratcliffe, D. Michaud & P. T. Boag, 1998. Do female black-capped chickadees prefer high-ranking males as extra-pair partners? Behavioural Ecology and Sociobiology 43: 25–36.

Prosser, M. R., P. J. Weatherhead, H. L. Gibbs & G. P. Brown, 2002. Genetic analysis of the mating system and opportunity for sexual selection in northern water snakes (*Nerodia sipedon*). Behavioural Ecology 13: 800–807.

Richardson, D. S. & T. Burke, 2001. Extrapair paternity and variance in reproductive success related to breeding density in Bullock's orioles. Animal Behaviour 62: 519–525.

Rios-Cardenas, O., 2005. Patterns of parental investment and sexual selection in teleost fishes: Do they support Bateman's principles? Integrative and Comparative Biology 45: 885–894.

Rossiter, S. J., R. D. Ransome, C. G. Faulkes, D. A. Dawson & G. Jones, 2006. Long-term paternity skew and the opportunity for selection in a mammal with reversed sexual size dimorphism. Molecular Ecology 15: 3035–3043.

Salzburger, W., T. Mack, E. Verheyen & A. Meyer, 2005. Out of Tanganyika: genesis, explosive speciation, key-innovations and phylogeography of the haplochromine cichlid fishes. BMC Evolutionary Biology 5: 17.

Salzburger, W., A. Meyer, S. Baric, E. Verheyen & C. Sturmbauer, 2002. Phylogeny of the Lake Tanganyika species flock and its relationship to the Central and East African haplochromine cichlid fish fauna. Systematic Biology 51: 113–135.

Salzburger, W., H. Niederstätter, A. Brandstätter, B. Berger, W. Parson, J. Snoeks & C. Sturmbauer, 2006. Colour-assortative mating among populations of *Tropheus moorii*, a cichlid fish from Lake Tanganyika, East Africa. Proceedings of the Royal Society of London Series B 273: 257–266.

Schulte-Hostedde, A. I., J. S. Millar & H. I. Gibbs, 2004. Sexual selection and mating patterns in a mammal with female-biased sexual size dimorphism. Behavioural Ecology 15: 351–356.

Schupke, P., 2003. Buntbarsche des Tanganjikasees. Die Arten der Gattung *Tropheus*. Aqualog, ACS GmbH, Germany.

Schürch, R. & B. Taborsky, 2005. The functional significance of buccal feeding in the mouthbrooding cichlid *Tropheus moorii*. Behaviour 142: 265–281.

Seehausen, O., 2006. African cichlid fish: a model system in adaptive radiation research. Proceedings of the Royal Society of London Series B 273: 1987–1998.

Seehausen, O. & J. J. M. van Alphen, 1999. Can sympatric speciation by disruptive sexual selection explain rapid evolution of cichlid diversity in Lake Victoria? Ecology Letters 2: 262–271.

Sefc, K. M., S. Baric, W. Salzburger & C. Sturmbauer, 2007. Species-specific population structure in rock-specialized sympatric cichlid species in Lake Tanganyika, East Africa. Journal of Molecular Evolution 64: 33–49.

Sheldon, B. C. & H. Ellegren, 1999. Sexual selection resulting from extrapair paternity in collared flycatchers. Animal Behaviour 57: 285–298.

Singer, A., C. Kvarnemo, K. Lindström & O. Svensson, 2006. Genetic mating patterns studied in pools with manipulated nest site availability in two populations of *Pomatoschistus minutus*. Journal of Evolutionary Biology 19: 1641–1650.

Storz, J. F., H. R. Bhat & T. H. Kunz, 2001. Genetic consequences of polygyny and social structure in an Indian fruit bat, *Cynopterus sphinx*. II. Variance in male mating success and effective population size. Evolution 55: 1224–1232.

Strutchbury, B. J. M., H. Piper, D. L. Neudorf, S. A. Tarof, J. M. Rhymer, G. Fuller & R. C. Fleischer, 1997. Correlates of extra-pair fertilization success in hooded warblers. Behavioural Ecology and Sociobiology 40: 119–126.

Sturmbauer, C. & R. Dallinger, 1995. Diurnal variation of spacing and foraging behaviour in *Tropheus moorii* (Cichlidae) in Lake Tanganyika, Eastern Africa. Netherlands Journal of Zoology 45: 386–401.

Sturmbauer, C., C. Fuchs, G. Harb, E. Damm, N. Duftner, M. Maderbacher, M. Koch & S. Koblmüller, 2008. Abundance, distribution, and territorial areas of rock-dwelling Lake Tanganyika cichlid fish species. Hydrobiologia (this issue). doi:10.1007/s10750-008-9557-z

Sturmbauer, C., U. Hainz, S. Baric, E. Verheyen & W. Salzburger, 2003. Evolution of the tribe Tropheini from Lake Tanganyika: synchronized explosive speciation producing multiple evolutionary parallelism. Hydrobiologia 500: 51–64.

Turner, G. F., O. Seehausen, M. E. Knight, C. Allender & R. L. Robinson, 2001. How many species of cichlid fishes are there in African lakes? Molecular Ecology 10: 793–806.

Wade, M. J. & S. J. Arnold, 1980. The intensity of sexual selection in relation to male sexual behaviour, female choice, and sperm precedence. Animal Behaviour 28: 446–461.

Wagner, R. H., M. D. Schug & E. S. Morton, 1996. Condition-dependent control of paternity by female purple martins: implications for coloniality. Behavioural Ecology and Sociobiology 38: 379–389.

Weatherhead, P. J. & P. T. Boag, 1997. Genetic estimate of annual and lifetime reproductive success in male red-winged blackbirds. Ecology 78: 884–896.

Webster, M. S., H. C. Chuang-Dobbs & R. T. Holmes, 2001. Microsatellite identification of extrapair sires in a socially monogamous warbler. Behavioural Ecology 12: 439–446.

Webster, M. S., S. Pruett-Jones, D. F. Westneat & S. J. Arnold, 1995. Measuring the effects of pairing success, extra-pair copulations and mate quality on the opportunity for sexual selection. Evolution 49. 1147–1157.

Werner, N. & A. Lotem, 2003. Choosy males in a haplochromine cichlid: first experimental evidence for male mate choice in a lekking species. Animal Behaviour 66: 293–298.

Whittingham, L. A. & P. O. Dunn, 2005. Effects of extra-pair and within-pair reproductive success on the opportunity for selection in birds. Behavioural Ecology 16: 138–144.

Whittingham, L. A. & J. T. Lifjeld, 1995. Extra-pair fertilizations increase the opportunity for sexual selection in the monogamous House Martin *Delichon urbica*. Journal of Avian Biology 26: 283–288.

Wickler, W., 1969. Zur Soziologie des Brabantbuntbarsches, *Tropheus moorei* (Pisces, Cichlidae). Zeitschrift für Tierpsychologie 26: 967–987.

Woolfenden, B. E., H. L. Gibbs & S. G. Sealy, 2002. High opportunity for sexual selection in both sexes of an obligate brood parasitic bird, the brown-headed cowbird (*Molothrus ater*). Behavioural Ecology and Sociobiology 52: 417–425.

Yanagisawa, Y. & M. Nishida, 1991. The social and mating system of the maternal mouthbrooder *Tropheus moorii* (Cichlidae) in Lake Tanganyika. Japanese Journal of Ichthyology 38: 271–282.

Yanagisawa, Y. & T. Sato, 1990. Active browsing by mouthbrooding females of *Tropheus duboisi* and *Tropheus moorii* (Cichlidae) to feed the young and/or themselves. Environmental Biology of Fishes 27: 43–50.

Yezerinac, S. M., P. J. Weatherhead & P. T. Boag, 1995. Extra-pair paternity and the opportunity for sexual selection in a socially monogamous bird (*Dendroica petechia*). Behavioural Ecology and Sociobiology 37: 179–188.

Hydrobiologia (2008) 615:37–48
DOI 10.1007/s10750-008-9564-0

SPECIATION IN ANCIENT LAKES

Assortative mating preferences between colour morphs of the endemic Lake Tanganyika cichlid genus *Tropheus*

Bernd Egger · Beate Obermüller · Eva Eigner · Christian Sturmbauer · Kristina M. Sefc

Published online: 10 September 2008

Abstract Female mate preferences effectuate reproductive isolation among and sexual selection within species. Both mechanisms have been associated with the diversification and speciation of cichlid species flocks of the East African Great Lakes. In Lake Tanganyika, the endemic genus *Tropheus* has diversified into >100 geographic colour morphs. Although distributed allopatrically at present, water level fluctuations have repeatedly displaced and merged the benthic, rock-dwelling populations. Tests for assortative mating were performed to explore the potential for reproductive isolation between morphs in secondary contact, and to assess the importance of sexual selection for the diversification of this group. In contrast to other haplochromine cichlids, *Tropheus* is a sexually monochromatic, territorial and maternally mouthbrooding fish, which establishes temporary pair bonds prior to spawning. Female mate preference trials involved two-way choices between a homotypic and a heterotypic male and were conducted on allopatric populations of red and blue morphs from the southern part of Lake Tanganyika. Female affiliation time near each male's compartment did not predict the mate preferences subsequently expressed in unrestrained interactions after removal of the compartment separators (spawning, pseudospawning and courtship). Consequently, mate preferences were inferred from unrestrained interactions with one test male at a time in replicate observation sessions. Of the 23 females tested, 13 courted, pseudospawned or spawned with the homotypic male, one blue female courted a red male, and nine females expressed no sexual motivation. The assortative mate preferences in the experiments ($P < 0.01$) suggest that colour differentiation between *Tropheus* populations can effectuate reproductive isolation, and is consistent with the notion that sexual selection contributed to the diversification of the genus.

Keywords Mate preference · Assortative mating · Cichlids · Sexual selection · *Tropheus* · Colour divergence

Guest editors: T. Wilke, R. Väinölä & F. Riedel
Patterns and Processes of Speciation in Ancient Lakes: Proceedings of the Fourth Symposium on Speciation in Ancient Lakes, Berlin, Germany, September 4–8, 2006

B. Egger · B. Obermüller · E. Eigner ·
C. Sturmbauer · K. M. Sefc (✉)
Department of Zoology, Karl-Franzens University of Graz, Universitätsplatz 2, Graz 8010, Austria
e-mail: kristina.sefc@uni-graz.at

Present Address:
B. Egger
School of Biological Sciences, University of Wales Bangor, Bangor, Gwynedd LL57 2UW, UK

Introduction

Experimental appraisal of female mate preferences provides a means to assess intra- and interspecies discrimination and to infer the intensity of sexual selection within groups (see e.g. Ludlow et al., 2001; Olsson, 2001; Devon & Slagsvold, 2005; Setchell, 2005) and the extent and the basis of pre-mating isolation between species or populations (e.g. Rolán-Alvarez et al., 1999; Summers et al., 1999; Payne et al., 2000; Kodric-Brown & Strecker, 2001; Ólafsdóttir et al., 2006). In species of the rapidly diversifying haplochromine fishes (family Cichlidae) of Lakes Victoria and Malawi, female choice exerts directional sexual selection on male nuptial colouration (Maan et al., 2004; Pauers et al., 2004) and induces assortative mating in experiments with sympatric (Seehausen, 1997; Seehausen & Van Alphen, 1998; Seehausen et al., 1998; Jordan et al., 2003) and allopatric species or colour morphs (Knight & Turner, 2004). Consistent with the differences in male colour patterns between morphologically undifferentiated, closely related species, visual cues were apparently sufficient for the recognition of conspecific males (Seehausen & Van Alphen, 1998; Couldridge & Alexander, 2002; Knight & Turner, 2004; Maan et al., 2004; Kidd et al., 2006).

In the cichlids of Lake Tanganyika, geographic colour variation is most strikingly realised in the over 100 colour variants of the genus *Tropheus* (Konings, 1998; Schupke, 2003), a member of the endemic tribe Tropheini (included in the 'modern haplochromines' by Salzburger et al., 2005). Most of the *Tropheus* variants are distributed allopatrically, but phylogeographic data indicate that populations underwent recurrent cycles of population fragmentation and secondary contact in the course of lake-level fluctuations (Sturmbauer & Meyer, 1992; Baric et al., 2003; Sturmbauer et al., 2005; Egger et al., 2007; Sefc et al., 2007). Given the recency of the latest major lake lowstand of about 300 m below present level (ending at 8,000–12,000 years BP; Scholz & Rosendahl, 1988), which must have displaced populations in shallower parts of the lake, the high number of allopatric colour morphs is difficult to explain. On the one hand, assortative mate choice could have enabled different morphs to persist in sympatry during secondary contact, but with rising water levels, the prezygotically isolated populations would have colonised the newly available habitat jointly. However, there are only a few instances of sympatry in the present colour morph distribution (Konings, 1998; Schupke, 2003). On the other hand, non-assortative mating would have homogenised morphs during secondary contact, and allopatric diversification must have proceeded very rapidly after colonisation of new habitats. Genetic analyses of accidentally admixed *Tropheus* morphs in Lake Tanganyika demonstrated colour-assortative mating (Salzburger et al., 2006).

The development of an experimental design for testing mate preferences in *Tropheus* has two objectives. Firstly, the assessment of reproductive isolation between populations is a prerequisite for the evaluation of possible scenarios of population interaction during historical periods of population admixture. Secondly, the potential for sexual selection as a driving force of colour divergence is affected by the presence of female mate preferences. The present paper is the first report on experimental investigations of mate preferences between two distinct, allopatric *Tropheus* colour morphs from the south-eastern shore of Lake Tanganyika.

Methods

Study species

The sexually monomorphic *Tropheus* are abundant in the shallow rocky habitat of Lake Tanganyika, where they graze on epilithic algae. Males and females defend separate territories, and females are fully included in the social interactions of the community and employ sex-independent colour signals in communication (Wickler, 1969; Sturmbauer & Dallinger, 1995). Female mate choice involves inspection visits to several territorial males before establishing a pair bond and moving into a mate's territory (Yanagisawa & Nishida, 1991). Apparently, females depend on the nutritional resources of a male's large territory for maturation of their ovaries, and forage actively under their mate's protection for two to three weeks prior to spawning (Yanagisawa & Nishida, 1991; Sturmbauer & Dallinger, 1995). Courtship and spawning behaviour in *Tropheus* is similar to that displayed by other haplochromines (McElroy & Kornfield, 1990), and includes lead swimming by the male, quivering by both sexes, release and snapping up of eggs by the female, and nuzzling of the male's anal fin by the female (Nelissen, 1976). Upon spawning, the female abandons the male's territory and settles in an unoccupied site to mouthbrood eggs and fry for up to six weeks (Yanagisawa & Nishida, 1991). Genetic analyses of *Tropheus* showed that broods are generally sired by a single male each (Egger et al., 2006).

The species-level taxonomy of the genus *Tropheus* is confounded by polyphyly and the existence of numerous geographic colour morphs within each of the six nominal species (Snoeks et al., 1994; Sturmbauer et al., 2003) and populations are generally referred to by locality rather than by species (e.g. Sturmbauer et al., 2005). In this study, mate preferences in a red colour morph, *Tropheus* "Moliro" (currently assigned to *Tropheus moorii kasabae* Nelissen), and a blue colour morph, *Tropheus* "Nakaku" (currently assigned to *Tropheus moorii* Boulenger), collected from Cape Chaitika and Nakaku (Fig. 1), were tested. The red and blue morphs belong to two distinct mitochondrial lineages (7F and 8G), but are grouped in the same clade (AFLP 4) by nuclear genetic markers (Egger et al., 2007).

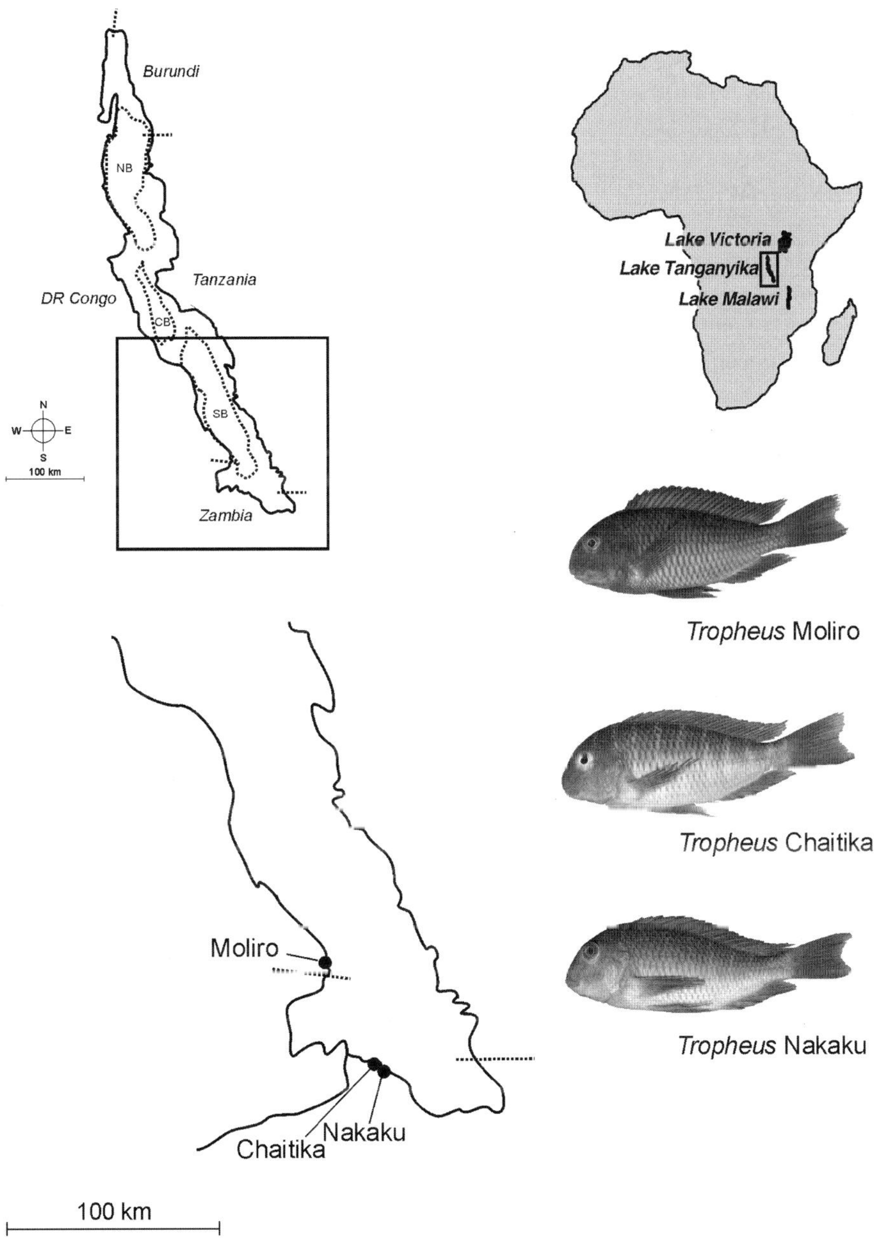

Fig. 1 *Tropheus* morphs used in our experiments and their location in Lake Tanganyika. The figure shows scans of anaesthetised fish. Note that colours intensify during social and sexual displays, but do not differ between sexes. Fish of the blue morph were collected from Cape Chaitika and Nakaku; these localities are separated only by a few kilometres and fish of the two populations are almost identical

Fish maintenance

All individuals were wild caught adults imported by an ornamental fish trader. Fish were housed in individual aquaria before trials (22 aquaria: 60 × 30 × 30 cm, two aquaria: 60 × 30 × 45 cm). After the trials were completed, fish were transferred into two stock tanks (100 × 60 × 60 cm). Both cultivation and experimental tanks (see below) were filtered with internal box filters, kept at 25–27°C by an internal heater and illuminated with an overhead white light on a 12:12 h light–dark cycle. One-third water changes were carried out at least every three weeks, and fish were fed 2 times daily with a mixture of pea and shrimp paste, pellets with high plant content and flake food.

Experimental protocol

Experimental tanks (150 × 70 × 50 cm) were divided by mesh partitions (mesh size 13 mm) into two outer compartments (45 cm) for males and one central compartment (60 cm) for the female. This set-up allowed fish to communicate with visual, chemical and auditory signals. Hollow bricks (clay bottle stands) were placed in each compartment (two in each of the small male compartments and six in the female's compartment) to serve as hiding place and territory focus and to provide a spawning surface. All fish were weighed and measured (total length) a week before introduction to the experimental tank. Series of three to five trials at a time were carried out in identically equipped tanks between August 2004 and October 2005, and involved a total of 10 females and 11 males of the red morph, six females and six males of the blue morph from Cape Chaitika, and seven females and five males of the blue morph collected at Nakaku.

Preliminary experiments had suggested that the inference of mate preferences from affiliation time and type of interaction across the barrier was unfeasible. Scoring of interactions on the partition proved to be difficult due to the fact that behaviour such as quivering and lateral displays are used both in courtship and aggression and cannot be fully expressed (see also Kidd et al., 2006). Moreover, in one trial, a female spawned across the barrier with the male she had spent significantly less time with. Therefore, an experimental design was developed, which allowed free interaction between the fish, taking into account the natural mating behaviour of *Tropheus*.

After introduction into the tank, fish were kept for one week with mesh partitions in place to allow the female to examine both males under standardised conditions, choose a mate and establish a pair-bond ('decision phase'). During the 'decision phase' of the first 12 trials (with 6 females from Moliro and 6 females from Cape Chaitika), affiliation time was recorded for comparison with subsequent courtship and mating behaviour; this was done in order to validate our decision to discard the simpler approach based on affiliation time. Fish were observed twice a day for 10 min, and the amount of time a female spent in a defined choice zone (25 cm in front of each male's compartment) and in the centre between choice zones and the duration of aggressive behaviour and courtship in choice zones were recorded. After the 'decision phase', females were permitted free access to one test male at a time ('sequential access phase'). 'Sequential access' test sessions were carried out twice a day, morning and afternoon, for 5 days. The partition between the female and one of the males was removed for 30 min, and the behaviour of the pair was observed. In the second session of the day, the female was allowed to interact with the other male. Fish were monitored continuously during the sequential access sessions, and showed no obvious response to the presence of the observer. When fish were in danger of being injured due to agonistic behaviour, the partition was restored before expiration of the 30-min observation period. Biting sometimes resulted in loss of one to two scales, but no serious injuries occurred in any of the trials. In most trials where females courted, pseudospawned or spawned with one of the males, agonistic behaviour was displayed against the other male, and those fish had to be separated in almost every open access session.

The order of access to the males was assigned randomly in each trial, but was preserved on consecutive days of the trial. The number of trials starting with access to the homomorphic male was approximately the same ($N = 11$) as the number of trials starting with the heteromorphic male ($N = 12$).

The interactions of the pair were recorded as spawning, pseudospawning, courtship, no interaction and aggression. Typical aggressive displays in *Tropheus* are mouth-fighting and circling (Nelissen, 1976). Pseudospawning employs the same behavioural sequences as spawning without release of eggs; see Nelissen (1976) and Wickler (1969) for a detailed description of the courtship and mating behaviour in *Tropheus*. When spawning occurred, the trial was considered completed and scored as 'spawning'. In the field, females mate with only one male per clutch (Egger et al., 2006), and spawning with one male apparently precludes the acceptance of an additional mate. When the female engaged in pseudospawning or intensive courtship with one of the males, but failed to interact or displayed agonistic behaviour with the alternative male consistently over the entire 'sequential access phase', trials were scored as 'pseudospawning' and 'courtship', respectively. Spawning of *Tropheus* pairs often spans several hours, and is frequently preceded by pseudospawning behaviour (personal observation). Lack of interaction and aggression between the female and both males were scored as 'negative'.

A total of 23 'sequential access trials' tested female mate preference in choices between the red and the blue morphs. Red females were confronted with males of their own morph and blue males from Cape Chaitika (6 trials) or Nakaku (4 trials), blue females from Cape Chaitika were given a choice between Cape Chaitika males and red males (6 trials), and blue females from Nakaku were tested against Nakaku males and red males (7 trials). Individual males were used in one to three ($X \pm SE = 1.82 \pm 0.72$) trials, but females were used only once.

The red and the blue morphs differed in total length (red: average $L_T = 10.2$ cm $\pm$ 0.9 cm, range 9.1–12.2 cm, $N = 21$; blue: average $L_T = 9.2$ cm $\pm$ 0.7 cm, range 8.2–11.5 cm, $N = 24$; $t_{43} = 4.454$, $P < 0.001$ in a two-tailed test), but there were no size differences between the two blue populations from Cape Chaitika ($N = 12$) and Nakaku ($N = 12$; $t_{22} = 0.338$, $P = 0.739$). Size ranges of males and females within each morph did not differ significantly (red males: average $L_T = 10.0$, $N = 11$, red females: average $L_T = 10.3$, $N = 10$, $t_{19} = 0.443$, $P = 0.663$; blue males: average $L_T = 9.3$, $N = 11$, blue females: average $L_T = 9.1$, $N = 13$, $t_{22} = 0.518$, $P = 0.610$).

Data analysis

The software JMating (Vers. 1.0.8, Carvajal-Rodriguez & Rolán-Alvarez, 2006) was used to test whether the frequencies of observed homo- and heteromorphic choices differed from a random expectation, employing indices of pairwise (PTI) and global sexual isolation (I_{PTI}) and indices of asymmetry between morphs (Rolán-Alvarez & Caballero, 2000; Coyne et al., 2005; Schwarz & McPheron, 2007). Significance of the deviations from random expectation was assessed on the basis of 10,000 bootstrapping replicates for pairwise isolation indices and asymmetry indices, and by a G-test for the global index. This analysis included all 23 trials, i.e. also those in which females had not expressed a preference for one of the males. To test whether the responsive females (14 out of 23, see Results) expressed preferences for homotypic males more often than expected under a null hypothesis of random mating between the morphs, we employed one-tailed binomial tests.

Two-tailed Wilcoxon signed rank tests were used to analyse female affiliation time recorded during the 'decision phase'.

Results

Mate preferences among colour morphs

Of the 23 trials under the 'sequential access' protocol, one resulted in 'spawning', eight were scored as 'pseudospawning' and five as 'courtship'; that is, 61% of our trials provided a measure of female preference (Table 1). Except in one case, spawning and pseudospawning were preceded by distinct courtship behaviour. In the 14 trials where females expressed a preference for one of the males, they either acted aggressively towards the other male (10 trials), experienced aggression by the other male (3 trials), or did not interact with the other male (1 trial). Females never reversed expressed mate preferences (pseudospawning and courtship) within trials. In 8 out of the 9 'negative' trials, both males behaved agonistically towards the female. In one negative trial, the female behaved agonistically towards both males. Details of the experiments are listed in Table 1.

Mate preference was assortative in 13 of the 14 positive trials (Fig. 2). Assortative preferences were expressed in all successful trials with red females ($N = 7$), and in six out of seven successful trials with blue females (Table 1). The only non-assortative case

Table 1 Preferred males, focal females and non-preferred males with scores of 'spawning', 'pseudospawning' and 'courtship' in 'sequential access' trials, and fish involved in negative trials

Spawning			Pseudospawning			Courtship			Negative		
Preferred male	Female	Non-preferred male	Preferred male	Female	Non-preferred male	Preferred male	Female	Non-preferred male	Homotypic male	Female	Heterotypic male
C10	C20	M16	N54	N60	M42	C8	C18	M15	C8	C19	M15
14.1 g,	14.9 g,	18.2 g,	12.0 g,	13.0 g,	14.9 g,	12.3 g,	16.0 g,	16.4 g,	12.3 g,	16.0 g,	16.4 g,
9.5 cm	9.4 cm	10.4 cm	8.9 cm	9.3 cm	9.9 cm	9.2 cm	9.7 cm	9.7 cm	9.2 cm	9.7 cm	9.7 cm
			N56	N62	M43	C8	C22	M14	C6	C21	M15
			10.0 g,	11.0 g,	15.0 g,	12.3 g,	16.5 g,	18.5 g,	16.7 g,	12.4 g,	16.4 g,
			8.9 cm	9.0 cm	9.3 cm	9.2 cm	9.6 cm	10.0 cm	10.0 cm	9.0 cm	9.7 cm
			M16	M25	C7	C10	C23	M14	N55	N52	M63
			18.2 g,	20.7 g,	13.2 g,	14.1 g,	15.0 g,	18.5 g,	11.9 g,	14.9 g,	21.5 g,
			10.4 cm	10.3 cm	8.8 cm	9.5 cm	9.2 cm	10.0 cm	9.2 cm	9 .8 cm	10.7 cm
			M12	M27	C7	*M43*	*N58*	*N53*	N56	N59	M45
			13.7 g,	16.2 g,	13.2 g,	*15.0* g,	*14.4* g,	*11.0* g,	10.0 g,	13.5 g,	26.0 g,
			9.1 cm	10.4 cm	8.8 cm	*9.3* cm	*8.8* cm	*8.8* cm	8.9 cm	8.9 cm	11.9 cm
			M17	M28	C6	M12	M26	C11	N56	N61	M45
			19.5 g,	24.7 g,	16.7 g,	13.7 g,	21.0 g,	11.1 g,	10.0 g,	10.8 g,	26.0 g,
			10.4 cm	11.1 cm	10.0 cm	9.1 cm	10.8 cm	8.6 cm	8.9 cm	8.9 cm	11.9 cm
			M45	M41	N51				N53	N64	M44
			26.0 g,	27.9 g,	23.4 g,				11.0 g,	12.5 g,	19.9 g,
			11.9 cm	12.2 cm	11.5 cm				8.8 cm	8.8 cm	10.4 cm
			M42	M46	N54				M12	M29	C9
			14.9 g,	12.6 g,	12.0 g,				13.7 g,	15.8 g,	11.8 g,
			9.9 cm	9.4 cm	8.9 cm				9.1 cm	9.2 cm	8.5 cm
			M43	M50	N53				M16	M24	C11
			15.0 g,	15.2 g,	11.0 g,				18.2 g,	23.7 g,	11.1 g,
			9.3 cm	9.3 cm	8.8 cm				10.4 cm	10.9 cm	8.6 cm
									M44	M48	N55
									19.9 g,	11.4 g,	11.9 g,
									10.4 cm	9.32 cm	9.2 cm

The trial resulting in non-assortative preference is highlighted by italic values. Identifiers of fish consist of a population code (blue morph: C, Chaitika and N, Nakaku; red morph: M, Moliro) and individual numbers; weight and size (total length) of individuals are given below their identifiers

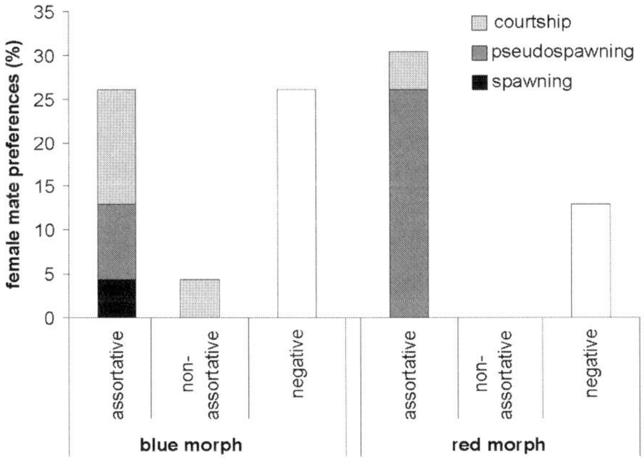

Fig. 2 Assortative and non-assortative mate preferences of blue and red females. Shaded bars show percentage of trials scored as spawning, pseudospawning and courtship; the open bars represent the negative trials, in which inference of female preference was not possible

consisted of courtship between a blue female and a red male, with agonistic behaviour towards the homotypic male (Table 1). The global estimator of sexual selection, based on all 23 trials, indicated significant positive assortative preferences with $I_{PTI} = 0.88$ ($G = 13.91$; $P < 0.01$). Pairwise indices suggested that the frequency of homotypic red pairs was higher (PTI = 2.3, $P = 0.005$) and the frequencies of mixed pairs were lower than expected under random mating (red ♀–blue ♂ pairs, PTI = 0, $P = 0.003$; blue ♀–red ♂ pairs, PTI = 0.25, $P = 0.005$). In contrast, the frequency of observed homotypic preferences in the blue morph did not deviate from the random expectation (PTI = 1.52, $P = 0.386$), which is probably due to the large proportion of negative trials (6 of the 13 tested blue females).

There were no differences in body sizes between females in successful and negative trials (red morph: $t_8 = 1.015$, $P = 0.340$; blue morph: $t_{11} = 0.483$, $P = 0.689$), or between the preferred males in successful trials and the males of the same morph as the test female in negative trials (red morph: $t_9 = -0.67$, $P = 0.948$; blue morph: $t_{10} = 0.158$, $P = 0.878$). The two morphs differ in body size (see Materials and methods), and female mate preferences could be size-assortative for a male of her own morph, and perhaps not detectable when the two alternative males are too similar in size or when both males are too dissimilar in size compared to the female. However, absolute body size differences were larger between alternative males offered in negative trials (mean difference = 1.54 cm) than between males offered in successful trials (mean difference = 0.66 cm; *t*-test for samples with unequal variances: $t_{7.919} = 2.262$, $P = 0.054$). Absolute body size differences between females and homotypic males were not smaller in successful trials than in negative ones (mean difference = 0.49 cm and 0.42 cm in successful and negative trials, respectively; $t_{21} = 0.356$, $P = 0.726$), and body size differences between females and heterotypic males were not larger in successful trials than in negative ones (mean difference = 0.86 cm and 1.37 cm in successful and negative trials, respectively; *t*-test for samples with unequal variances: $t_{11.069} = -1.201$, $P = 0.255$). Hence, body sizes of males and females in a test trio do not appear to determine failure or success to detect female preferences. Rather, we assume that the females in the negative trials were not gravid at the time of the experiment, which would prevent them from establishing a pair bond and rather elicit a territorial response from both of the candidate males (see Discussion). If this was indeed the case, analyses should be restricted to the successful trials only. In fact, the results were not affected by the exclusion of the nine unsuccessful trials, as the responsive females expressed preferences for homotypic males more often than expected under a null hypothesis of random mating between the morphs (binomial test: $P = 0.00092$). Analysis of results for each morph separately showed the significant assortative preferences in the red females ($N = 7$, binomial test: $P = 0.008$), and a trend to assortative mating in blue females (binomial test: $P = 0.062$ for six assortative choices in seven experiments). However, the difference in the ratio of assortative to non-assortative preferences between red and blue morphs was not significant ($\chi_1^2 = 1.07$, $P = 0.299$). Likewise, the asymmetry indices calculated in JMating based on all 23 trials indicated no asymmetry in sexual isolation between the morphs (P values >0.3).

There is no indication that female behaviour was influenced by the order in which the partitions were removed, as in half of the successful trials, the females chose the males on the side of the tank where the partition was removed first. The observed spawning took place on the fourth day of the 'sequential access' phase, and observed pseudospawning started on day 1 in three trials, and on days 2, 3, 4 and 5 in one, one, two and one experiment, respectively.

Affiliation time and mate preference

In 12 of the mate preference trials, the amount of time a female spent close to each male was scored during the 'decision phase' (with individuals permanently separated by partitions). In seven of the 12 trials, the females spent significantly more time with one of the males than with the other (Wilcoxon signed-ranks tests, $P < 0.05$ in each of the seven trials). In only four trials, the majority of time was spent with the homomorphic male (one red and three blue females), whereas one blue female spent more time with the red male, and two red females associated preferentially with blue males. In contrast to the assortative mate preferences indicated by interactions and spawning in the 'sequential access' phase, affiliation time did not indicate a significant preference for homomorphic males ($N = 6$ blue females, 6 red females, $I_{PTI} = 0.14$, $G = 1.53$, $P > 0.1$). Affiliation preferences were not correlated with male size (Wilcoxon signed-ranks test for paired samples on sizes of preferred versus rejected males, $T = -1.7$, $N = 7$, $P = 0.091$).

Affiliation time predicted mate preferences (as inferred from the 'sequential access' phase) in only three trials. In contrast, two females spent significantly less time with the male they later pseudospawned and courted with; in three trials, females spent equal amounts of time with each male, but later expressed a preference (one spawning, one pseudospawning and one courtship score); and in two further trials, the amount of time females spent near each male differed significantly, but no preference was evident in the 'sequential access' phase. In the remaining two trials, neither affiliation time nor subsequent interactions indicated any preference.

Discussion

Assortative mating between *Tropheus* morphs

This study demonstrated positive assortative mating preferences in experiments with two allopatric *Tropheus* colour morphs. Females of the red morph chose a homomorphic male in each of seven successful trials, and blue females preferred the homomorphic male in six of seven successful trials. However, in 9 out of the 23 trials (~40%), no preference of the female for one of the males could be detected. We believe that the high proportion of negative trials was due to the fact that test females were not assessed for ripeness before trials. In subsequent experiments (not presented here), care was taken to use gravid females, and the rate of successful experiments was raised to nearly 100% (Egger et al., in prep.). In the majority of negative trials (8 of 9), both males behaved agonistically towards the female. Intruding non-gravid females are probably recognised as territorial rivals and therefore attacked and chased away by the male. Most chemical signalling during reproduction in fish appears to originate from females, and indicates reproductive state or receptivity to the males (Sargent et al., 1998). Furthermore, fluctuations in the hormone levels throughout the female reproductive cycle have been shown to correspond with the presence and absence of affiliative preferences for territorial males in the African cichlid *Astatotilapia* (*Haplochromis*) *burtoni* (Clement et al., 2004; Martin, 2004). Both male territoriality against non-gravid females and these females' lack of sexual motivation may have prevented the detection of mating preferences in our negative experiments.

A tendency for colour-assortative mating in the field has recently been detected in parentage and assignment studies in *Tropheus* from Mpulungu Bay, Lake Tanganyika, where adults of four different allopatric morphs had been released by fish traders in 1998. Several years after the incident, the introduced morphs were still present; 80% of the genetically reconstructed parent-offspring relationships were within morphs, and mixed mating had occurred between similarly coloured morphs only (Salzburger et al., 2006). Both the mate preference tests in the laboratory (this study) and the genetic analysis of the field populations (Salzburger et al., 2006) imply that phenotypically strongly differentiated populations of *Tropheus* experience a considerable degree of prezygotic reproductive isolation.

Male colouration has been shown to determine female choice in several cichlids of Lake Malawi (Knight & Turner, 1999; Jordan et al., 2003; Kidd et al., 2006) and Victoria (Seehausen & Van Alphen, 1998; Maan et al., 2004). It is tempting to relate the mate preference patterns observed in *Tropheus* to the colour differences between populations, particularly as reproductive isolation under natural conditions also depended on the extent of phenotypic distinctness in the admixed populations of Mpulungu Bay (Salzburger et al., 2006). Similarly, mating was highly assortative among distinct colour forms of the *Pseudotropheus zebra* complex (Lake Malawi), but not between two populations with similar male breeding colours (Knight & Turner, 2004). Such associations between female preferences and male courtship traits in allopatric populations are expected when courtship signals diverged in response to female preferences (Summers et al., 1999; Knight & Turner, 2004). In contrast, olfactory rather than visual cues determined discrimination of female *P. emmiltos* against males of a sympatric species with different breeding colours (*P. fainzilberi*; Plenderleith et al., 2005), which makes it possible that the divergence of male nuptial colouration in other allopatric populations may not be caused by direct inter-sexual selection on colour (Arnegard & Kondrashov, 2004; Knight & Turner, 2004; Seehausen & Schluter, 2004; Plenderleith et al., 2005). Several other fish species also demonstrate olfactory mate recognition (McLennan & Ryan, 1999; Kodric-Brown & Strecker, 2001; Wong et al., 2005). In the freshwater isopod *Asellus aquaticus*, assortative mating of two populations differing in pigmentation and body size was based on size rather than colouration, and rapid divergence in pigmentation was attributed to natural selection rather than divergent mating preferences (Hargeby & Erlandsson, 2006). Our trials permitted fish to use multiple communication channels, and colour morph and body size were fully correlated in the males used in our experiments (Table 1). Morph-assortative preferences could therefore be due to colour, body size, population specific odour or acoustic signals, and different experimental designs will be needed to discriminate between cues responsible for assortative mating in these populations.

Males of different species have been shown to discriminate between con- and heterospecific females (Wiernasz, 1995; Knight & Turner, 1999; McLennan, 2004; Wong et al., 2004) and among conspecific females (Amundsen & Forsgren, 2001; Werner & Lotem, 2003) and allocate their courtship efforts accordingly; females, on their part, are influenced by the intensity of male courtship (e.g. Forsgren, 1997; Parri et al., 1997). In our trials, male behaviour might have influenced female responses both during the 'decision' and during the 'access' phase, and mate preferences observed during the 'sequential access phase' may actually have been determined by both female and male preferences. Although the repeated use of certain males could introduce a bias to the experiment, the strong pattern of colour-assortative preferences observed in our experiment makes it unlikely that the result was substantially affected by individual male traits. Furthermore, the same males were preferred by females of their own morph and rejected by females of the other morph.

Tropheus employ colour signals in different contexts to indicate motivation and social status to individuals of the same and of the other sex, and both genders possess concurrent repertoires of population-specific 'neutral', 'dominant' and 'inferiority' colour patterns (Wickler, 1969; Nelissen, 1976; Sturmbauer & Dallinger, 1995). Hence, signal evolution is likely to be affected by social interactions such as those involved in status, territory acquisition and defence (West-Eberhart, 1983; Seehausen & Schluter, 2004; Benson & Basolo, 2006). Since territoriality is a prerequisite to reproductive success for male *Tropheus*, sexual and social selection might be tightly linked in this system (see also Kraaijeveld et al., 2004). At this point, evidence for and against sexual selection as a strong diversifying force in *Tropheus* is mixed: the rapid diversification into numerous colour races after the latest lake level rise, high maternal investment in brood care and colour-assortative mating contrast with sexual monomorphism, pair-bonding and social and genetic monogamy within spawning events (Egger et al., 2006; Sefc, 2008). Therefore, and as long as the exact mechanism of mate recognition remains unknown, colour differentiation among *Tropheus* populations cannot be positively ascribed to direct male preference, although observations of assortative mating are certainly, but not exclusively, consistent with divergent sexual selection (Andersson, 1994).

No connection between affiliation time and mate preference

The amount of time a female spends near a male's compartment under experimental conditions does not necessarily reflect a mating preference for that male (Gabor, 1999; Aspbury & Basolo, 2002), although it predicted subsequent mate choice in some studies (White & Galef, 1999; Couldridge & Alexander, 2001; Santangelo, 2005). The lack of connection between association preferences during the 'decision phase' and subsequent courtship and mating patterns in the 'sequential access' phase in our trials with *Tropheus* may be due to various reasons. Mating preferences are not the only motivation to associate with a particular male or spend time in certain parts of the aquarium. Poeciliid sailfin mollies, for example, displayed a universal preference to associate with large conspecifics independent of the sex of the chosen individual (Gabor, 1999). In *Tropheus*, male body size did not influence association preferences, but other factors, such as activities related to territory defence, may determine how a female divides her time between different sections of her compartment.

Conclusions and further considerations

The described design for mate choice experiments in the monogamous, pair-bonding *Tropheus* is time consuming, but provides direct (spawning) or indirect (pseudospawning and intense courtship) evidence of mate choice. As one would expect, the rate of positive trials (yielding evidence of mate preference) increases when only gravid females are used. Individuals of two distinct populations exhibited highly assortative mating preferences, suggesting that differentiated morphs could persist sympatrically. Moreover, discriminative mating is a prerequisite to divergent sexual selection, and could therefore contribute to the observed variety of colour races. Now that assortative mating among two morphs has been demonstrated, many questions emerge with regard to mate choice processes in *Tropheus*. What are the decisive cues? How different must populations be in order to mate assortatively? Does genetic relatedness between populations influence the strength of assortative mating? A further issue for predictions of mating decisions in secondary contact is the influence of territory size and quality on female choice (Candolin & Reynolds, 2001; Ludlow et al., 2001; Maan et al., 2004; Kohda et al., 2005), since differences between the territories of potential mates could potentially override the females' otherwise assortative preferences in the field.

Acknowledgements We thank the organisers of SIAL IV (Berlin, 2006) for an outstanding symposium and the editors for their efforts in compiling this special issue. We also thank two anonymous reviewers for their suggestions on data analysis and presentation. The work was supported by the Austrian Science Foundation (Grant P17380-B06).

References

Amundsen, T. & E. Forsgren, 2001. Male mate choice selects for female colouration in a fish. Proceedings of the National Academy of Sciences USA 98: 13155–13160.

Andersson, M., 1994. Sexual Selection. Princeton University Press, Princeton.

Arnegard, M. E. & A. S. Kondrashov, 2004. Sympatric speciation by sexual selection alone is unlikely. Evolution 58: 222–237.

Aspbury, A. & A. Basolo, 2002. Repeatable female preferences, mating order and mating success in the poeciliid fish, *Heterandria formosa*. Behavioral Ecology and Sociobiology 51: 238–244.

Baric, S., W. Salzburger & C. Sturmbauer, 2003. Phylogeography and evolution of the Tanganyikan cichlid genus *Tropheus* based upon mitochondrial DNA sequences. Journal of Molecular Evolution 56: 54–68.

Benson, K. E. & A. L. Basolo, 2006. Male–male competition and the sword in male swordtails, Xiphophorus helleri. Animal Behaviour 71: 129–134.

Candolin, U. & J. D. Reynolds, 2001. Sexual signaling in the European bitterling: females learn the truth by direct inspection of the resource. Behavioral Ecology 12: 407–411.

Carvajal-Rodriguez, A. & E. Rolán-Alvarez, 2006. JMATING: a software for the analysis of sexual selection and sexual isolation effects from mating frequency data. BMC Evolutionary Biology 6: 40.

Clement, T. S., K. E. Grens & R. D. Fernald, 2004. Female affiliative preference depends on reproductive state in the African cichlid fish, *Astatotilapia burtoni*. Behavioral Ecology 16: 83–88.

Couldridge, V. C. K. & G. J. Alexander, 2001. Does the time spent near a male predict female mate choice in a Malawian cichlid? Journal of Fish Biology 59: 667–672.

Couldridge, V. C. K. & G. J. Alexander, 2002. Colour patterns and species recognition in four closely related species of Lake Malawi cichlid. Behavioral Ecology 13: 59–64.

Coyne, J. A., S. Elwyn & E. Rolán-Alvarez, 2005. Impact of experimental design on *Drosophila* sexual isolation studies: direct effects and comparison to field hybridization data. Evolution 59: 2588–2601.

Devon, T. & T. Slagsvold, 2005. When and from whom do female pied flycatchers (*Ficedula hypoleuca*) solicit copulations. Behaviour 142: 1059–1076.

Egger, B., B. Obermüller, H. Phiri, C. Sturmbauer & K. M. Sefc, 2006. Monogamy in the maternally mouthbrooding Lake Tanganyika cichlid fish *Tropheus moorii*. Proceedings of the Royal Society of London. Series B 273: 1797–1802.

Egger, B., S. Koblmüller, C. Sturmbauer & K. M. Sefc, 2007. Nuclear and mitochondrial data reveal different evolutionary processes in the Lake Tanganyika cichlid genus *Tropheus*. BMC Evolutionary Biology 7: 137.

Forsgren, E., 1997. Mate sampling in a population of sand gobies. Animal Behaviour 53: 267–276.

Gabor, C., 1999. Association patterns of sailfin mollies (*Poecilia latipinna*): alternative hypotheses. Behavioural Ecology and Evolution 46: 333–340.

Hargeby, A. & J. Erlandsson, 2006. Is size-assortative mating important for rapid pigment differentiation in a freshwater isopod? Journal of Evolutionary Biology 19: 1911–1919.

Jordan, R., K. Kellogg, F. Juanes & J. J. Stauffer, 2003. Evaluation of female mate choice cues in a group of Lake Malawi Mbuna (Cichlidae). Copeia 1: 181–186.

Kidd, M. R., P. D. Danley & T. D. Kocher, 2006. A direct assay of female choice in cichlids: all the eggs in one basket. Journal of Fish Biology 68: 373–384.

Knight, M. E. & G. F. Turner, 1999. Reproductive isolation among closely related Lake Malawi cichlids: can males recognize conspecific females by visual cues? Animal Behaviour 58: 761–768.

Knight, M. E. & G. F. Turner, 2004. Laboratory mating trials indicate incipient speciation by sexual selection among populations of the cichlid fish *Pseudotropheus zebra* from Lake Malawi. Proceedings of the Royal Society of London Series B 271: 675–680.

Kodric-Brown, A. & U. Strecker, 2001. Responses of *Cyprinodon maya* and *C labiosus* females to visual and olfactory cues of conspecific and heterospecific males. Biological Journal of the Linnean Society 74: 541–548.

Kohda, M., K. Sasaki, Y. Sakai, N. Ohnishi & K. Matsumoto, 2005. Preliminary study on female choice of mates versus sites in a wrasse, *Cirrhilabrus temminckii*. Ichthyological Research 52: 406–409.

Konings, A., 1998. Tanganyika Cichlids in Their Natural Habitat. Cichlid Press, El Paso.

Kraaijeveld, K., J. Gregurke, C. Hall, J. Komdeur & R. A. Mulder, 2004. Mutual ornamentation, sexual selection, and social dominance in the black swan. Behavioral Ecology 15: 380–389.

Ludlow, A. M., M. Itzkowitz & D. R. Baird, 2001. Male mating success and female preferences for multiple traits in the twoline pupfish (*Cyprinodon bifasciatus*). Behaviour 138: 1303–1318.

Maan, M. E., O. Seehausen, L. Söderberg, L. Johnson, E. A. P. Ripmeester, H. D. J. Mrosso, M. I. Taylor, T. J. M. Van Dooren & J. J. M. Van Alphen, 2004. Intraspecific sexual selection on a speciation trait, male colouration, in the Lake Victoria cichlid *Pundamilia nyererei*. Proceedings of the Royal Society of London, Series B 271: 2445–2452.

Martin, J., 2004. Hormonal and physiological profiles of female *Haplochromis burtoni* as it relates to affiliative behavior. Stanford Undergraduate Research Journal 3: 55–61.

McElroy, D. M. & I. Kornfield, 1990. Sexual selection, reproductive behavior, and speciation in the mbuna species flock of Lake Malawi (Pisces: Cichlidae). Environmental Biology of Fishes 28: 273–284.

McLennan, D. A., 2004. Male brook sticklebacks' (*Culaea inconstans*) response to olfactory cues. Behaviour 141: 1411–1424.

McLennan, D. A. & M. J. Ryan, 1999. Interspecific recognition and discrimination based upon olfactory cues in northern swordtails. Evolution 53: 880–888.

Nelissen, M., 1976. Contribution to the ethology of *Tropheus moorii* Boulenger (Pisces, Cichlidae) and a discussion of the significance of its colour pattern. Revue de Zoologie Africaine 90: 17–29.

Ólafsdóttir, G. Á., M. G. Ritchie & S. S. Snorrason, 2006. Positive assortative mating between recently described sympatric morphs of Icelandic sticklebacks. Biology Letters 2: 250–252.

Olsson, M., 2001. No female choice in Mallee dragon lizards, *Ctenophorus fordi*. Evolutionary Ecology 15: 129–141.

Parri, S., R. V. Alatalo, J. Kotiaho & J. Mappes, 1997. Female choice for male drumming in the wolf spider *Hygrolycosa rubrofasciata*. Animal Behaviour 53: 305–312.

Pauers, M. J., J. S. McKinnon & T. J. Ehlinger, 2004. Directional sexual selection on chroma and within-pattern colour contrast in *Labeotropheus fuelleborni*. Proceedings of the Royal Society of London, Series B 271: 444–447.

Payne, R. B., L. L. Payne, J. L. Woods & M. D. Sorenson, 2000. Imprinting and the origin of parasite-host species associations in brood parasitic indigobirds *Vidua chalybeata*. Animal Behaviour 59: 69–81.

Plenderleith, M., C. Van Oosterhout, R. L. Robinson & G. F. Turner, 2005. Female preference for conspecific males based on olfactory cues in a Lake Malawi cichlid fish. Biology Letters 1: 411–414.

Rolán-Alvarez, E., J. Erlandsson, K. Johannesson & R. Cruz, 1999. Mechanisms of incomplete prezygotic reproductive isolation in an intertidal snail: testing behavioural models in wild populations. Journal of Evolutionary Biology 12: 879–890.

Rolán-Alvarez, E. & A. Caballero, 2000. Estimating sexual selection and sexual isolation effects from mating frequencies. Evolution 54: 30–36.

Salzburger, W., T. Mack, E. Verheyen & A. Meyer, 2005. Out of Tanganyika: genesis, explosive speciation, key-innovations and phylogeography of the haplochromine cichlid fishes. BMC Evolutionary Biology 5: 17.

Salzburger, W., H. Niederstätter, A. Brandstätter, B. Berger, W. Parson, J. Snoeks & C. Sturmbauer, 2006. Colour-assortative mating among populations of *Tropheus moorii*, a cichlid fish from Lake Tanganyika, East Africa. Proceedings of the Royal Society of London Series B 273: 257–266.

Santangelo, N., 2005. Courtship in the monogamous convict cichlid: what are individuals saying to rejected and selected mates? Animal Behaviour 69: 143–149.

Sargent, R. C., V. N. Rush, B. D. Wisenden & H. Y. Yan, 1998. Courtship and mate choice in fishes: integrating behavioral and sensory ecology. American Zoologist 38: 82–96.

Scholz, C. & B. Rosendahl, 1988. Low lake stands in Lakes Malawi and Tanganyika, East Africa, delineated with multifold seismic data. Science 240: 1645–1648.

Schupke, P., 2003. African Cichlids II: Tanganyika I: *Tropheus*. Aqualog A. C. S Gmbh, Rodgau.

Schwarz, D. & B. A. McPheron, 2007. When ecological isolation breaks down: sexual isolation is an incomplete barrier to hybridization between *Rhagoletis* species. Evolutionary Ecology Research 9: 829–841.

Seehausen, O., 1997. Distribution of and reproductive isolation among colour morphs of a rock-dwelling Lake Victoria

cichlid (*Haplochromis nyererei*). Ecology of Freshwater Fish 6: 59–66.
Seehausen, O. & J. J. M. Van Alphen, 1998. The effect of male colouration on female mate choice in closely related Lake Victoria cichlids (*Haplochromis nyererei* complex). Behavioural Ecology and Sociobiology 42: 1–8.
Seehausen, O. & D. Schluter, 2004. Male-male competition and nuptial-colour displacement as a diversifying force in Lake Victoria cichlid fishes. Proceedings of the Royal Society of London, Series B 271: 1345–1353.
Seehausen, O., F. Witte, J. J. M. Van Alphen & N. Bouton, 1998. Direct mate choice maintains diversity among sympatric cichlids in Lake Victoria. Journal of Fish Biology 53: 37–55.
Sefc, K. M., 2008. Variance in reproductive success and the opportunity for selection in a serially monogamous species: simulations of the mating system of *Tropheus* (Teleostei: Cichlidae). Hydrobiologia. doi:10.1007/s10750-008-9563-1.
Sefc, K. M., S. Baric, W. Salzburger & C. Sturmbauer, 2007. Species-specific population structure in rock-specialized sympatric cichlid species in Lake Tanganyika, East Africa. Journal of Molecular Evolution 64: 33–49.
Setchell, J. M., 2005. Do female mandrills prefer brightly coloured males? International Journal of Primatology 26: 715–735.
Snoeks, J., L. Rüber & E. Verheyen, 1994. The Tanganyika problem: comments on the taxonomy and distribution patterns of its cichlid fauna. Archiv für Hydrobiologie Beiheft Ergebnisse der Limnologie 44: 355–372.
Sturmbauer, C. & A. Meyer, 1992. Genetic divergence, speciation and morphological stasis in a lineage of African cichlid fishes. Nature 358: 578–581.
Sturmbauer, C. & R. Dallinger, 1995. Diurnal variation of spacing and foraging behaviour in *Tropheus moorii* (Cichlidae) in Lake Tanganyika, Eastern Africa. Netherlands Journal of Zoology 45: 386–401.
Sturmbauer, C., U. Hainz, S. Baric, E. Verheyen & W. Salzburger, 2003. Evolution of the tribe Tropheini from Lake Tanganyika: synchronized explosive speciation producing multiple evolutionary parallelism. Hydrobiologia 500: 51–64.
Sturmbauer, C., S. Koblmüller, K. M. Sefc & N. Duftner, 2005. Phylogeographic history of the genus *Tropheus*, a lineage of rock-dwelling cichlid fishes endemic to Lake Tanganyika. Hydrobiologia 542: 335–366.
Summers, K., R. Symula, M. Clough & T. Cronin, 1999. Visual mate choice in poison frogs. Proceedings of the Royal Society of London, Series B 266: 2141–2145.
Werner, N. Y. & A. Lotem, 2003. Choosy males in a haplochromine cichlid: first experimental evidence for male mate choice in a lekking species. Animal Behaviour 66: 293–298.
West-Eberhart, M. J., 1983. Sexual selection, social competition, and speciation. The Quarterly Review of Biology 58: 155–183.
White, D. J. & B. G. J. Galef, 1999. Affiliative preferences are stable and predict mate choices in both sexes of Japanese quail, *Coturnix japonica*. Animal Behaviour 58: 865–871.
Wickler, W., 1969. Zur Soziologie des Brabantbuntbarsches, *Tropheus moorei* (Pisces, Cichlidae). Zeitschrift für Tierpsychologie 26: 967–987.
Wiernasz, D. C., 1995. Male choice on the basis of female melanin pattern in Pieris butterflies. Animal Behaviour 49: 45–51.
Wong, B. B. M., J. S. Keogh & M. D. Jennions, 2004. Mate recognition in a freshwater fish: geographical distance, genetic differentiation, and variation in female preference for local over foreign males. Journal of Evolutionary Biology 17: 701–708.
Wong, B. B. M., H. S. Fisher & G. G. Rosenthal, 2005. Species recognition by male swordtails via chemical cues. Behavioral Ecology 16: 818–822.
Yanagisawa, Y. & M. Nishida, 1991. The social and mating system of the maternal mouthbrooder *Tropheus moorii* (Cichlidae) in Lake Tanganyika. Japanese Journal of Ichthyology 38: 271–282.

Hydrobiologia (2008) 615:49–56
DOI 10.1007/s10750-008-9567-x

SPECIATION IN ANCIENT LAKES

Variation of territory size and defense behavior in breeding pairs of the endemic Lake Tanganyika cichlid fish *Variabilichromis moorii*

Christian Sturmbauer · Christoph Hahn · Stephan Koblmüller · Lisbeth Postl · Danny Sinyinza · Kristina M. Sefc

Published online: 11 September 2008

Abstract Lake Tanganyika harbors the ecologically, morphologically, and behaviorally most diverse species flock of cichlid fishes. It is comprised by substrate breeding and mouthbrooding species, most of which live in littoral habitats. Species communities are characterized by complex behavioral and trophic interactions, resulting in a dense pattern of partially overlapping territories, depending on the degree of ecological distinctness. We studied territorial behavior of breeding pairs in a substrate breeding species, with respect to territory size and defense behavior. The study species *Variabilichromis moorii* belongs to the tribe Lamprologini, the most species rich tribe of cichlids in Lake Tanganyika. Our study shows that breeding pairs of *V. moorii* can have highly complex territories, in which both parents hold separate subterritories which are shifted slightly according to the movements of the fry, but the outer borders are conjointly defended. The size of the total defended territory varied from <1 to almost 4 m^2, averaging at about 2 m^2. Depending on presence of competitors or fry-predators evoking agonistic interactions, the territory size varied quite substantially over the day. Attack rates and size of the defended area decreased with water depth. Agonistic behavior was observed toward heterospecifics as well as conspecifics, with heterospecific attacks mostly concerning territorial neighbors and potential fry-predators in about equal frequencies.

Keywords Lamprologini · Ecology · Fish community · Territorial behavior · Eastern Africa

Guest editors: T. Wilke, R. Väinölä & F. Riedel
Patterns and Processes of Speciation in Ancient Lakes: Proceedings of the Fourth Symposium on Speciation in Ancient Lakes, Berlin, Germany, September 4–8, 2006

Electronic supplementary material The online version of this article (doi:10.1007/s10750-008-9567-x) contains supplementary material, which is available to authorized users.

C. Sturmbauer (✉) · C. Hahn · S. Koblmüller · L. Postl · K. M. Sefc
Department of Zoology, University of Graz, Universitätsplatz 2, 8010 Graz, Austria
e-mail: Christian.sturmbauer@uni-graz.at

D. Sinyinza
Department of Research and Specialist Services, Fisheries Research Division, Ministry of Agriculture, Food and Fisheries, P.O. Box 55, Mpulungu, Zambia

Introduction

It was at the turn of the last century when the ichthyologist George Albert Boulenger (1898) gave the first scientific account of the incredible cichlid fauna in the three East African Lakes Malawi, Tanganyika, and Victoria. With an age of 9–12 MY and the establishment of a real lacustrine habitat with deep water conditions about 5–6 MYA Lake Tanganyika is the oldest of the three East African Great Lakes (Cohen et al., 1993, 1997). With about 250 species classified into 12 (Poll, 1986)—or alternatively 16 (Takahashi, 2003)—tribes, the cichlid fishes of Lake Tanganyika comprise the most diverse species flock in terms of morphology, ecology, and behavior (Poll, 1986; Snoeks, 2000). The structure of the shore lines of Lake Tanganyika varies among sandy, muddy, and rocky sections, interrupted by river estuaries. It has long been suggested that the type of shore restricts the dispersal of many species specialized to particular habitats by forming ecological barriers (Fryer & Iles, 1972), so that allopatric speciation, driven in part by lake level fluctuations, accounted for the formation of several species (Sturmbauer, 1998; Sturmbauer et al., 2001). Each continuous stretch of shore is inhabited by arrays of species which might differ from that of the next shore section, especially after habitat discontinuities.

The rocky shores of Lake Tanganyika harbor the most species-rich community of the lake, formed by several lineages of mouthbrooding and substrate breeding species which coexist by finely tuned interactions. Such densely packed communities enforce competition for space and resources and their members exhibit complex interactions in a territorial framework (Kawanabe, 1981; Takamura, 1983, 1984; Kuwamura, 1987, 1992; Kohda & Yanagisawa, 1992; Ochi, 1993; Yanagisawa, 1993; Kohda, 1995; Sturmbauer & Dallinger, 1995; Kohda & Tanida, 1996; Matsumoto & Kohda, 2004, 2007; Ota & Kohda, 2006). Territories were found to be partially overlapping among species occupying different trophic niches, implying that aggression toward other individuals and species is differential, with attack rates depending on the risk of intruders snatching the crucial resources within the defender's territory (e.g., Nakano & Nagoshi, 1990). Differential agonistic behavior toward intruders has been suggested to be beneficial to the territory owner, since costs accompanying territorial defense should be reduced by not attacking all fishes approaching the territory with equal intensity (Nakano & Nagoshi, 1990). Aside of resource defense, motivation for territoriality relates to reproduction and brood care. Breeding space is an issue, especially for substrate-breeding species who guard and defend eggs and offspring (Keenleyside, 1991). Moreover, knowledge about territorial behavior and territory sizes not only provides information about community structure and ecological interaction but also helps to understand evolutionary trajectories of the involved species.

Our study focuses on aspects of territorial behavior in the context of brood care. The target species *Variabilichromis moorii* (formerly named *Neolamprologus moorii*) represents a monotypic genus forming one of several ancient lineages of the exclusively substrate breeding tribe Lamprologini, the most species rich lineage of cichlids in Lake Tanganyika (Sturmbauer et al., 1994; Day et al., 2007). The species is endemic to the lake and restricted to the southern lake basin. Only little, if any geographic variation is found in terms of color, while genetic differentiation is high among populations (Duftner et al., 2006). *V. moorii* lives above rocks in the upper littoral (Brichard, 1989) and feeds on small invertebrates and aufwuchs, i.e., epilithic algae and organisms therein. It is a biparental substrate-breeder, which breeds underneath and between rocks (Karino, 1997). The purpose of the present study was to add to the knowledge on the territoriality of this species (Karino, 1997, 1998), by focusing on the territorial behavior of breeding pairs of *V. moorii* to elucidate temporal and spatial stability and frequencies of agonistic behavior against con- and hetero-specific competitors and fry predators.

Materials and methods

Study area

The study transect was established about 2 km south of the Kalambo River estuary with its southernmost near-shore point at S08°37′21.9″ and E31°12′0.8″. Data was collected by snorkeling in March 2006. The habitat was a cobble shore with patches of sand and vegetation in the deeper areas and had an extension of 400 m^2. The area was subdivided into grids of two by two meters with coordinates marked on flat stones lettered from A-0 to K-10. Water depth was measured at each coordinate mark and a substrate map was drawn on underwater tablets by snorkeling and a map of the whole transect was constructed from these sketches (Fig. 1).

Strategy of the behavioral observations

First, the overall number of *V. moorii* in the study transect was counted by swimming through the transect in parallel to the shore along the coordinate

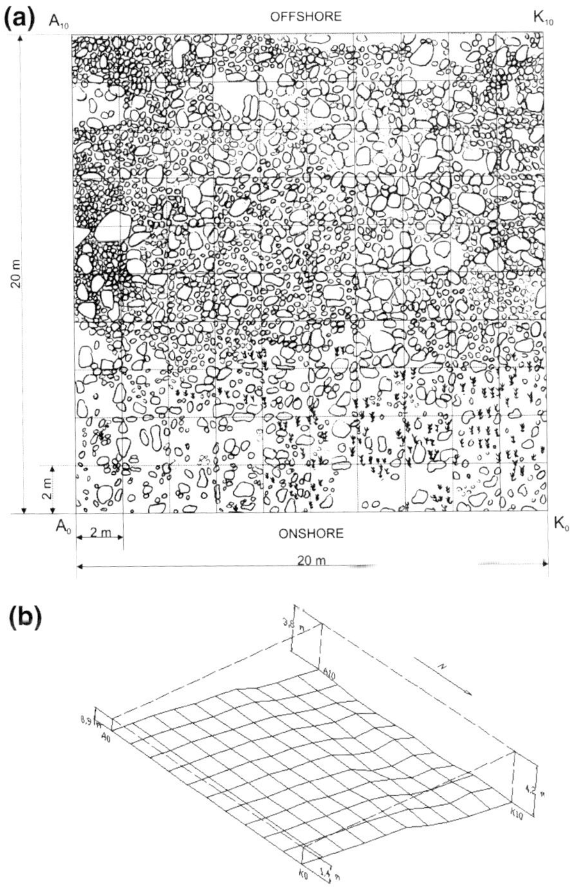

Fig. 1 Research transect. (**a**) Substrate map, (**b**) depth profile

rows of the transect (e.g., from row A_0–A_1 until K_0–K_1). Each count was repeated three times per row, averaged, and the density of individuals per row per m^2 was calculated. Then 12 breeding pairs with small (<2 cm) offspring were selected for observation, spread evenly over the research transect. Two daily periods were defined, one from 8.00 h to 12.00 h (morning) and from 14.00 h to 18.00 h (afternoon). For each breeding pair three 15-min observations were carried out for each daily period. The location of each parent was recorded every 30 s on underwater writing tablets. These regular recordings were taken as estimate for the territory core area in which an individual spent most of the time. In addition, all short-term movements for territorial defense, divided into conspecific and heterospecific attacks, were recorded as an estimate of the total defended territory. Both the territory core and the total defended territory were quantified as maximum convex polygon after transferring the data to millimeter grid paper and digitizing them. The areas were averaged for each daily period or breeding pair to obtain average territory core areas and total defended areas. By overlaying the area sketches of observation units per breeding pair (six observation units) and daily period (three observation units), we calculated the "superimposed area" for both the territory core and the total defended territory. Through these measures, termed superimposed core area and superimposed total area, we integrated observed territory sizes over time.

It was not possible to distinguish male and female individuals of the breeding pairs. While females were found to be significantly larger than males (Karino, 1997), it was not possible to sex individuals under water by snorkeling, because the size difference of <1 cm was not possible to determine. For this reason the sizes of the sub core-territories of males and females cannot be discussed in this study.

Concerning heterospecific attacks, the target species were also recorded for each agonistic action. The target species were categorized into resource competitors, fry-predators and piscivorous species considering species characterizations of Brichard (1989) and Konings (1998). Conspecific and

Table 1 Frequency of attacks per hour for each of the 12 surveyed breeding pairs in the morning, afternoon and over the whole day

Pair no.	Morning			Afternoon			Whole day		
	CA/h	HA/h	TA/h	CA/h	HA/h	TA/h	CA/h	HA/h	TA/h
1	13.3	44.0	57.3	18.7	53.3	72.0	16.0	48.7	64.7
2	22.7	44.0	66.7	21.3	56.0	77.3	22.0	50.0	72.0
3	5.3	30.7	36.0	10.7	50.7	61.3	8.0	40.7	48.7
4	12.0	66.7	78.7	9.3	88.0	97.3	10.7	77.3	88.0
5	8.0	38.7	46.7	10.7	40.0	50.7	9.3	39.3	48.7
6	12.0	21.3	33.3	10.7	28.0	38.7	11.3	24.7	36.0
7	30.7	32.0	62.7	18.7	26.7	45.3	24.7	29.3	54.0
8	6.0	17.3	41.3	20.0	36.0	56.0	13.0	26.7	48.7
9	20.0	24.0	44.0	18.7	33.3	52.0	19.3	28.7	48.0
10	8.0	18.7	26.7	2.7	17.3	20.0	5.3	18.0	23.3
11	13.3	29.3	42.7	10.7	20.0	30.7	12.0	24.7	36.7
12	21.3	10.7	32.0	13.3	13.3	26.7	17.3	12.0	29.3

Attack rates were averages calculated from three 15 min observations in the morning and three observations in the afternoon. CA/h: Conspecific attacks per hour; HA/h: heterospecific attacks per hour; TA/h: total number of attacks per hour heterospecific attacks were counted and quantified in attacks per hour. Statistical comparisons were carried out in the SigmaStat software package (SPSS Inc.). Data were tested for normal distribution to carry out the appropriate tests.

Results

Ecological parameters of the study habitat

Water temperature at the study area at Kalambo was constant at 28°C. The average water depth of the inshore border of the transect was 1.18 m and the average depth of the outermost border was 4.15 m, resulting in an inclination of 15%. The substrate consisted of sand (white areas in Fig. 1) and rocks with little macrophyte vegetation between (*Elodea* sp. and *Vallisneria* sp.). *Variabilichromis moorii* was the most abundant cichlid species in the transect counting 339 adult and sub-adult individuals in the whole transect, translating into 31% of the total (adult and sub-adult) cichlid fish population and a density of 0.8 fish per m^2.

Defense behavior

There were significantly more heterospecific attacks than conspecific attacks, both considering morning and afternoon data separately (Table 1; *t*-test for independent samples; $P = 0.002$) and combining morning and afternoon data ($P = 0.005$). A *t*-test for paired samples showed significantly more heterospecific attacks in the afternoon than in the morning ($P = 0.031$), whereas there was no significant difference in conspecific attacks ($P = 0.177$) and when combining con- and heterospecific attacks ($P = 0.210$). Heterospecific attacks were observed toward at least 29 different species (Table 2). Some intruders were not identified to species level, e.g., *Mastacembelus* spp., *Synodontis* spp., *Petrochromis* spp. The species against which the most attacks were directed was *Ophthalmotilapia ventralis*, followed by *Tropheus moorii*, *Lepidiolamprologus elongatus*, *Neolamprologus fasciatus*, and *Chalinochromis brichardi*. As only five among the top-ten targets were fry-eaters or piscivorous species, there was no increased aggressiveness toward fry-predators and piscivorous species evident. Fry-predators are indicated by an asterisk and piscivorous species by a double-asterisk in Table 2. However, the most potent predators *L. elongatus* and *N. fasciatus*, were rapidly recognized and attacked over long distances by all breeding pairs. The heterospecific attack rate decreased with depth (Pearson $r = -0.703$, $P = 0.011$), whereas the conspecific attack rate showed no correlation with depth.

Table 2 Attacked species and total number of observed attacks

Attacked species	Number of attacks
Ophthalmotilapia ventralis	91
Tropheus moorii	52
*Lepidiolamprologus elongatus***	45
*Neolamprologus fasciatus***	39
*Chalinochromis brichardi**	34
*Gnathochromis pfefferi**	21
Mastacembelus spp.**	21
Xenotilapia spiloptera	21
Lamprichthys tanganicanus	20
Petrochromis spp.	14
Aulonocranus dewindti	11
Telmatochromis temporalis	11
Perissodus microlepis	10
Telmatochromis vittatus	9
Eretmodus cyanostictus	8
*Lamprologus lemairii***	8
Ectodus descampsi	6
Lobochilotes labiatus	6
Neolamprologus caudopunctatus	6
Neolamprologus tetracanthus	5
Synodontis multipunctatus	5
*Altolamprologus compressiceps**	4
*Ctenochromis horei**	2
Julidochromis ornatus	2
*Boulengerochromis microlepis***	1
Cyathopharynx foae	1
Simochromis sp.	1
Synodontis sp.	1

Asterisks indicate fry-predators (*) or piscivorous species (**)

Territory areas

The graphs of all six daily observations are shown in Fig. 2 for breeding pair 1 only. Those for the remaining breeding pairs are shown in the

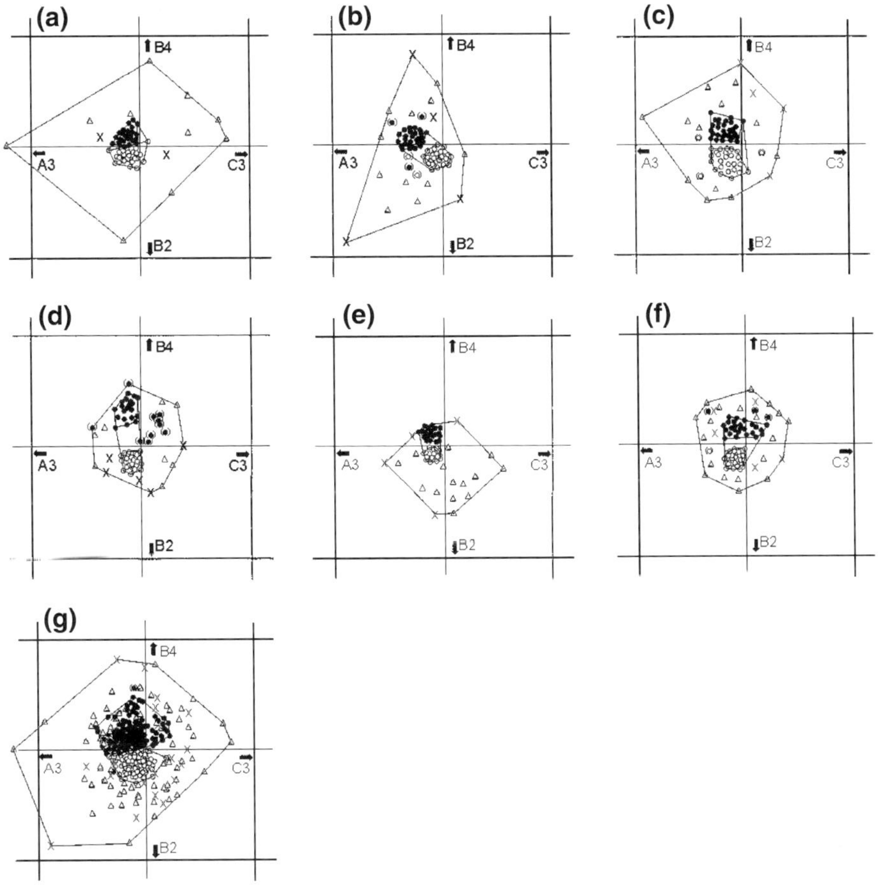

Fig. 2 Territorial area of breeding pair 1, derived from 15-min behavioral observations. Territory core areas for both parents, as well as total defended territories are shown. (**a**)–(**c**) show observations in the morning (9:00–12:00); (**d**)–(**f**) show afternoon-observations (13:00–18:00); (**g**) shows the circa-diurnal summary of all observations superimposed into a single graph, termed superimposed core area and superimposed total area in the text. Symbols: open circle: regular 30 s census of fish No.1 of the breeding pair; filled circle: regular 30 s census of fish No. 2 of the breeding pair; cross: conspecific attack; triangle: heterospecific attack; Symbols in parentheses were considered outliners and not used for the calculation of the territorial core area

Table 3 Summary of total defended territory sizes and territorial core areas of 12 breeding pairs

	Minimum	Maximum	Average across pairs (s.d.)
Total defended territory, averaged across observations	0.23	2.48	0.81 (0.25)
Total defended territory, superimposed across observations	0.93	3.95	2.12 (0.82)
Joint core area of both adult fish, averaged across observations	0.06	0.34	0.18 (0.03)
Joint core area of both adult fish, superimposed across observations	0.26	0.80	0.44 (0.17)

Area sizes were calculated from six 15 min observations per pair. Area sizes are given in m^2; s.d. stands for standard deviation

Supplementary material—Appendix 2. It turned out that the core area of the territories in which the adults and their brood stayed, were relatively small in relation to the total area defended. However, both areas varied from observation to observation (Supplementary material—Appendix 1). The breeding pair divided the territory area between them with little overlap. The territory core areas of the two adults shifted slightly from observation to observation. Total defended territories highly depended on the presence of intruders or predators and thus varied considerably (see Supplementary material—Appendix 2). Total defended territories, averaged across the six observations, ranged from 0.23 to 2.48 m^2 per breeding pair, with an average of 0.81 m^2 (s.d. 0.252 m^2) across all pairs. Territory core areas ranged from 0.06 to 0.34 m^2 per pair, with an average of 0.18 m^2 (s.d. 0.031 m^2) across all pairs (Table 3).

By superimposing all six observations into one single graph for every pair, the smallest superimposed total area of a breeding pair measured 0.93 m^2, and the largest area was 3.95 m^2, amounting to an average of 2.12 m^2 (s.d. 0.821 m^2) for all 12 pairs (Table 3). All territory areas obtained from single observations are shown in Supplementary material—Appendix 1.

We also tested for correlation of territory area and water depth (see Supplementary material—Appendix 1) and found negative correlation with average total defended territory of the breeding pairs (Pearson $r = -0.622$, $P = 0.031$), but not with superimposed total defended territory (Pearson $r = -0.52$, $P = 0.08$). No correlation between territory core areas and water depth was observed.

Discussion

Depending on the resources to be defended and the objects to be repelled, the territories of cichlids have been classified into four types: feeding, mating, nest-guarding, and brood guarding (Yuma & Kondo, 1997). In the present study, fry-guarding pairs of *V. moorii* were investigated, indicating that the observed territories functioned as a combination of both a brood guarding and feeding territory.

Perhaps the most remarkable observation about territory areas of breeding pairs of *V. moorii* was that each parent had its own sub-territory which seldom overlapped with that of the other parent (see Fig. 2; Supplementary material—Appendix 2). The fry was always located in between the two adults. In addition, both parents shifted their sub-territories over the day. Interestingly, the superimposed sub-territories, defined by defense activities, hardly overlapped as well. The seemingly equal investment of both parents in brood defense is remarkable, given that genetic analyses revealed a high degree of extra-pair paternity in *V. moorii* broods from the same location (Sefc et al. 2008). The slight but steady movements of the breeding pairs are the reason why the superimposed (sub-) territories are often much larger than the areas scored for a single 15-min observation (see Supplementary material—Appendix 1).

The finding that the total territory areas decreased with water depth can be explained by the method used to define the total area, namely from the area defended from intruders. Attack rates on heterospecific intruders also decreased with depth (see above data), and attack rates and total territory size were positively correlated ($r = 0.662$, $P = 0.019$). In a study of parental investment by breeding pairs of *V. moorii*, Karino (1997) found that brood defense, defined as the attack rate against fry-eaters adjusted for the intruder's preference for fry, was predominantly determined by brood size, and not by fry size. While that study did not consider water depth as additional factor influencing attack rates, our study did not distinguish between the diet preferences of the heterospecific intruders and between brood and fry sizes of the defended broods. Taken together, the fact that the depth-dependence of heterospecific attack rates was not cancelled by the correlation between brood-size and attacks against fry-eaters in our study, and vice versa in Karino (1997), indicates that both factors may be relevant in determining brood defense rates by the guarding pair. Whether predator density is the decisive factor for the decrease of heterospecific attacks with increasing water depth, must be left as open question for the future.

A similar, negative relationship between water depth and attack rates has been observed by Karino (1998) in a study of solitary, non-breeding *V. moorii*. This previous study further reported an increase in territory size with water depth, but in that case, territory size was defined by the sites where fish picked algae, and would correspond to our definition of core area rather than total defended area. In the present study on breeding pairs, there was no significant increase of core territory size with water depth. Karino (1998) explained the increase of territory size with water depth by the necessity to compensate reduced algal growth in deeper regions to satisfy nutritional needs of *V. moorii*. When not breeding, fish can expand the territory they use as food resource. However, during breeding, the highest priority lies in defending the fry, and individuals probably cannot sustain such a large area, but remain closer to their offspring. Territory sizes of non-breeding *V. moorii* were studied in Sturmbauer et al. (2008) and turned out to be similar.

The fact that heterospecific attack rates were significantly higher than conspecific attack rates is probably explained by differences in the provocation by intruders. Conspecific attacks are directed against individuals of *V. moorii* that have their territory in the nearer ambiance, and the attack rate is determined, and also somewhat limited, by the number of conspecifics in the "neighborhood". Frostman & Sherman (2004) demonstrated that another lamprologine cichlid species, *Neolamprologus pulcher*, exhibits strikingly different behavioral responses to familiar and unfamiliar neighbors in that the aggression level toward unfamiliar neighbors was significantly higher. Whether *V. moorii* also has the ability to distinguish between familiar and unfamiliar conspecifics and consequently exhibits differential intensities of agonistic behavior against conspecifics remains to be tested.

Heterospecific attacks, on the other hand, were not restricted to species with a nearby or overlapping territory, and were often directed against migrating fish such as non-territorial juveniles, females or predator species which do not necessarily have stable territories. It is interesting that there were significantly more heterospecific attacks in the afternoon. One might speculate that some fish are more mobile in the afternoon, or potential fry predators utilize the increased wave action and water turbidity during afternoon hours to ambush their prey. Among the top four species attacked, there are two non-piscivorous territorial neighbors and their non-territorial juveniles, *Ophthalmotilapia ventralis* and *Tropheus moorii*, and two piscivorous predators (*Lepidiolamprologus elongatus*, *Neolamprologus fasciatus*). In agreement with Karino (1998), the predators *L. elongatus* and *N. fasciatus* where rapidly recognized, and attacked over long distances by all breeding pairs. This is also consistent with observations on other brood-guarding species, that predatory fishes presenting a high risk for broods are attacked from greater distances than other species (Perrone, 1978; McKaye, 1984; Nakano & Nagoshi, 1990; Kohda, 1997; Karino, 1998; Haley & Muller, 2002). Furthermore, aggression intensity of territory owners toward intruders is often positively related to the degree of overlap of the food resources contested between them (e.g., Low, 1971; Myrberg, 1972; Ebersole, 1977; Kohda, 1981; Hori, 1991; Matsumoto & Kohda, 2004), which implies that territory owners have to be able to discriminate potential intruders based on their food habits (see also Thresher, 1976; Ebersole, 1977; Kohda, 1997). The high frequency of attacks toward *Ophthalmotilapia ventralis* and *T. moorii* may thus be explained by food competition (Takamura, 1984; Karino, 1998), but this is a less likely explanation for the frequent attacks on *Chalinochromis brichardi*, which is considered an invertebrate feeder and opportunistic fry-predator. Furthermore, *C. brichardi* occurred at much lower numbers than the highly abundant *T. moorii* and *O. ventralis* in the study transect (2% vs. 10 and 12%, respectively; Sturmbauer et al. 2008), such that encounter rates fail to explain species-specific attack rates. There is certainly need and great potential for studies clarifying the perception of co-distributed species as competitors and predators within a species community.

Acknowledgments We thank P. Ngalande and the team at the Mpulungu Station of the Ministry of Agriculture and Cooperatives, Republic of Zambia, as well as J. Knapp and N. Feiner for their cooperation during fieldwork. This study was financially supported by the Austrian Science Foundation (Grants P 17680 to CS and P17380 to KMS).

References

Brichard, P., 1989. Book of cichlids and all the other fishes of Lake Tanganyika. T.F.H. Publications, Neptune City, New Jersey, USA.

Cohen, A. S., K. E. Lezzar, J. J. Tiercelin & M. Soreghan, 1997. New palaeogeographic and lake-level reconstructions of Lake Tanganyika: implications for tectonic, climatic and biological evolution in a rift lake. Basin Research 9: 107–132.

Cohen, A. S., M. J. Soreghan & C. A. Scholz, 1993. Estimating the age of formation of lakes: an example from Lake Tanganyika, East-African Rift system. Geology 21: 511–514.

Day, J. J., S. Santini & J. Garcia-Moreno, 2007. Phylogenetic relationships of the Lake Tanganyika cichlid tribe Lamprologini: The story from mitochondrial DNA. Molecular Phylogenetics and Evolution 45: 629–642.

Duftner, N., K. M. Sefc, S. Koblmüller, B. Nevado, E. Verheyen, H. Phiri & C. Sturmbauer, 2006. Distinct population structure in a phenotypically homogeneous rock-dwelling cichlid fish from Lake Tanganyika. Molecular Ecology 15: 2381–2396.

Ebersole, J. P., 1977. The adaptive significance of interspecific territoriality in the reef fish *Eupomacentrus leucostictus*. Ecology 58: 914–920.

Fryer, G. & T. D. Iles, 1972. The cichlid fishes of the Great Lakes of Africa: their biology and evolution. T.F.H. Publications, Neptune City, New Jersey, USA.

Frostman, P. & P. T. Sherman, 2004. Behavioral response to familiar and unfamiliar neighbors in a territorial cichlid, *Neolamprologus pulcher*. Ichthyological Research 51: 283–285.

Haley, M. & C. R. Muller, 2002. Territorial behaviour of beaugregory damselfish (*Stegastes leucostictus*) in response to egg predators. Journal of Experimental Marine Biology and Ecology 273: 151–159.

Hori, M., 1991. Feeding relationships among cichlid fishes in Lake Tanganyika: effects of intra- and interspecific variations of feeding behavior on their coexistence. Ecology International Bulletin 19: 89–101.

Karino, K., 1997. Influence of brood size and offspring size on parental investment in a biparental cichlid fish, *Neolamprologus moorii*. Journal of Ethology 15: 39–43.

Karino, K., 1998. Depth-related differences in territory size and defence in the herbivorous cichlid, *Neolamprologus moorii*, in Lake Tanganyika. Ichthyological Research 45: 89–94.

Kawanabe, H., 1981. Territorial behaviour of *Tropheus moorii* (Osteichthyes, Cichlidae) with a preliminary consideration of territorial forms in animals. African Study Monographs 1: 101–106.

Keenleyside, H. M. A., 1991. Parental care. In Keenleyside, H. M. A. (ed.), Cichlid Fishes—Behaviour, Ecology and Evolution. Chapman & Hall, London: 191–208.

Kohda, M., 1981. Interspecific territoriality and agonistic behavior of a temperate pomacentrid fish, *Eupomacentrus altus* (Pisces: Pomacentridae). Zeitschrift für Tierpsychologie 56: 205–216.

Kohda, M., 1995. Territoriality of male cichlid fishes in Lake Tanganyika. Ecology of Freshwater Fish 4: 180–184.

Kohda, M., 1997. Interspecific society among herbivorous cichlid fishes. In Kawanabe, H., M. Hori & M. Nagoshi (eds), Fish Communities in Lake Tanganyika. Kyoto University Press, Kyoto: 107–120.

Kohda, M. & K. Tanida, 1996. Overlapping territory of the benthophagous cichlid fish, *Lobochilotes labiatus*, in Lake Tanganyika. Environmental Biology of Fishes 45: 13–20.

Kohda, M. & Y. Yanagisawa, 1992. Vertical distribution as of two herbivorous fishes of the genus *Tropheus* in Lake Tanganyika. Ecology of Freshwater Fish 1: 99–103.

Konings, A., 1998. Tanganyika cichlids in their natural habitat. Cichlid Press, El Paso, TX.

Kuwamura, T., 1987. Distribution of fishes in relation to the depth and substrate at Myako, east-middle coast of Lake Tanganyika. African Study Monographs 7: 1–14.

Kuwamura, T., 1992. Overlapping territories of *Pseudosimochromis curvifrons* males and other herbivorous cichlid fishes in Lake Tanganyika. Ecological Research 7: 43–53.

Low, R. W., 1971. Interspecific territoriality in a pomacentrid fish, *Pomacentrus flavicauda* Whitley. Ecology 52: 648–654.

Matsumoto, K. & M. Kohda, 2004. Territorial defense against various food competitors in the Tanganyikan benthophagous cichlid *Neolamprologus tetracanthus*. Ichthyological Research 51: 354–359.

Matsumoto, K. & M. Kohda, 2007. Male foraging avoidance in female feeding territories in a harem polygynous cichlid in Lake Tanganyika. Journal of Ethology 25: 21–27.

McKaye, K. R., 1984. Patterns of territoriality and brood defense in Central American substratum spawners and African mouth brooders. In Potts, G. W. & R. J. Wootton (eds), Fish Reproduction: Strategy and Tactics. Academic, London: 245–273.

Myrberg, A. A., 1972. Social dominance and territoriality in the bicolor damselfish, *Eupomacentrus partitus* (Poey) (Pisces: Pomacentridae). Behaviour 41: 207–231.

Nakano, S. & M. Nagoshi, 1990. Brood defense and parental roles in a biparental cichlid fish *Lamprologus toae* in Lake Tanganyika. Japanese Journal of Ichthyology 36: 468–476.

Ochi, H., 1993. Maintenance of separate territories for mating and feeding by males of a maternal mouthbrooding cichlid, *Gnathochromis pfefferi* in Lake Tanganyika. Japanese Journal of Ichthyology 40: 173–182.

Ota, K. & M. Kohda, 2006. Description of alternative male reproductive tactics in a shell-brooding cichlid, *Telmatochromis vittatus*, in Lake Tanganyika. Journal of Ethology 24: 9–15.

Perrone, M., 1978. The economy of brood defense by parental cichlid fishes, *Cichlasoma maculicauda*. Oikos 31: 137–141.

Poll, M., 1986. Classification des cichlidae du lac Tanganika. Tribus, genres et espèces. Academie Royale de Belgique Memoires de la Classe des Sciences 45: 1–163.

Sefc, K. M., K. Mattersdorfer, C. Sturmbauer, & S. Koblmüller, 2008. High frequency of multiple paternity in broods of a socially monogamous cichlid fish with biparental nest defence. Molecular Ecology 17: 2531–2543.

Snoeks, J., 2000. How well known is the ichthyodiversity of the large East African lakes? Advances in Ecological Research 31: 17–38.

Sturmbauer, C., 1998. Explosive speciation in cichlid fishes of the African Great Lakes: a dynamic model of adaptive radiation. Journal of Fish Biology 53(Supplement A): 18–36.

Sturmbauer, C., S. Baric, W. Salzburger, L. Rüber & E. Verheyen, 2001. Lake level fluctuations synchronize genetic divergences of cichlid fishes in African Lakes. Molecular Biology and Evolution 18: 144–154.

Sturmbauer, C. & R. Dallinger, 1995. Diurnal variation of spacing and foraging behaviour in *Tropheus moorii* (Cichlidae) in Lake Tanganyika, eastern Africa. Netherlands Journal of Zoology 45: 386–401.

Sturmbauer, C., E. Verheyen & A. Meyer, 1994. Mitochondrial phylogeny of the Lamprologini, the major substrate spawning lineage of cichlid fishes from Lake Tanganyika in Eastern Africa. Molecular Biology and Evolution 11: 691–703.

Sturmbauer, C., C. Fuchs, G. Harb, E. Damm, N. Duftner, M. Maderbacher, M. Koch & S. Koblmüller, 2008. Abundance, distribution, and territory areas of rock-dwelling Lake Tanganyika cichlid fish species. Hydrobiologia. doi:10.1007/s10750-008-9557-z.

Takahashi, T., 2003. Systematics of Tanganyikan cichlid fishes (Teleostei: Perciformes). Ichthyological Research 50: 367–382.

Takamura, K., 1983. Interspecific relationship between two Aufwuchs-eaters *Petrochromis polyodon* and *Tropheus moorii* (Pisces, Cichlidae) of Lake Tanganyika, with a discussion on the evolution and function of a symbiontic relationship. Physiology and Ecology Japan 20: 59–69.

Takamura, K., 1984. Interspecific relationships of Aufwuchs-eating fishes in Lake Tanganyika. Environmental Biology of Fishes 10: 225–241.

Thresher, R. E., 1976. Field analysis of the territoriality of the three spot damselfish, *Eupomacentrus planifrons* (Pomacentridae). Copeia 1976: 266–276.

Yanagisawa, Y., 1993. Long-term territory maintenance by female *Tropheus duboisi* (Cichlidae) involving foraging during the mouth-brooding period. Ecology of Freshwater Fish 2: 1–7.

Yuma, M. & T. Kondo, 1997. Interspecific relationships and habitat utilization among benthivorous cichlids. In Kawanabe, H., M. Hori & M. Nagoshi (eds), Fish Communities in Lake Tanganyika. Kyoto University Press, Kyoto: 87–103.

Hydrobiologia (2008) 615:57–68
DOI 10.1007/s10750-008-9557-z

SPECIATION IN ANCIENT LAKES

Abundance, distribution, and territory areas of rock-dwelling Lake Tanganyika cichlid fish species

Christian Sturmbauer · Corinna Fuchs · Georg Harb · Elisabeth Damm · Nina Duftner · Michaela Maderbacher · Martin Koch · Stephan Koblmüller

Published online: 11 September 2008

Abstract Lake Tanganyika, the second-oldest and second-deepest lake in the world, harbors an impressive cichlid fish fauna counting about 250 endemic species that are characterized by a great level of ecological, morphological, and behavioral specialization. This study describes and compares cichlid fish communities at two rocky shores with differential human impact in the south of Lake Tanganyika. Species inventories and depth-dependent abundances were elaborated. About 41 and 46 sympatric cichlid species were recorded in the two study sites, respectively. *Variabilichromis moorii* was the most abundant species (29–60% of total number of fishes), followed by *Aulonocranus dewindti* (3–19%), *Tropheus moorii* (12%), *Ophthalmotilapia ventralis* (4–10%), *Eretmodus cyanostictus* (6–11%), and *Cyathopharynx furcifer* (0.01–9%). All other species had abundances below 5%. It further emerged that large cichlids such as *Petrochromis* species, *Cyathopharynx furcifer*, and *Lobochilotes labiatus* were very rare at one location, with frequencies of 0.55% or less. Territorial sizes of three particularly abundant species, *Variabilichromis moorii*, *Aulonocranus dewindti*, and *Tropheus moorii*, were assessed by behavioral observations. We distinguished between territorial core areas and total defended area, yielding average core areas between 0.4 (*V. moorii*) and 1.6 m^2 (*T. moorii*), and total defended areas averaging for each species between 1.6 (*V. moorii*) and 5.0 m^2 (*A. dewindti*) with no significant differences between the two study sites. The data on individual densities are also relevant for evolutionary studies, in that they allow more accurate calculations of effective population sizes.

Keywords *Aulonocranus dewindti* · *Tropheus moorii* · *Variabilichromis moorii* · Ecology · Fish community · Territorial behavior · Eastern Africa

Guest editors: T. Wilke, R. Väinölä & F. Riedel
Patterns and Processes of Speciation in Ancient Lakes: Proceedings of the Fourth Symposium on Speciation in Ancient Lakes, Berlin, Germany, September 4–8, 2006

C. Sturmbauer (✉) · C. Fuchs · G. Harb · E. Damm · N. Duftner · M. Maderbacher · M. Koch · S. Koblmüller
Department of Zoology, University of Graz, Universitätsplatz 2, 8010 Graz, Austria
e-mail: christian.sturmbauer@uni-graz.at

N. Duftner
Section of Integrative Biology, University of Texas at Austin, 1 University Station #C0930, Austin, TX 78712, USA

Introduction

With an age of 9–12 million years (Cohen et al., 1993, 1997) and a maximum depth of 1,470 m, Lake Tanganyika is the second-oldest and second-deepest lake in the world. The old age of the ecosystem translates into a highly complex organismic community with a high degree of endemism, great levels of ecological, morphological, and behavioral specialization. Lake Tanganyika harbors an impressive cichlid fish fauna, counting 200 valid species plus several more awaiting scientific description (Snoeks, 2000; Koblmüller et al., 2008). They are currently classified into 12 (Poll, 1986), or alternatively 16 (Takahashi, 2003), tribes. Beside cichlid fishes, the lake also contains considerable species flocks of catfish, spiny eels, Nile perches, crustaceans, and gastropods (Coulter, 1991). The shorelines add up to a length of about 2,000 km and are structured into stretches of rocks, cobbles, pebbles, sand, or mud, interleaved by river estuaries. Depending on the kind and degree of specialization to particular habitats, many species are subdivided into populations. Most of the lacustrine cichlids live along the shoreline, where they are specialized to a variety of habitat types and dietary resources. Each shore type accommodates a different species community of cichlids, comprised by mostly specialized and thus stenotopic species, with only few ubiquitous species (Fryer & Iles, 1972). The littoral fish communities at rocky shores are the most species-rich communities of the lake, dominated by epilithic algae eaters and invertebrate feeders. Several species coexist by niche partitioning and complex interactions in a territorial framework, with territories partially overlapping between species utilizing different food resources, implying differential aggression levels toward other species and conspecific individuals. In order to derive evolutionary conclusions for these complex communities, it is important to collect descriptive data on abundances, community structure, and behavior of these fishes. Quite a substantial number of studies targeted territorial behavior and territory size of Lake Tanganyika cichlids (Kawanabe, 1981; Takamura, 1983, 1984; Kuwamura, 1987, 1992; Kohda & Yanagisawa, 1992; Ochi, 1993; Kohda, 1995; Sturmbauer & Dallinger, 1995; Kohda & Tanida, 1996; Matsumoto & Kohda, 2004), providing important information that might serve as a basis for further ecological and evolutionary investigations.

The present study aims to add ecological data on littoral species and describes the cichlid fish communities at two rocky shores with a differential human impact in the south of Lake Tanganyika. Species inventories were elaborated for both sites, depth-dependent abundances were calculated for the four most abundant species. Territorial sizes of three particularly abundant species, *Variabilichromis moorii*, *Aulonocranus dewindti*, and *Tropheus moorii*, were estimated by behavioral observations.

Variabilichromis moorii is a member of the substrate breeding tribe Lamprologini, the most species-rich sub-assemblages of Tanganyika cichlids (Salzburger et al., 2002). It is restricted to the southern section of the lake and represents a monotypic, quite distinct lineage within the tribe Lamprologini (Sturmbauer et al., 1994; Day et al., 2007). It was also suggested to have an ancestral morphological configuration of supra-orbital bones, and was therefore placed at the base of the phylogenetic tree by Colombe & Allgayer (1985) and Stiassny (1997). Colombe & Allgayer (1985) erected the new genus *Variabilichromis*, which was rejected by Poll (1986), but corroborated by Stiassny (1997). Despite high genetic differentiation among populations (Duftner et al., 2006), hardly any geographic color variation is found. The species feeds on small invertebrates and is a bi-parental substrate-breeder, which tends to hide eggs and fry underneath rock crevices (Karino, 1997). Because of high predation pressure, *Variabilichromis* has high parental investment and both parents intensively attack potential fry-predators (Karino, 1997; Sturmbauer et al., 2008).

Ophthalmotilapia ventralis and *Aulonocranus dewindti* belong to an exclusively maternally mouthbrooding lineage within the tribe Ectodini (Koblmüller et al., 2004). Both species occur over cobbles, where *O. ventralis* inhabits deeper water and *Aulonocranus dewindti* mainly occurs in the shallow zone of cobble and rock habitats, which may contain sandy patches. *Ophthalmotilapia ventralis* feeds on plankton and loose Aufwuchs. *Aulonocranus dewindti* mostly feeds on insect larvae and crustaceans and shares its adaptations for this type of feeding preference with the Lake Malawi genus *Aulonocara*. Both *Aulocranus* and *Aulonocara* have enlarged sensory pores on the head, enabling them to detect prey in the sand. Territorial males of *O. ventralis* and *A. dewindti* construct small nests in the form of sand craters with a diameter of about 30 cm and a distance of 2–3 m to neighboring males. Males form breeding colonies and decorate their spawning sites with pebbles. During the day, females hover in groups about half a meter above the substrate, while males try to attract them to their nests where they spawn. Females brood up to 35 eggs in their buccal cavity to release them after 20 days in nursery schools in shallow water (Konings, 1998).

The evolutionary history of *Tropheus moorii*, the third target species for estimating territory sizes, has been studied extensively (Sturmbauer & Meyer, 1992; Baric et al., 2003; Sturmbauer et al., 2005; Egger et al., 2007). Although more than 120 distinctly colored "races" are recognized in the genus (Konings, 1998; Schupke, 2003), species delineation is highly problematic (see Egger et al., 2007). *Tropheus moorii* is confined to rocky habitat for foraging and mating and has a limited capacity for dispersal across open water, and therefore, discontinuities in rocky habitat constitute barriers for dispersal (Sefc et al., 2007). In *Tropheus moorii*, both sexes are equally colorful, defend territories, and live in a social system (Yanagisawa & Nishida, 1991; Sturmbauer & Dallinger, 1995). The species is an epilithic algae feeder specializing on filamentous algae (Sturmbauer et al., 1992).

Materials and methods

Study area

The behavioral section of this study was carried out during 3 weeks in September and October 2004. Missing data on fish abundance were completed in 2006. Two rock-cobble shores were surveyed, one about 2 km south of the Kalambo mouth (8°37′36″S, 31°12′02″E) and the second at Kasakalawe (8°47′24″S, 31°04′40″E), both at the southern lake shores in the Republic of Zambia. At Kalambo, the substrate was formed by larger cobbles, usually in several layers, with small sandy patches in between (Fig. 1A). At Kasakalawe, the bottom was dominated by homogeneous smaller cobbles, often in just one layer, interrupted by pebbles (Fig. 1B). Due to the flat inclination of the shore at Kasakalawe, the study transect started 20 m offshore. The study transects had an extension of 400 m^2 and were subdivided into squares of two by two meters with coordinates marked on flat stones lettered from A-0 to K-10. Water depth was measured at each coordinate mark. A substrate map was drawn on underwater tablets by snorkeling and a map of the whole transect was constructed from these sketches.

During the 3-week study period, water temperature ranged between 27°C and 28°C. The inclination was 15% at Kalambo and 5.5% at Kasakalawe. The average depth of the inshore borders of the Kalambo transect was 0.57 m. The outermost borders had an average depth of 3.57 m. At Kasakalawe, the most inshore borders of the study transect were located at an average depth of 1.75 m and the most offshore borders at 2.85 m (Fig. 1).

Fish census

The total number of cichlid species and their depth-distribution was determined for both the study sites. For each species, censuses were carried out by snorkeling in parallel, one person over each transect row (e.g., from A0/A1 to K0/K1), so that the fish were not counted twice in different sectors. One person was noting down the data onshore. Each transect row was censused in triplicate and the average number was taken. For density analysis, the most abundant species of both habitats were individually plotted; rare species or species restricted to only one study site were pooled in the category "rare species."

Behavioral observations

Territorial sizes of three species *Variabilichromis moorii*, *Aulonocranus dewindti*, and *Tropheus moorii* were determined from observations of the locomotory activity of randomly chosen adult individuals within the transect. Three twenty-minute observations each were carried out per species and transect in each of the following daily periods: dawn (06.30–07.00), morning (09.00–12.00), afternoon (14.00–17.00), and dusk (17.30–18.00). A maximum of five investigators worked in the transect at a time to avoid cross-disturbances. After random choice, adult territorial individuals were observed for 2 min to allow the fish to adapt to the observer who stayed almost motionless in close vicinity. Then, the timed 20-min observation was carried out. Position records were taken every minute by recording the current coordinate on an underwater tablet. These regular records reflect the core area of the territory (core territory, territorial center) in which the fish spent most of the time. As a second type of locomotory activity, short-term movements were recorded to account for territorial defense action of

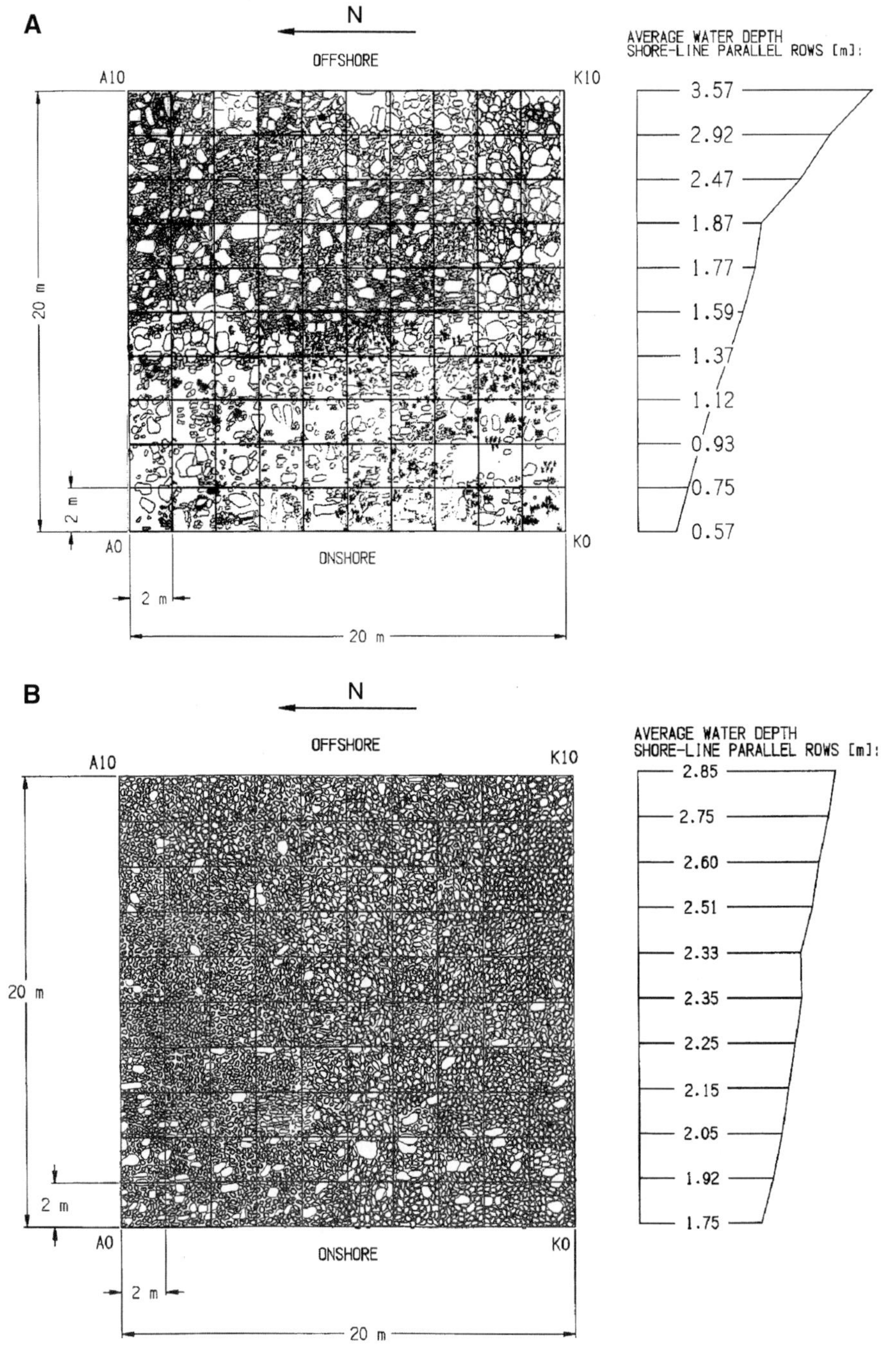

Fig. 1 Transect layout and the depth profile of the cobble shores at Kalambo (**A**) and Kasakalawe (**B**). White spaces in between cobble outlines indicate sand patches

the fish. The outermost coordinates of such burst activities marked the territorial borders. Observation records were transferred to a millimeter grid paper to be scanned and digitalized using the computer program Photoshop. In this way, the total territorial area was estimated in the form of the minimum convex polygon. Then, the total territorial area and core area were averaged over all recordings per study site, by excluding multiple records of the same individuals (inferred from the position of the territory in the transect). Minimum, maximum, average, standard deviation, and median of the territorial area were determined for each species. In order to test for differences in territory size between study sites, ANOVA (in case values were normally distributed) and Kruskal–Wallis tests (in case values were not normally distributed) were performed in SigmaStat 3.0.

Distribution in relation to water depth

The distribution in relation to water depth was calculated for the four most abundant species, *Variabilichromis moorii*, *Aulonocranus dewindti*, *Ophthalmotilapia ventralis*, and *Tropheus moorii*, based upon three counts per depth zone. Therefore, the average of the depth measurements at the closest corner points was calculated for each fish.

Results

Fish abundance

At Kalambo, a total of 46 cichlid species was recorded in the 400 m^2 study area (Table 1). The highest density among all cichlid species in the study area was found for *Variabilichromis moorii* with almost 28.9% of the total cichlid population. A total of 255 individuals were counted in the transect area, which can be translated to 0.64 individuals/m^2 or 1,275 individuals/100 m coast line. Adult *Aulonocranus dewindti* (males and females) were counted with an abundance of 166 (18.9% or 830 individuals/100 m coast line). There was also a great abundance of adult *Tropheus moorii* (12.3% or 108 individuals in the study transect, or 540 individuals/100 m coast line) and adult male and female *Cyathopharynx furcifer* counted 75 individuals (1.1 + 7.4 = 8.5% or 375 individuals/100 m coast line). Adult *Ophthalmotilapia ventralis* (territorial males) counted 52 individuals (5.9% or 208 individuals/100 m coast line), and females and juveniles counted 36 individuals (4.0%), adding up to 9.9%. Fifty individuals were found of *Eretmodus cyanostictus*, translating into an abundance of 5.7% or 250 individuals/100 m coast line. Adult individuals of *Lobochilotes labiatus* (22 individuals, 2.5%, 110 individuals/100 m coast line) and *Petrochromis* spp. (25 individuals, 2.8%, 125 individuals/100 m coast line) were quite frequent at Kalambo. The remaining species had abundances of less than 3% (Table 1), with several more rare species with a density of less than 0.01 fish per m^2.

At Kasakalawe, a total of 41 species were recorded (Table 1). Again, *Variabilichromis moorii* showed the highest density, with 552 individuals (60.4% or 2,760 individuals/100 m coast line). Adult *Tropheus moorii* followed with 107 individuals (11.6% or 535 individuals/100 m coast line), 100 individuals were counted of *Eretmodus cyanostictus* (11% or 500 individuals/100 m coast line), and 38 individuals for (adult male) *Ophthalmotilapia ventralis* (4.1% or 190 individuals/100 m coast line). Adult male *Aulonocranus dewindti* were found with an abundance of 3% (26 individuals in the whole transect or 130 individuals/100 m coast line). All remaining species made less than 3%, with several additional species with an abundance of less than 0.01%, equivalent to four fish in the entire study area of 400 m^2. A much lower density than that at Kalambo was observed for adult individuals of *Lobochilotes labiatus* with only five individuals (0.55%); adult specimens of *Petrochromis* spp. were also very rare (five individuals, 0.5%) and adult *Cyathopharynx furcifer* were almost absent (Table 1).

Territory areas

Estimates of territorial sizes of *Variabilichromis moorii*, *Aulonocranus dewindti*, and *Tropheus moorii* by delineating minimum convex polygons are shown in Fig. 2. The three mouthbrooding species generally had larger total territorial areas. In both study habitats, the largest average core territories were determined for *Tropheus moorii*, amounting to 1.60 and 1.30 m^2 in Kalambo and Kasakalawe, respectively. With 1.37 and 1.24 m^2, *Aulonocranus dewindti* had the second-largest core territories and *Variabilichromis moorii* the smallest (0.52 and 0.37 m^2, respectively). The largest total territory areas were obtained for *Aulonocranus dewindti* (4.16 and 5.03 m^2, in Kalambo versus Kasakalawe, respectively) and *Tropheus moorii* (4.31 and 4.69 m^2, respectively). *Variabilichromis moorii* consistently had the smallest total territories (1.72 and 1.60 m^2, respectively). The highest standard deviation was calculated for the average total territory size of *Tropheus moorii* at Kalambo with 2.73 m^2 (see Tables 2, 3). No significant differences in the sizes of core and total territory areas could be detected among sites and among daily periods.

Table 1 Absolute and relative abundance of cichlid fish species at the study transects (400 m^2) at Kalambo and Kasakalawe

	Kalambo		Kasakalawe	
	Absolute	Relative (%)	Absolute	Relative (%)
Altolamprologus compressiceps	5.0	0.6	6.0	0.6
Aulonocranus dewindti (adult territorial males)	21.3*	2.5*	26.0	2.8
Aulonocranus dewindti (females)	144.7*	16.4*	?	
Boulengerochromis microlepis	+		+	
Chalinochromis brichardi	19.5	2.2	–	
Callochromis macrops	+		–	
Cyathopharynx furcifer (adult territorial males)	10.0	1.1	+	
Cyathopharynx furcifer (females and juveniles)	65.3	7.4	?	
Eretmodus cyanostictus	50.1*	5.7*	100.0	11.0
Gnathochromis pfefferi	+		+	
Haplotaxodon sp.	+		+	
Julidochromis marlieri	+		–	
Julidochromis ornatus	10.0	1.1	16.3	1.8
Lamprologus callipterus	+		+	
Lamprologus lemairii	+		+	
Lepidiolamprologus elongatus	8.0*	0.9*	6.0	0.7
Lepidiolamprologus attenuatus	+		+	
Limnotilapia dardennii	+		+	
Lobochilotes labiatus	22.0	2.5	5.0	0.6
Neolamprologus modestus	3.6	0.4	+	
Neolamprologus caudopunctatus	+		24.6	2.7
Neolamprologus cylindricus	+		–	
Neolamprologus fasciatus	+		5.6	0.6
Neolamprologus furcifer	+		+	
Neolamprologus sexfasciatus	–		3.0	0.3
Neolamprologus tetracanthus	16.5*	1.9*	+	
Ophthalmotilapia nasuta	+		–	
Ophthalmotilapia ventralis (adult territorial males)	51.5*	5.9*	37.7	4.1
Ophthalmotilapia ventralis (females and juveniles)	35.5*	4.0*	?	
Oreochromis tanganicae	+		+	
Perissodus microlepis	+		+	
Petrochromis (adults, 5 spp.)	25.0*	2.8*	4.6	0.5
Plecodus straeleni	+		+	
Pseudosimochromis curvifrons	+		+	
Simochromis (adults, 3 spp.)	30.0	3.4	21.3	2.3
Ctenochromis horei	+		+	
Telmatochromis temporalis	+		+	
Telmatochromis vittatus	+		+	
Tropheus moorii	108.0*	12.3*	106.7	11.6
Tylochromis polylepis	+		+	
Variabilichromis moorii	255.0*	28.9*	552.3	60.4
Xenotilapia boulengeri	+		+	
Xenotilaphia spiloptera	+		+	
Sum	881.0	100.0	915.1	100.0

Symbols: "+" abundance of less than four individuals in the transect (0.01 individuals per m^2; not counted in % estimates); "–" absence at study site; "*" average abundance of two independent counts (2004 and 2006); "?" symbolize missing data

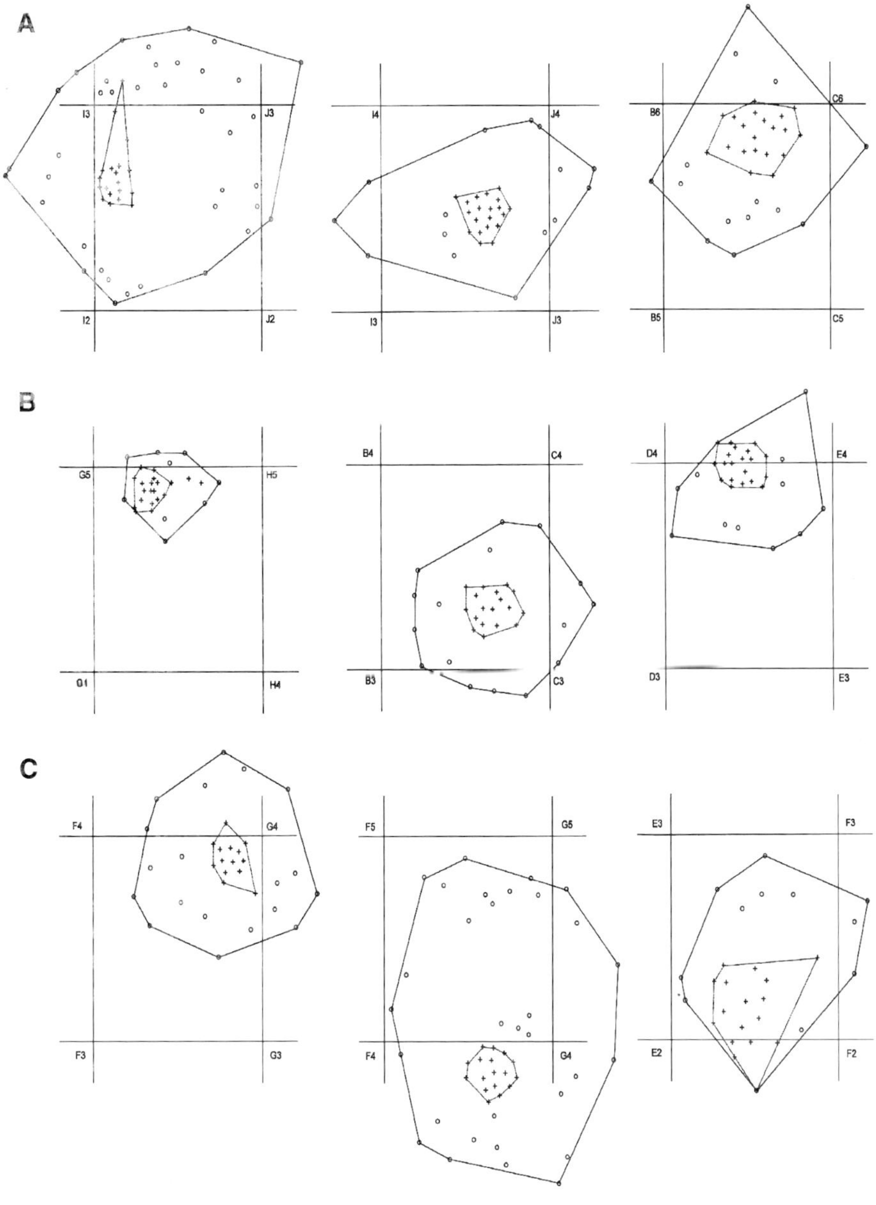

Fig. 2 Examples for territorial areas in (**A**) *Aulonocranus dewindti*, (**B**) *Variabilichromis moorii*, and (**C**) *Tropheus moorii*. The grid size was 2 × 2 m. Cross symbols indicate regular censuses per minute (core territory), and circle symbols indicate short-term locomotory activity (total territory)

Distribution of the four most abundant species in relation to water depth

At Kalambo, *Variabilichromis moorii* preferred depths between 1.40 and 3.50 m, with a maximum density of 0.8 individuals per m^2 at a depth around 2.6 m. Its average density in the 400 m^2 transect amounted to 0.64 individuals per m^2, the highest of all cichlid fish. *Tropheus moorii* had its peak in density at 1.5 m, counting 0.6 individuals per m^2. At greater depths, density of *T. moorii* slightly declined. The average density was 0.27 individuals per m^2 (Fig. 3A). *Ophthalmotilapia ventralis* (territorial males) started to occur at a water depth of about 1.3 m and reached densities of 0.2 individuals per m^2. Females moved about and had an average abundance of 0.09 individuals per m^2. *Aulonocranus dewindti* males did not exhibit depth-dependent density, having an average density of 0.05 individuals per m^2. Females of this species were moving about in swarms and appeared patchily (average density of 0.36 individuals per m^2).

At the second study transect at Kasakalawe, which started at a greater water depth, *Variabilichromis moorii*

Table 2 Territory sizes of *Aulonocranus dewindti*, *Variabilichromis moorii*, and *Tropheus moorii* at Kalambo, including average territorial areas, standard deviation, minimum, maximum, and median

	Aulonocranus dewindti		*Variabilichromis moorii*		*Tropheus moorii*	
	Core territory	Total territory	Core territory	Total territory	Core territory	Total territory
Average (m^2)	1.37	4.16	0.52	1.72	1.60	4.31
Standard deviation	0.66	1.74	0.24	1.08	1.31	2.73
Median	1.69	4.52	0.5	1.64	1.16	3.62
Minimum	0.26	1.67	0.18	0.61	0.14	1.51
Maximum	2.12	6.51	0.85	4.59	4.09	9.46

Table 3 Territory sizes of *Aulonocranus dewindti*, *Variabilichromis moorii*, and *Tropheus moorii* at Kasakalawe, including average territorial areas, standard deviation, minimum, maximum, and median

	Aulonocranus dewindti		*Variabilichromis moorii*		*Tropheus moorii*	
	Core territory	Total territory	Core territory	Total territory	Core territory	Total territory
Average (m^2)	1.24	5.03	0.37	1.60	1.30	4.69
Standard deviation	0.98	2.29	0.20	0.80	0.73	2.40
Median	0.84	5.88	0.32	1.71	1.11	3.10
Minimum	0.22	2.45	0.16	0.47	0.27	2.68
Maximum	2.92	9.59	0.76	2.92	2.51	9.46

was the most abundant cichlid fish species (1.38 individuals per m^2), with a maximum of more than 1.5 individuals per m^2. The species did not show depth-dependent density in the depth range of the study transect. A rather homogeneous density pattern of *Tropheus moorii* was found, with densities between 0.2 and 0.4 individuals per m^2 (averaging at 0.27 individuals per m^2, Fig. 3B). *Ophthalmotilapia ventralis* (territorial males) started to occur at a water depth of about 2 m and reached densities of 0.2 individuals per m^2. Females moved around and were not present in countable numbers in the study period. The density of *Aulonocranus dewindti* at Kasakalawe was lower than that at Kalambo and never exceeded 0.2 individuals per m^2, averaging at 0.27 individuals per m^2 as in Kalambo.

Discussion

Commonalities and differences in the cichlid communities

Great diversity in a species community is often accompanied by a high structural diversity of habitat. Thus, one would expect a higher number of species in the rocky habitat compared to the sandy one. This is also true for Lake Tanganyika, with about 65% of its cichlids species restricted to the rocky littoral habitat (Kuwamura, 1987). Therefore, it is not astounding that both study sites comprise more than 20% of the currently described 200 cichlid species (see Koblmüller et al., submitted). At both study sites, the cichlid community was dominated by Aufwuchs-eaters and zooplanktivorous species: *Variabilichromis moorii*, *Tropheus moorii*, *Eretmodus cyanostictus*, *Aulonocranus dewindti*, and *Ophthalmotilapia ventralis*. While the total number of observed species was similar at both study sites (Kalambo, 46 species; Kasakalawe, 41 species), slight differences in the community structure became evident. The absence of *Chalinochromis brichardi*, *Julidochromis marlieri*, *Neolamprologus cylindricus*, and *Ophthalmotilapia nasuta* at Kasakalawe is not surprising given their species-specific distributions in the Zambian part of Lake Tanganyika (Kohda et al., 1996). In the southwestern part of Lake Tanganyika, *Chalinochromis brichardi* and *Ophthalmotilapia nasuta* are only reported northwest of Mbete Bay, which is formed at the mouth of the Izi River and was shown to constitute an important barrier for gene flow between

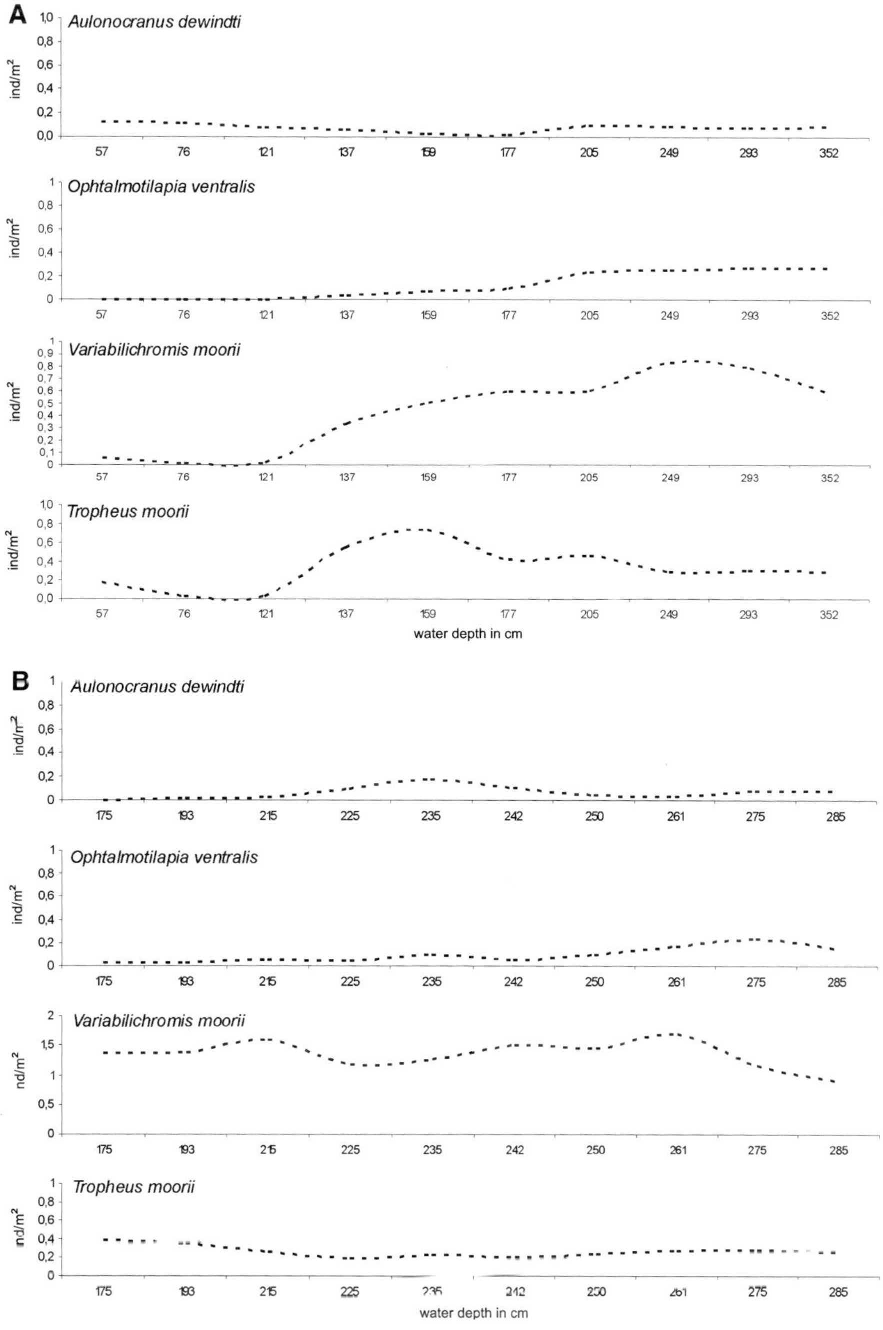

Fig. 3 Depth distribution of the four most abundant cichlid species at (**A**) Kalambo and (**B**) Kasakalawe. The *x*-axis represents the horizontal distance from the shore with the grid lines of the study transect at 2-m intervals indicated. For each grid line interval, the water depth (cm; obtained as the average of the depth at the two grid lines) is shown below the axes. The *y*-axis shows fish densities. The density curves were fitted to the observed average densities by eye. The water depth of each territory was calculated by averaging the depth of the closest coordinate points

populations of rock-dwelling cichlid species distributed on both sides of this bay (Duftner et al., 2006; Koblmüller et al., 2007; Sefc et al., 2007). In Zambian waters, *Julidochromis marlieri* and *Neolamprologus cylindricus* are only reported from the southeastern coast, with Kalambo as the southernmost record of *J. marlieri* (pers. observation). The absence of *Callochromis macrops* from the study site at Kasakalawe might be explained by its specific habitat requirements. *Callochromis macrops* is a species of the very shallow sandy habitat, and although both study sites consisted of mainly rocky bottom, the transect at Kalambo was bordered by a stretch of sandy bottom in the very shallow water.

However, one should keep in mind that species that are very rare and/or only occasionally enter the study site might not have been recorded in our survey.

Besides slight differences in species composition, it became evident that adult individuals of *Cyathopharynx furcifer*, *Lobochilotes labiatus*, and of species of the genus *Petrochromis* were much rarer at Kasakalawe (Table 1). These differences in the abundance of species with the largest body size might be explained by the different habitat structures of the two localities. While there were plenty of larger rock crevices at Kalambo, such potential hiding places for these comparatively large cichlids were almost absent in the relatively homogeneous cobble shore at Kasakalawe. Alternatively, fishing pressure might be higher at Kasakalawe than at Kalambo, where the study area was private property and where fishing was prohibited. Concerning the second-abundant territorial species *Tropheus moorii*, the observed densities of 108 and 107 individuals in 400 m^2, respectively, seem higher than those found for other localities at Lake Tanganyika. In fact, only few studies investigated fish densities of *Tropheus* in relation to habitat steepness or its social system (Yanagisawa & Nishida, 1991; Kohda & Yanagisawa 1992). About 60 adult *Tropheus* individuals were reported from Yanagisawa & Nishida (1991) at Bemba, Lake Tanganyika within a 20 m × 30 m (600 m^2) observation quadrant. Similar fish densities were reported by Kohda & Yanagisawa (1992) for two *Tropheus* populations (Bemba and Luhanga) situated at the northern end of Lake Tanganyika. Within the investigated transect of 18 m × 20 m (360 m^2), 36 *T. moorii* "Bemba" and 38 *T. duboisi* "Luhanga" were reported (Kohda & Yanagisawa 1992).

Territorial areas

Four types of territories—feeding, mating, nest-guarding, and brood guarding—have been defined for cichlid fish, depending on the resources to be defended and the objects to be repelled from the territories (Yuma & Kondo, 1997). In the present study, territory size has been determined for solitary *Variabilichromis moorii* and *Tropheus moorii* (we did not distinguish between males and females) and territorial male *Aulonocranus dewindti*. In all species, a clear distinction into core and total area became evident. While the fish spent most of their time close to their shelter (*V. moorii*, *T. moorii*) or mating site (*A. dewindti*), the outer border of the territory was marked by attacks against potential competitors for food (con- and heterospecific; see also, Takamura, 1983, 1984; Kuwamura, 1992; Kohda & Takemon, 1996; Karino, 1998) or, in the case of sexually active males, conspecific males (see also, Kohda, 1995) and sexually inactive females (see also, Yanagisawa & Nishida, 1991; Kohda, 1995). It is well documented that territory size in Lake Tanganyika cichlids usually increases with depth (Kohda, 1991; Kohda & Yanagisawa, 1992; Karino, 1998). This has been explained, at least in the case of herbivorous species, by lower costs of territory defense due to reduced competitor density in deeper water (Hixon, 1980; Kuwamura & Karino, 1991), and by the need to maintain greater food resources to compensate for lower algal productivity (Karino, 1998). Unfortunately, our study sites had very moderate inclinations and were not suited for testing this hypothesis, as the depth range covered was too narrow.

Acknowledgements We thank P. Ngalande and the team at the Mpulungu Station of the Ministry of Agriculture and Cooperative, Republic of Zambia, for their cooperation during fieldwork. This study was financed by the Austrian Science Foundation (P17680 to C.S.). N.D. was supported by an Erwin Schrödinger Fellowship from the Austrian Science Foundation.

References

Baric, S., W. Salzburger & C. Sturmbauer, 2003. Phylogeography and evolution of the Tanganyikan cichlid genus *Tropheus* based upon mitochondrial DNA sequences. Journal of Molecular Evolution 56: 54–68.

Cohen, A. S., K. E. Lezzar, J. J. Tiercelin & M. Soreghan, 1997. New palaeogeographic and lake-level reconstructions of Lake Tanganyika: Implications for tectonic, climatic and biological evolution in a rift lake. Basin Research 9: 107–132.

Cohen, A. S., M. J. Soreghan & C. A. Scholz, 1993. Estimating the age of formation of lakes: An example from Lake Tanganyika, East African rift system. Geology 21: 511–514.

Colombe, J. & R. Allgayer, 1985. Déscription de *Variabilichromis*, *Neolamprologus* et *Paleolamprologus*, genres nouveaux du lac Tanganyika, avec redescription des genres *Lamprologus* Schilthuis 1891 et de *Lepidiolamprologus* Pellegrin 1904. Revue Francaise des Cichlidophiles 49: 9–28.

Coulter, G. W., 1991. Lake Tanganyika and its Life. Oxford University Press, London, New York.

Day, J. J., S. Santini & J. Garcia-Moreno, 2007. Phylogenetic relationships of the Lake Tanganyika cichlid tribe Lamprologini: The story from mitochondrial DNA. Molecular Phylogenetics and Evolution 45: 629–642.

Duftner, N., K. M. Sefc, S. Koblmüller, B. Nevado, E. Verheyen, H. Phiri & C. Sturmbauer, 2006. Distinct population structure in a phenotypically homogeneous rock-dwelling cichlid fish from Lake Tanganyika. Molecular Ecology 15: 2381–2396.

Egger, B., S. Koblmüller, C. Sturmbauer & K. M. Sefc, 2007. Nuclear and mitochondrial data reveal different evolutionary processes in the Lake Tanganyika cichlid genus *Tropheus*. BMC Evolutionary Biology 7: 137.

Fryer, G. & T. D. Iles, 1972. The Cichlid Fishes of the Great Lakes of Africa. THF Publ, Neptune City, NJ.

Hixon, M. A., 1980. Food production and competitor density as the determinants of feeding territory size. American Naturalist 115: 510–530.

Karino, K., 1997. Influence of brood size on parental investment in a biparental cichlid fish, *Neolamprologus moorii*. Journal of Ethology 15: 39–43.

Karino, K., 1998. Depth-related differences in territory size and defense in the herbivorous cichlid, *Neolamprologus moorii*, in Lake Tanganyika. Ichthyological Research 45: 89–94.

Kawanabe, H., 1981. Territorial behaviour of *Tropheus moorii* (Osteichthyes, Cichlidae) with a preliminary consideration of territorial forms in animals. African Study Monographs 1: 101–106.

Koblmüller, S., W. Salzburger & C. Sturmbauer, 2004. Evolutionary relationships in the sand-dwelling cichlid lineage of Lake Tanganyika suggest multiple colonization of rocky habitats and convergent origin of biparental mouthbrooding. Journal of Molecular Evolution 58: 79–96.

Koblmüller, S., K. M. Sefc, N. Duftner, M. Warum & C. Sturmbauer, 2007. Genetic population structure as indirect measure of dispersal ability in a Lake Tanganyika cichlid. Genetica 130: 121–131.

Koblmüller, S., K. M. Sefc & C. Sturmbauer, 2008. The Lake Tanganyika cichlid species assemblage: Recent advances in molecular phylogenetics. Hydrobiologia (this volume).

Kohda, M., 1991. Intra- and interspecific social organization among three herbivorous cichlid fishes in Lake Tanganyika. Japanese Journal of Ichthyology 38: 147–163.

Kohda, M., 1995. Territoriality of male cichlid fishes in Lake Tanganyika. Ecology of Freshwater Fish 4: 180–184.

Kohda, M. & Y. Takemon, 1996. Group foraging by the herbivorous cichlid fish, *Petrochromis fasciolatus*, in Lake Tanganyika. Ichthyological Research 43: 55–63.

Kohda, M. & K. Tanida, 1996. Overlapping territory of the benthophagous cichlid fish, *Lobochilotes labiatus*, in Lake Tanganyika. Environmental Biology of Fishes 45: 13–20.

Kohda, M. & Y. Yanagisawa, 1992. Vertical distributions of two herbivorous fishes of the genus *Tropheus* in Lake Tanganyika. Ecology of Freshwater Fish 1: 99–103.

Kohda, M., Y. Yanagisawa, T. Sato, K. Nakaya, Y. Niimura, K. Matsumoto & H. Ochi, 1996. Geographical colour variation in cichlid fishes at the southern end of Lake Tanganyika. Environmental Biology of Fishes 45: 237–248.

Konings, A., 1998. Tanganyika Cichlids in their Natural Habitat. Cichlid Press, El Paso, TX.

Kuwamura, T., 1987. Distribution of fishes in relation to the depth and substrate at Myako, east-middle coast of Lake Tanganyika. African Study Monographs 7: 1–14.

Kuwamura, T., 1992. Overlapping territories of *Pseudosimochromis curvifrons* males and other herbivorous cichlid fishes in Lake Tanganyika. Ecological Research 7: 43–53.

Kuwamura, T. & K. Karino, 1991. Vertical distribution of fishes on the rocky shore of Kasenga, southerm Lake Tanganyika. Ecology and Limnology of Lake Tanganyika 7: 18–20.

Matsumoto, K. & M. Kohda, 2004. Territorial defense against various food competitors in the Tanganyikan benthophagous cichlid *Neolamprologus tetracanthus*. Ichthyological Research 51: 354–359.

Ochi, H., 1993. Maintenance of separate territories for mating and feeding by males of a maternal mouthbrooding cichlid, *Gnathochromis pfefferi* in Lake Tanganyika. Japanese Journal of Ichthyology 40: 173–182.

Poll, M., 1986. Classification des Cichlidae du lac Tanganika: tribus, genre et epèces. Academie Royale de Belgique Memoires de la Classe des Sciences 45: 1–163.

Salzburger, W., A. Meyer, S. Baric, E. Verheyen & C. Sturmbauer, 2002. Phylogeny of the Lake Tanganyika cichlid species flock and its relationship to the central and East African haplochromine cichlid fish faunas. Systematic Biology 51: 113–135.

Schupke, P., 2003. African Cichlids II: Tanganyika I: Tropheus. Aqualog, A. C. S. Gmbh, Rodgau, Germany.

Sefc, K. M., S. Baric, W. Salzburger & C. Sturmbauer, 2007. Species specific population structure in rock-specialized sympatric cichlid species in Lake Tanganyika, East Africa. Journal of Molecular Evolution 64: 33–49.

Snoeks, J., 2000. How well known is the ichthyodiversity of the large East African lakes? Advances in Ecological Research 31: 17–38.

Stiassny, M., 1997. A phylogenetic overview of the lamprologine cichlids of Africa (Teleostei, Cichlidae): A morphological perspective. South African Journal of Science 93: 513–523.

Sturmbauer, C. & R. Dallinger, 1995. Diurnal variation of spacing and foraging behaviour in *Tropheus moorii* (Cichlidae) in Lake Tanganyika, Eastern Africa. Netherlands Journal of Zoology 45: 386–401.

Sturmbauer, C., C. Hahn, S. Koblmüller, H. Phiri, D. Sinyinza & K. M. Sefc, 2008. Variation of territorial size and defense behavior in breeding pairs of the endemic Lake Tanganyika cichlid fish *Variabilichromis moorii*. Hydrobiologia (this volume).

Sturmbauer, C., S. Koblmüller, K. M. Sefc & N. Duftner, 2005. Phylogeographic history of the genus *Tropheus*, a lineage of rock-dwelling cichlid fishes endemic to Lake Tanganyika. Hydrobiologia 542: 335–366.

Sturmbauer, C., W. Mark & R. Dallinger, 1992. Ecophysiology in Aufwuchs-eating cichlids in Lake Tanganyika: Niche separation by trophic specialization. Environmental Biology of Fishes 35: 283–290.

Sturmbauer, C. & A. Meyer, 1992. Genetic divergence, speciation and morphological stasis in a lineage of African cichlid fishes. Nature 358: 578–581.

Sturmbauer, C., E. Verheyen & A. Meyer, 1994. Mitochondrial phylogeny of the Lamprologini, the major substrate spawning lineage of cichlid fishes from Lake Tanganyika in Eastern Africa. Molecular Biology and Evolution 11: 691–703.

Takahashi, T., 2003. Systematics of Tanganyikan cichlid fishes (Teleostei: Perciformes). Ichthyological Research 50: 367–382.

Takamura, K., 1983. Interspecific relationship between two Aufwuchs-eaters *Petrochromis polyodon* and *Tropheus moorii* (Pisces, Cichlidae) of Lake Tanganyika, with a discussion on the evolution and function of a symbiontic relationship. Physiology and Ecology Japan 20: 59–69.

Takamura, K., 1984. Interspecific relationships of Aufwuchs-eating fishes in Lake Tanganyika. Environmental Biology of Fishes 10: 225–241.

Yanagisawa, Y. & M. Nishida, 1991. The social and mating system of the maternal mouthbrooder *Tropheus moorii* (Cichlidae) in Lake Tanganyika. Japanese Journal of Ichthyology 38: 271–282.

Yuma, M. & T. Kondo, 1997. Interspecific relationships and habitat utilization among benthivorous cichlids. In Kawanabe, H., M. Hori & M. Nagoshi (eds), Fish Communities in Lake Tanganyika. Kyoto University Press, Kyoto: 87–103.

Hydrobiologia (2008) 615:69–79
DOI 10.1007/s10750-008-9565-z

SPECIATION IN ANCIENT LAKES

Subtle population structure and male-biased dispersal in two *Copadichromis* species (Teleostei, Cichlidae) from Lake Malawi, East Africa

Dieter Anseeuw · Gregory E. Maes · Paul Busselen · Dries Knapen · Jos Snoeks · Erik Verheyen

Published online: 10 September 2008

Abstract Various attributes of cichlid biology have been suggested to drive their propensity for rapid speciation, including population substructuring over short geographic distances. While this seems especially true for the rock-dwelling Mbuna species from Lake Malawi, the present study shows that geographic or habitat barriers are not sufficient to explain population substructuring in the less substrate-bound Utaka (non-Mbuna) species. We found similar levels of subtle population structure in the rock-dwelling *Copadichromis quadrimaculatus* and in the sand-dwelling *C.* sp. 'virginalis kajose' ($F_{ST} < 0.01$ in both species) without a discernable geographical pattern. We suggest that aspects of the reproductive strategy, by which seasonal aggregation alternates with more free-ranging stages, may facilitate the establishment of small population differences in Utaka. This hypothesis agrees with our finding that in these cichlids dispersal appears to be male biased.

Keywords Cichlids · *Copadichromis* · Population structure · Sex-biased dispersal · Habitat preference · Mating strategy

Guest editors: T. Wilke, R. Väinölä & F. Riedel
Patterns and Processes of Speciation in Ancient Lakes: Proceedings of the Fourth Symposium on Speciation in Ancient Lakes, Berlin, Germany, September 4–8, 2006

D. Anseeuw (✉) · P. Busselen
Interdisciplinary Research Centre, Katholieke Universiteit Leuven Campus Kortrijk, Etienne Sabbelaan 53, 8500 Kortrijk, Belgium
e-mail: dieter.anseeuw@kuleuven-kortrijk.be

G. E. Maes
Laboratory of Aquatic Ecology and Evolutionary Biology, Katholieke Universiteit Leuven, Ch. de Bériotstraat 32, 3000 Leuven, Belgium

D. Knapen
Laboratory for Ecophysiology, Biochemistry and Toxicology, University of Antwerp, Groenenborgerlaan 171, 2020 Antwerpen, Belgium

J. Snoeks
Zoology Department, Royal Museum for Central Africa, Leuvensesteenweg 13, 3080 Tervuren, Belgium

J. Snoeks
Laboratory of Animal Diversity and Systematics, Katholieke Universiteit Leuven, Ch. de Bériotstraat 32, 3000 Leuven, Belgium

E. Verheyen
Vertebrate Department, Royal Belgian Institute of Natural Sciences, Vautierstraat 29, 1000 Brussels, Belgium

Introduction

The remarkable diversity, extraordinary degree of endemism and high speciation rates of the cichlids in the Great Lakes of East Africa have been of longstanding interest to evolutionary biologists. Despite decades of research, there are still many uncertainties concerning the various factors leading to the accumulation of genetic differences among conspecific populations and how these factors may influence the rate and modes of speciation in these fishes (Salzburger & Meyer, 2004). In this context, the most extensively studied cichlids from Lake Malawi are the rock-dwelling "Mbuna", a group of fishes that comprises about 240–300 of the estimated 800 endemic haplochromine species in this lake (Snoeks, 2004). Earlier studies demonstrated that an unsuitable habitat of even a few hundred meters between Mbuna populations may significantly reduce their genetic exchange. Hence, areas of deep water and sandy stretches appear to be important barriers for the dispersal in stenotopic (i.e. exhibiting high site fidelity) rock-dwelling species. This feature in combination with the patchy distribution of rock habitats may at least partly explain why certain rock-dwelling cichlids have speciated so rapidly (van Oppen et al., 1997; Arnegard et al., 1999; Rico & Turner, 2002).

However, most of the endemic haplochromines from Lake Malawi do not belong to the Mbuna group. Many of these taxa are not confined to littoral rocky habitats and they often display a more vagile lifestyle. Interestingly, the high number of non-Mbuna species and their young evolutionary age suggest that they too have speciated in an explosive way, whereas the most likely explanation for hyperspeciation in the Mbuna (i.e. allopatry due to barriers to dispersal) may not be sufficient for this group. To date, only a handful of studies have investigated factors that may affect the accumulation of genetic differences among non-Mbuna populations.

Essentially pelagic non-Mbuna species such as various *Diplotaxodon* species, which occupy an apparently continuous habitat without physical barriers to migration, do not show any genetic population subdivision (Shaw et al., 2000). Littoral non-Mbuna, including the genus *Copadichromis*, have habitat preferences intermediate to those of the highly stenotopics (Mbuna) and the free-ranging pelagics. These species live near-shore either over sandy, rocky or intermediate substrate. Among five populations of a sand-dwelling species (*Copadichromis* sp. 'virginalis kajose'), Taylor & Verheyen (2001) found weak but significant levels of genetic differentiation with no discernible geographical pattern. In another study on three species of the benthic genus *Protomelas*, Pereyra et al. (2004) found different patterns of population substructuring which could be related to habitat preferences: the rock-dwelling species *P. taeniolatus* Trewavas showed relatively high levels of population structure over distances less than 1 km, whereas the sandy-shore species *P. similis* Regan did not show any significant structure over distances up to 21 km, and *P. fenestratus* Trewavas, living at the interface between rocks and sand, showed intermediate levels of population substructuring.

Habitat preference is but one possible factor affecting the degree of population subdivision in cichlids. Mating strategies may also influence the population structure through reproductive bias in one or both sexes, breeding territoriality, assortative mating and sexual selection (van Oppen et al., 1998; Perrin & Mazalov, 2000; Avise et al., 2002; Knight & Turner, 2004).

The present study investigates the population structure in two seldom-studied Utaka species, *Copadichromis* sp. 'virginalis kajose' and *C. quadrimaculatus* Regan. Utaka are relatively substrate-bound cichlids, although for feeding they gather in schools to forage on zooplankton in the water column. During the breeding season, they typically aggregate in large groups on in-shore breeding grounds (so-called 'leks') where the males aggressively defend a spawning site. *Copadichromis* sp. 'virginalis kajose' is a typical sand-dweller with a lake-wide distribution. Preliminary data suggest considerable morphological and colour variation but only weak genetic population substructuring in this species (Turner, 1996; Taylor & Verheyen, 2001; Snoeks, 2004). Our second target species, *C. quadrimaculatus*, is usually found in rocky habitats where adults gather for breeding and males actively defend a rock or crevice for spawning. Occasionally, adults may display a truly pelagic behaviour as well (Iles, 1960; Thompson et al., 1995; Duponchelle et al., 2000). *C. quadrimaculatus* also shows some variation in colour and body form, which seems unrelated to geographic distribution (Turner, 1996; Chisambo, 2000).

Our study extends the earlier work by Taylor & Verheyen (2001) by including the rock-breeding *C. quadrimaculatus*, increasing the number of populations sampled per species and studying sexes separately. We also included a population from Lake Malombe, a satellite lake which is connected to Lake Malawi by the Shire River. We examined whether differences in habitat selection between the two

Copadichromis species relate to the amount of gene flow among their populations. It may be expected that the rock-dwelling species is more constrained in dispersal and, in view of the patchy distribution of rock habitats, will therefore exhibit a higher degree of population substructuring. Furthermore, we investigated whether the population subdivision accommodates to a model of isolation-by-distance in these lake-wide distributed species. Finally, since it has been suggested that some cichlids with high population substructuring show sex-biased dispersal (Knight et al., 1999; Pereyra, 2003), we also investigated the relative contribution of sex dependent dispersal to the population substructuring in our two non-Mbuna species.

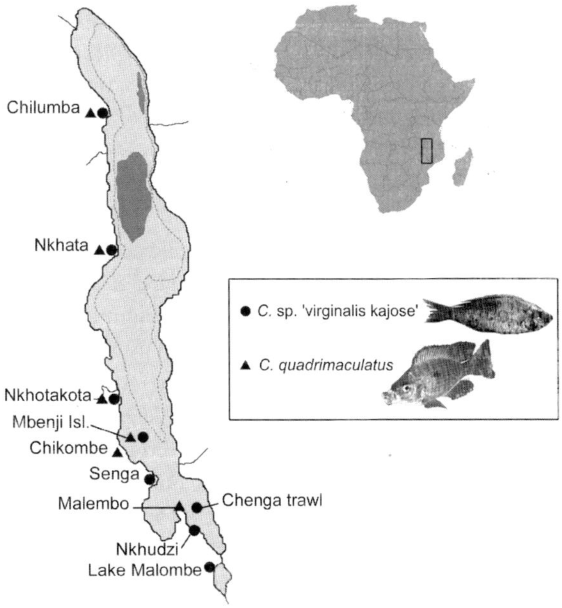

Fig. 1 Map of Lake Malawi, East Africa, showing sampling locations along the Western shore. Abbreviations used: Chilumba (CH), Nkhata Bay (NK), Nkhotakota (NO), Mbenji Island (MI), Chikombe (CK), Senga Bay (SB), Malembo Bay (MA), Chenga trawl (CT), Nkhudzi Bay (NKU), Lake Malombe (LM)

Materials and methods

Sample collection

In May 2002, a total of 427 *Copadichromis* sp. 'virginalis kajose' individuals were collected from seven localities in Lake Malawi and one in Lake Malombe (Fig. 1). Additionally, 252 specimens of *C. quadrimaculatus* were collected from six localities in Lake Malawi (Fig. 1). All fish were collected with open water seines (chirimila nets) and beach seines by artisanal fishermen, except for one catch of *C.* sp. 'virginalis kajose' that was collected by a commercial trawler. Sample size per locality ranged from 40 to 63 individuals in *C.* sp. 'virginalis kajose' and from 30 to 63 in *C. quadrimaculatus*. Pelvic fin clips were preserved in 100% ethanol and stored at room temperature. The sampled specimens were fixed in 10% formalin and are curated in the Royal Museum for Central Africa in Tervuren, Belgium.

DNA preparation and amplification

Total genomic DNA was extracted using Proteinase K digestion and salt precipitation, according to Aljanabi & Martinez (1997). DNA extracts were resuspended in 100 µl of autoclaved ddH_2O. All samples were analysed for genetic variation at six microsatellite markers: Pzeb1, Pzeb3, Pzeb4, Pzeb5 (van Oppen et al., 1997), UNH002 (Kellogg et al., 1995) and TmoM11 (Zardoya et al., 1996). PCRs were performed under the following conditions: 94°C for 120 s, followed by 5 cycles of 94°C for 45 s; 55°C for 45 s; 72°C for 45 s, followed by 30 cycles of 90°C for 30 s; 55°C for 30 s; 72°C for 30 s, followed by 72°C for 10 min. Ten microliter of reaction mixes included 1 µl template DNA, 0.5 µM of each primer, 200 µM of each dNTP, 0.26 units Taq polymerase (Sigma Aldrich, Germany), 1 µl 10× reaction buffer (Sigma Aldrich). PCR amplification products were run on 6% denaturing polyacrylamide gels using an ALF Express DNA Sequencer (Amersham Pharmacia Biotech). Fragment sizes were scored with ALFWin Fragment Analyser v1.0 (Amersham Pharmacia Biotech), using M13mp8 DNA standards as external references, following van Oppen et al. (1997).

For a selected subset of *C.* sp. 'virginalis kajose' specimens (see below), we examined the mitochondrial DNA (mtDNA) sequence variation in a 331 bp fragment of the control region (D-loop). PCR amplification was carried out in a 25 µl buffered reaction mixture, containing 5 µl template DNA, 0.5 µM of each primer, H16498 and L15995 (Meyer et al., 1990), 200 µM of each dNTP, 2.5 µl of 10× buffer (1 mM $MgCl_2$) and 0.65 units of Red Taq Polymerase

(Sigma Aldrich). PCR conditions consisted of an initial denaturation for 2 min at 94°C, followed by 35 cycles of denaturation at 94°C for 60 s, annealing at 52°C for 60 s, and extension at 72°C for 120 s and was terminated by a final extension at 72°C for 10 min. The success of amplification was checked by electrophoresis on 1% agarose gel. The PCR products were purified following the TMQiaquick PCR purification kit protocol, prior to being added as template for chain termination sequencing, carried out in 10 μl reaction volumes, using 2 μl purified DNA template, 1 μl 5× Sequencing Buffer (Applied Biosystems), 2 μl Ready Reaction Mix (Applied Biosystems) and 2 μl primer (L15995, 2 μM). The sequencing program consisted of an initial denaturation step of 1 min at 96°C, followed by 25 cycles of 10 s at 96°C, 5 s at 50°C, and 4 min at 60°C. DNA fragments were purified by ethanol precipitation and subsequently visualised on a 3130 capillary sequencer (Applied Biosystems).

Data analyses

Genetic diversity estimates, such as mean number of alleles, and observed and unbiased expected number of heterozygotes at each locus, were calculated using GENETIX 4.05 (Belkhir, 2000). Allelic richness (*AR*) was calculated using FSTAT 2.9.3 (Goudet, 2001). Departures from Hardy–Weinberg equilibrium and linkage disequilibrium were tested using exact tests as implemented by GENEPOP 3.3 (Raymond & Rousset, 1995). *P*-values were obtained by the Markov Chain method using 5,000 dememorisations, 500 batches and 2,000 iterations per batch. Deviations from Hardy–Weinberg equilibrium expectations (mostly heterozygote deficits) can have a biological reason (Wahlund effect, inbreeding, selection, recent population expansion), but can also be due to technical artefacts (degraded DNA, large allele dropouts, null alleles) or a combination of these events (Garcia De Léon et al., 1997). While biological causes should affect all loci, null alleles or allele dropouts may affect just one or a few loci. MICRO-CHECKER (van Oosterhout et al., 2004) was used to identify probable genotyping errors. When null alleles were the most likely cause of deviation, allele and genotype frequencies that did not conform to the Hardy–Weinberg equilibrium were adjusted to accommodate for the presence of a null allele. The null allele frequencies were estimated following the Expectation Maximization algorithm of Dempster et al. (1977) and unbiased F_{ST}'s (Weir, 1996) were estimated following the Excluding Null Alleles (ENA) method, recently described in Chapuis & Estoup (2007). The calculations were performed in FreeNA (Chapuis & Estoup, 2007).

Estimates of global population differentiation using Wright's F_{ST} (Wright, 1969) and pairwise F_{ST} values estimated using θ (Weir & Cockerham, 1984) were calculated in GENETIX. *P*-values of F_{ST} values were computed using a permutation approach (2,000 iterations) as implemented in GENETIX. To test whether population differentiation followed an isolation-by-distance model, we regressed the pairwise $\theta/(1-\theta)$ estimates against the Log-transformed geographic distances. Both the shortest straight-line distances between sampling localities, and the distance along the shoreline were used. Mantel tests were conducted with 2,000 permutations as implemented in GENETIX.

In addition, we used a model-based approach implemented in the software STRUCTURE 2.2 (Pritchard et al., 2000; Falush et al., 2007) to infer the population substructure from multi-locus genotype data. For each species, the number of populations represented in our data was estimated by calculating the posterior probability for different numbers of putative populations (*K* from one to ten populations). Burn-in was set at 10,000 steps followed by 100,000 MCMC iterations at each *K*. Simulations were run five times for each *K* to test convergence of the priors and the appropriateness of the chosen burn-in and simulation length. We performed clustering under the admixture model without prior population information and with correlated allele frequencies between populations. To determine the most likely number of clusters, the rate of change in the log probability of data between successive *K* values was evaluated (Evanno et al., 2005).

For each species, evidence for sex-biased dispersal was tested only between male and female populations from the same localities and using two different approaches. The population differentiation for each sex was estimated using Wright's F_{ST}. In the case of sex-biased dispersal, we expect that the dispersing sex should be less genetically structured (Goudet et al., 2002). Queller & Goodnight's (1989) relatedness (*r*) was used as a further estimate of sex-biased dispersal, calculated as the average relatedness of individuals within samples compared to the whole population. Genotypes from the dispersing sex should on average be less likely (less related) in the population in which they were sampled compared to genotypes from the philopatric sex (Goudet et al., 2002). Differences in F_{ST} and r between sexes were tested using a randomisation approach implemented in FSTAT (Goudet, 2001).

The mtDNA sequences were visually aligned, after manual correction in Chromas 1.45. The sequences have been submitted to GenBank and the accession numbers are listed in Table 5. Genetic homogeneity among populations was tested through analysis of variance (AMOVA), using ARLEQUIN 3.0 (Excoffier et al., 2005). P-values were obtained using 1,000 permutations.

Results

Genetic diversity and Hardy–Weinberg equilibrium

All six microsatellite loci surveyed were highly polymorphic. Levels of genetic variability and allele size ranges were consistent between samples within species, among species, as well as with respect to previous results by Taylor & Verheyen (2001). The total number of alleles per locus per population varied from four in Pzeb5 to 38 in Pzeb1. Observed heterozygosity per locus ranged from 0.10 in Pzeb5 to 0.94 in UNH002. Across loci, within-sample observed heterozygosities varied between 0.59 (Nkhudzi Bay) and 0.71 (Senga Bay) for *C.* sp. 'virginalis kajose', and between 0.56 (Nkhata Bay) and 0.73 (Nkhotakota) for *C. quadrimaculatus* (Table 1).

In *C.* sp. 'virginalis kajose', 25 of 56 single-locus tests, and in *C. quadrimaculatus,* 16 of 42 single-locus tests for deviation from Hardy–Weinberg equilibrium were significant at a table wide Bonferroni corrected alpha level ($\alpha_{\text{corr}} = 0.001$). Most deviations occurred across populations at loci Pzeb1, UNH002 and TmoM11, all involving heterozygote deficits. Loci Pzeb3, Pzeb4 and Pzeb5 were in agreement with Hardy–Weinberg equilibrium expectations in eleven or more of the fourteen populations. Linkage disequilibrium tests across all loci and all populations showed non-significant in both *C.* sp. 'virginalis kajose' and *C. quadrimaculatus* (all $P > 0.05$).

Table 1 Mean sample size (n), allelic richness (AR; based on $n = 40$ for *Copadichromis* sp. 'virginalis kajose' and $n = 30$ for *C. quadrimaculatus*), observed (Ho) and expected heterozygosities (He) and inbreeding coefficients F_{IS} (all P-values <0.01)

Population	n	AR	He	Ho	F_{IS}
Copadichromis sp. 'virginalis kajose'					
CH	51	16.46	0.7493	0.6830	0.241
NK	54	16.89	0.7541	0.6821	0.105
NO	48	18.88	0.7614	0.7326	0.105
MI	57	18.87	0.7648	0.6871	0.110
SB	63	18.17	0.7854	0.7090	0.048
CT	40	19	0.7770	0.7083	0.101
NKU	61	17.59	0.7729	0.5929	0.098
LM	53	16.85	0.7375	0.6195	0.169
Copadichromis quadrimaculatus					
CH	35	16.48	0.8159	0.7095	0.145
NK	38	16.64	0.7595	0.5614	0.273
NO	30	17.5	0.8009	0.7333	0.101
MI	55	18.21	0.7929	0.7212	0.099
CK	31	15.96	0.7784	0.6613	0.166
MA	63	16.59	0.7872	0.6402	0.194

Measures for microsatellites are the mean over six loci. Populations are listed from northernmost to southernmost (see Fig. 1)

Genetic population structure

In each species, low but significant population structuring was detected. The F_{ST} estimate for population subdivision across all samples was $\theta = 0.006$ ($P < 0.001$) in *C.* sp. 'virginalis kajose' and $\theta = 0.007$ ($P < 0.001$) in *C. quadrimaculatus*.

Genetic differentiation was further examined using pairwise F_{ST} estimates. In total, respectively 20 (71.4%) pairwise values proved significant for *C.* sp. 'virginalis kajose' (Table 2). All adjacent populations differed significantly from each other, while geographically more distant samples did not always differ. The northernmost population (Chilumba) differed significantly from all other populations in F_{ST}. The Nkhata population appears to be least differentiated from other populations with only significant F_{ST} values when compared to Lake Malombe, Nkhotakota and Chilumba. For *C. quadrimaculatus*

Table 2 Pairwise F_{ST} (θ) values for *Copadichromis* sp. 'virginalis kajose'

	LM	CT	NKU	SB	MI	NO	NK	CH
LM	–							
CT	0.009*$^{(-)}$	–						
NKU	0.005	0.007*	–					
SB	0.007**	0.001	0.007**	–				
MI	0.01**	0.002	0.006*	0.005*$^{(-)}$	–			
NO	0.003	0.011**	0.011**	0.007**	0.011**	–		
NK	0.006**	0.007	0.004$^{(+)}$	0.001	0.002	0.008**	–	
CH	0.011**	0.01**	0.015**	0.01**	0.008**	0.013**	0.009**	–

Localities encoded as in Fig. 1. Significance codes: * $P < 0.05$; ** $P < 0.001$. Pairwise F_{ST} values based on genotypes corrected for the presence of null alleles that differed from F_{ST} values based on non-corrected genotypes are indicated by $^{(+)}$ (significant) and $^{(-)}$ (not significant)

Table 3 Pairwise F_{ST} (θ) values for *Copadichromis quadrimaculatus*

	MA	CK	MI	NO	NK	CH
MA	–					
CK	0.034**	–				
MI	0.01**	0.007*	–			
NO	0.006	0.004	0.005	–		
NK	0.004	0.013**	0.009**	0.011*	–	
CH	0.013**	0.016**	0.007*	0.008	0.02**	–

Localities encoded as in Fig. 1. Significance codes: * $P < 0.05$; ** $P < 0.001$

fixation indices displayed 10 (66.7%) significant values (Table 3). Again, the pairwise comparisons demonstrated no obvious geographic patterning. Estimation of population differentiation using genotype frequencies corrected for the presence of null alleles returned identical results for the pairwise comparisons in *C. quadrimaculatus*, but reduced the number of significant pairwise F_{ST}'s by one for *C.* sp. 'virginalis kajose' (see Table 2). However, comparison of the F_{ST} estimates revealed no significant differences in values between F_{ST}'s calculated for observed or corrected genotype frequencies (*t*-test, $P = 0.19$) for the two species. This indicates that, although three loci showed deviations from Hardy–Weinberg expectations, the estimates of population differentiation return a similar pattern with or without correction.

Isolation-by-distance was investigated using both "lake shore distance" and "shortest distance" between the localities. In both species, neither of the two Mantel tests gave a significant correlation (*P*-values from 0.24 to 0.27). Thus, based on these data, no isolation-by-distance pattern could be detected.

The model-based clustering approach implemented in the STRUCTURE software yielded estimated Ln probabilities for $1 \leq K \leq 10$ ranging from -12355 to -13714 for *C.* sp. 'virginalis kajose' and from -5237 to -6301 for *C. quadrimaculatus* with the highest posterior probability given for $K = 1$ in both species. Additionally, assignment of individuals to putative clusters was highly symmetrical, indicating little or no genetic structure. Therefore, no underlying genetic structure could be inferred despite the small but significant F_{ST} estimates found among populations.

Sex-biased dispersal

Tests of sex-biased dispersal were carried out for the localities from where both male and female specimens were collected. The populations and sample sizes (between brackets) were, respectively, Nkhudzi Bay (♂: 29; ♀: 32), Senga Bay (♂: 29; ♀: 30), Mbenji Islands (♂: 29; ♀: 28), Nkhata Bay (♂: 29; ♀: 24) and Chilumba (♂: 25; ♀: 26) for *C.* sp. 'virginalis kajose' and Mbenji Islands (♂: 25; ♀: 30) and Nkhata Bay (♂: 20; ♀: 18) for *C. quadrimaculatus*.

In *C.* sp. 'virginalis kajose', overall fixation indices (F_{ST}) and relatedness (*r*) did not differ significantly between sexes (*P*-values between 0.12 and 0.63). For *C. quadrimaculatus*, pairwise differentiation θ ($P < 0.01$) and relatedness ($P = 0.036$) were significantly higher in the female populations.

Table 4 Comparison of F_{ST} and relatedness within males and females and in both sexes combined for *Copadichromis* sp. 'virginalis kajose' and *C. quadrimaculatus*

	Copadichromis sp. ' virginalis kajose'		*Copadichromis quadrimaculatus*	
	Males	Females	Males	Females
F_{ST} (θ)	0.005	0.010**	0.003	0.023**
	0.007**		0.008**	
Relatedness (r)	0.009	0.018**	0.005	0.041**
	0.013*		0.012**	

Significance notation: * $P < 0.05$; ** $P < 0.005$

Although, for *C.* sp. 'virginalis kajose', the differences in F_{ST} and r between males and females were not significant; however, the estimates for the sexes were consistent with the general trend that males disperse more than females (Table 4): Allele frequencies (pairwise F_{ST}) differed significantly among the female populations, but not among the male populations; and relatedness in female populations was significantly higher than expected by random mating, while the null hypothesis could not be rejected for the male populations.

As the evidence for male-biased dispersal was less clear-cut in *C.* sp. 'virginalis kajose', we further verified this trend for this species by a supplementary analysis using mtDNA sequence data. We examined the mtDNA variability among male and female populations in *C.* sp. 'virginalis kajose' from three different localities (Table 5). AMOVA showed that, when all populations were treated as a single group, 80.9% ($P < 0.01$) of the genetic variance was found within populations, whereas 19.1% ($P < 0.01$) was attributable to differences among populations. An AMOVA on males and females separately showed that the genetic variance across male and female populations comprised 17.5% ($P = 0.001$) and 27.6% ($P < 0.001$), respectively. To test if the genetic differentiation in females significantly differed from that in males, we grouped male and female populations by geographic locality (three groups) and found that 77.2% ($P < 0.01$) of the variance was due to differences within populations, 19.4% ($P < 0.05$) was due to differences between geographic groups and a significant 3.4% ($P < 0.01$) was due to differences between (male and female) populations within the groups.

Table 5 Populations sampled, sample sizes (n), number of haplotypes (h) and haplotype diversity (Hd) for mitochondrial DNA sequences in *C.* sp. 'virginalis kajose'

Locality	n	h	Hd	GenBank Accession #
Chilumba				
Males	20	4	0.805	EF211848–EF211867
Females	20	8	0.742	EF211905–EF911924
Total	40	10	0.81	
Senga Bay				
Males	16	7	0.525	EF211832–EF211847
Females	18	5	0.696	EF211887–EF211904
Total	34	7	0.674	
Nkhudzi Bay				
Males	19	6	0.795	EF211868–EF211886
Females	22	8	0.84	EF211925–EF211946
Total	41	11	0.815	

Discussion

Genetic diversity and Hardy–Weinberg equilibrium

In each of the samples, all six microsatellite loci examined displayed high levels of polymorphism. However, it is important to note that 25 out of 56 single-locus tests in *C.* sp. 'virginalis kajose' and 16 of 42 in *C. quadrimaculatus* showed a deficit in heterozygotes. Most of the observed Hardy–Weinberg disequilibria were detected at three loci (Pzeb1, UNH002 and TmoM11). Because the levels of differentiation between the populations were far less than the mean F_{IS} values, the Wahlund effect hypothesis can be rejected. It is indeed difficult to suppose that subunits within our samples would create F_{ST} values 10–20 times greater than those of samples obtained from separated localities. Analyses of homozygote size classes in MICROCHECKER did not point to the occurrence of large allele dropouts by possibly degraded DNA as a cause for the observed heterozygote deficiencies. The presence of non-amplifying null alleles in Pzeb1 and UNH002 was previously described by van Oppen et al. (1998), Shaw et al. (2000), Rüber et al. (2001) and Rico & Turner (2002). It is the most likely explanation for the heterozygote deficits that we detected at the three loci. Null alleles could lead to both over- or underestimation of levels of population differentiation (van Oosterhout et al., 2004). In this study, correction of allele frequencies for the presence of null alleles resulted in similar population structure patterns and suggests that our results are not significantly biased.

Population structure in the two *Copadichromis* species

Previous studies on cichlids of Lake Malawi showed that species can exhibit very different patterns of population structure depending on their substrate preference. Both *Copadichromis* species examined display a degree of population substructuring intermediate to that of the highly structured rock-dwelling Mbuna and the panmictic pelagic cichlids. We found very low but significant degrees of genetic differentiation among populations in both species: overall $F_{ST} = 0.006$ ($P < 0.001$) for *C.* sp. 'virginalis kajose' and overall $F_{ST} = 0.007$ ($P < 0.001$) for *C. quadrimaculatus*. We did not find a significant difference in the degree of population substructuring between the rock-frequenting *C. quadrimaculatus* and the sand-dwelling *C.* sp. 'virginalis kajose'. This finding contrasts with previous observations by Pereyra et al. (2004) that, within three species of the non-Mbuna genus *Protomelas*, the level of population structure varied with substrate preference.

Although most pairwise F_{ST} values between populations were significant, the Bayesian clustering approach was unable to identify more than a single genetic cluster within each of the two Utaka species. However, when $F_{ST} < 0.05$ the ability of this method to detect population structure may become poor (Pearse & Crandall, 2004; J. K. Pritchard, pers. comm. GEM). Therefore, caution should be taken when interpreting low values of population differentiation (Hedrick, 1999). The small F_{ST} values and the absence of geographic structuring indicate that habitat discontinuities apparently do not impose restrictions to dispersal in these taxa. Moreover, the Lake Malombe population of *C.* sp. 'virginalis kajose', in spite of its geographically isolated position, is similarly differentiated from the Lake Malawi populations as the Lake Malawi populations from each other. These observations may imply that at least some Utaka disperse over long distances and do not refrain from migrating through the Shire River, the only connection between Lake Malombe and Lake Malawi.

On the other hand, despite the absence of apparent migration barriers, both species do tend to develop some genetic substructuring. Additional preliminary data on *Mchenga eucinostomus* (Regan) and *C. chrysonotus* (Boulenger) (Anseeuw et al., unpublished data), two other Utaka species from the sandy and the rocky habitat, respectively, also indicate levels of genetic substructuring similar to those in *C.* sp. 'virginalis kajose' and *C. quadrimaculatus*. We suggest that the formation of subtle population substructuring in Utaka may be facilitated by aspects of their mating strategy. Sexually active individuals of lek-breeding cichlids, such as the Utaka, periodically congregate on inshore breeding arenas (leks) to mate (McKaye, 1983; Kidd et al., 2006). Non-breeding individuals are not territorial and have a more free-ranging lifestyle (McKaye, 1983). Such sequence of events, in which temporal territoriality and aggregating on specific mating grounds during the reproductive phase alternates with non-territorial more free-ranging life stages, is likely to induce low population substructuring irrespective of the substrate preference of the species.

Sex-biased dispersal

Inspection of the microsatellite data for the sexes separately suggests male-biased dispersal in *C. quadrimaculatus*. *F*-statistics and relatedness both show that female populations are more genetically structured than male populations, which indicates that male *C. quadrimaculatus* display a higher dispersal behaviour. For *C.* sp. 'virginalis kajose', the difference in F_{ST} and r between males and females turned out statistically non-significant. Nonetheless, overall F_{ST} and r values were only significant for the female populations, suggesting potential weak asymmetry in dispersal between sexes. We therefore performed a supplementary analysis on a set of three *C.* sp. 'virginalis kajose' populations for which we examined the mtDNA sequence variability. By this approach, we found that the genetic structuring is significantly higher in female populations than in male populations. Since mtDNA is strictly maternally inherited, the degree of mtDNA differentiation among male populations can only be reduced through recent dispersal. In contrast, a female can pass on her haplotype to the next generation. Hence, a female could leave copies of it in different populations as she migrates and those offspring could also disperse to other populations. Consequently, female dispersal should have a far greater effect on homogenizing mtDNA variation than does male dispersal (O'Corry-Crowe et al., 1997). Therefore, if mtDNA results indicate that between-population differentiation is higher among females than males, as in our case, the most conservative hypothesis is that dispersal is likely to be biased towards males.

Male-biased dispersal was already detected in the highly structured Mbuna from Lake Malawi (Knight et al., 1999). According to a theoretical model by Perrin & Mazalov (2000), the tendency towards male-biased dispersal may be induced in polygynous and promiscuous mating systems with strong male–male competition for females, insofar as breeding resources are not limiting female reproductive output. In Utaka, breeding males seasonally congregate on leks where they defend an individual spawning territory to compete for females and to mate. All parental care is provided by the females, which leave the arena with the eggs (McKaye, 1983). The mouthbrooding females aggregate to form separate schools (McKaye, 1983, 1984) and release their fry together on nursery grounds (McKaye, 1984). The juveniles form mixed schools of females and non-territorial males (Konings, 2001) in which the surviving females may remain for a considerably longer time than male individuals, additionally increasing the chance for male-biased dispersal.

Conclusion

The Utaka in this study demonstrated low genetic population structure without an obvious spatial patterning. This suggests that these species experience little or no geographical restrictions to dispersal. Unlike in the strictly rock-dwelling Mbuna or members of the non-Mbuna genus *Protomelas*, substrate preference seems to have little effect on the degree of population structure in these Utaka. We have suggested that aspects of the reproductive strategy, such as periodic lekking, may be responsible for the development of low levels of genetic substructuring in the Utaka. The importance of mating behaviour on small-scale genetic substructuring seems further confirmed by the occurrence of male-biased dispersal.

Acknowledgements The field work was supported by the Interdisciplinary Research Centre of the KULeuven Campus Kortrijk, a grant from the King Leopold III Fund for Nature Exploration and Conservation to DA and grants from the Fund for Scientific Research (FWO) and the 'Stichting tot bevordering van het wetenschappelijk onderzoek in Afrika' to JS. G.E. Maes is a post-doctoral fellow of the Fund for Scientific Research (FWO). We wish to thank the Malawi Fisheries Department, and especially Sam Mapila and Orton Kachinjika, for permission to conduct fieldwork in Malawi, the Department for International Development (DFID) of the British High Commission in Malawi, and especially George Turner (Bangor University, UK), for putting a 4 × 4 vehicle at our disposal for the sampling trip. We thank the late Davis Mandere for his valuable assistance in the field, the Fisheries Research Unit in Monkey Bay and Nkhata Bay for their assistance and the late Stuart Grant for his hospitality and for providing fish samples.

References

Aljanabi, S. M. & I. Martinez, 1997. Universal and rapid salt-extraction of high quality genomic DNA for PCR-based techniques. Nucleic Acids Research 25: 4692–4693.

Arnegard, M. E., J. A. Markert, P. D. Danley, J. R. Stauffer Jr., A. J. Ambali & T. D. Kocher, 1999. Population structure and colour variation of the cichlid fish *Labeotropheus fuelleborni* Ahl along a recently formed archipelago of rocky habitat patches in Southern Lake Malawi. Proceedings of the Royal Society B: Biological Sciences 266: 119–130.

Avise, J. C., A. G. Jones, D. Walker, J. A. Dewoody, et al., 2002. Genetic mating systems and reproductive natural histories of fishes: Lessons for ecology and evolution. Annual Review of Genetics 36: 19–45.

Belkhir, K., 2000. GENETIX, Version 4.05. Laboratoire Genome, Populations, Interactions. CNRS UPR 9060, Montpellier, France.

Chapuis, M. P. & A. Estoup, 2007. Microsatellite null alleles and estimation of population differentiation. Molecular Biology and Evolution 24(3): 621–631.

Chisambo, J., 2000. A taxonomic revision of the deep-bodied spotted *Copadichromis* spp. (Teleostei, Cichlidae) from Lake Malawi/Nyasa Africa. Masters' Thesis, University of Waterloo, Ontario, Canada.

Dempster, A. P., N. M. Laird & D. B. Rubin, 1977. Maximum likelihood from incomplete data via the EM algorithm. Journal of the Royal Statistical Society, Series B 34: 1–38.

Duponchelle, F., A. J. Ribbink, A. Msukwa, J. Mafuka & D. Mandere, 2000. Depth distribution and breeding patterns of the demersal species most commonly caught by trawling in the South West Arm of Lake Malawi. In Duponchelle, F. & A. J. Ribbink (eds), Fish Ecology Report. Lake Malawi/Nyasa/Niassa Biodiversity Conservation Project. Southern African Development Community, Gaborone. Botswana/Global Environmental Facility, Washington, DC: 15–168.

Evanno, G., S. Regnaut & J. Goudet, 2005. Detecting the number of clusters of individuals using the software STRUCTURE: A simulation study. Molecular Ecology 14: 2611–2620.

Excoffier, L., G. Laval & S. Schneider, 2005. Arlequin version 3.0: An integrated software package for population genetics data analysis. Evolutionary Bioinformatics Online 1: 47–50.

Falush, D., M. Stephens & J. K. Pritchard, 2007. Inference of population structure using multilocus genotype data: Dominant markers and null alleles. Molecular Ecology Notes 7: 574–578.

Garcia De Léon, F. J., L. Chikhi & L. Bonhomme, 1997. Microsatellite polymorphism and population subdivision in natural populations of European sea bass *Dicentrarchus labrax* (Linnaeus, 1758). Molecular Ecology 6: 51–62.

Goudet, J. 2001. FSTAT, A Program to Estimate and Test Gene Diversities and Fixation Indices (Version 2.9.3). Available from http://www.unil.ch/izea/softwares/fstat.html.

Goudet, J., N. Perrin & P. Waser, 2002. Tests for sex-biased dispersal using bi-parentally inherited genetic markers. Molecular Ecology 11: 1103–1114.

Hedrick, P. W., 1999. Perspective: Highly variable loci and their interpretation in evolution and conservation. Evolution 53: 313–318.

Iles, T. D., 1960. A group of zooplankton feeders of the genus *Haplochromis* (Cichlidae) in Lake Nyasa. Annals and Magazine of Natural History 13: 257–280.

Kellogg, K., J. A. Markert, J. R. Stauffer Jr. & T. D. Kocher, 1995. Microsatellite variation demonstrates multiple paternity in lekking cichlid fishes from Lake Malawi, Africa. Proceedings of the Royal Society B: Biological Sciences 260: 79–84.

Kidd, M. R., C. E. Kidd & T. D. Kocher, 2006. Axes of differentiation in the bower-building cichlids of Lake Malawi. Molecular Ecology 15: 459–478.

Knight, M. E. & G. F. Turner, 2004. Laboratory mating trials indicate incipient speciation by sexual selection among populations of the cichlid fish *Pseudotropheus zebra* from Lake Malawi. Proceedings of the Royal Society B: Biological Sciences 271: 675–680.

Knight, M. E., M. J. H. van Oppen, C. Rico, G. M. Hewitt & G. F. Turner, 1999. Evidence for male-biased dispersal in Lake Malawi cichlids from microsatellites. Molecular Ecology 8: 1521–1527.

Konings, A., 2001. Malawi Cichlids in their Natural Habitat, 3rd ed. Cichlid Press, El Paso: 352.

McKaye, K. R., 1983. Ecology and breeding behavior of a cichlid fish, *Cyrtocara eucinostomus*, on a large lek in Lake Malawi, Africa. Environmental Biology of Fishes 8: 81–96.

McKaye, K. R., 1984. Behavioural aspects of cichlid reproductive strategies: Patterns of territoriality and brood defence in Central American substratum spawners and African mouth brooders. In Potts, G. W. & R. J. Wootton (eds), Fish Reproduction: Strategies and Tactics. Academic Press, London: 245–273.

Meyer, A., T. D. Kocher, P. Basasibwaki & A. C. Wilson, 1990. Monophyletic origin of Lake Victoria cichlid fishes suggested by mitochondrial DNA sequences. Nature 347: 550–553.

O'Corry-Crowe, G. M., R. S. Suydam, A. Rosenberg, K. J. Frost & A. E. Dizon, 1997. Phylogeography, population structure and dispersal patterns of the beluga whale *Delphinapterus leucas* in the western Nearctic revealed by mitochondrial DNA. Molecular Ecology 6: 955–970.

Pearse, D. E. & K. A. Crandall, 2004. Beyond F_{ST}: Analysis of population genetic data for conservation. Conservation Genetics 5: 585–602.

Pereyra, R., 2003. Population structure, genetic diversity and dispersal of cichlid fishes from Lake Malawi. PhD Thesis, University of East Anglia, Norwich, UK.

Pereyra, R., M. I. Taylor, G. F. Turner & C. Rico, 2004. Variation in habitat preference and population structure among three species of the Lake Malawi cichlid genus Protomelas. Molecular Ecology 13: 2691–2697.

Perrin, N. & V. Mazalov, 2000. Local competition, inbreeding and the evolution of sex-biased dispersal. American Naturalist 155: 116–127.

Pritchard, J. K., M. Stephens & P. Donnelly, 2000. Inference of population structure using multilocus genotype data. Genetics 155: 945–959.

Raymond, M. & F. Rousset, 1995. GENEPOP version 1.2: Populations genetics software for exact tests and ecumenicism. Journal of Heredity 86: 248–249.

Rico, C. & G. F. Turner, 2002. Extreme microallopatric divergence in a cichlid species from Lake Malawi. Molecular Ecology 11: 1585–1590.

Rüber, L., A. Meyer, C. Sturmbauer & E. Verheyen, 2001. Population structure in two sympatric species of the Lake Tanganyika cichlid tribe Eretmodini: Evidence for introgression. Molecular Ecology 10: 1207–1225.

Queller, D. C. & K. F. Goodnight, 1989. Estimating relatedness using molecular markers. Evolution 43: 258–275.

Salzburger, W. & A. Meyer, 2004. The species flocks of East African cichlid fishes: Recent advances in molecular phylogenetics and population genetics. Naturwissenschaften 91: 277–290.

Shaw, P. W., G. F. Turner, M. R. Idid, R. L. Robinson & G. R. Carvalho, 2000. Genetic population structure indicates sympatric speciation of Lake Malawi pelagic cichlids. Proceedings of the Royal Society B: Biological Sciences 267: 2273–2280.

Snoeks, J., 2004. The Cichlid Diversity of Lake Malawi/Nyasa/Niassa: Identification, Distribution and Taxonomy. Cichlid Press, El Paso, Texas: 360.

Taylor, M. I. & E. Verheyen, 2001. Microsatellite data reveals weak population substructuring in *Copadichromis* sp. 'virginalis kajose', a demersal cichlid from Lake Malawi, Africa. Journal of Fish Biology 59: 593–604.

Thompson, A. B., E. H. Allison & B. P. Ngatunga, 1995. Spatial and temporal distribution of fish in the pelagic waters. In Menz, A. (ed.), The Fishery Potential and Productivity of the Pelagic Zone of Lake Malawi/Niassa. Natural Resources Institute, Chatham, UK: 201–232.

Turner, G. F., 1996. Offshore Cichlids of Lake Malawi. Cichlid Press, Lauenau, Germany: 240.

van Oosterhout, C., W. F. Hutchinson, D. P. M. Wills & P. Shipley, 2004. Micro-checker: Software for identifying and correcting genotyping errors in microsatellite data. Molecular Ecology Notes 4: 535–538.

van Oppen, M. J. H., C. Rico, J. C. Deutsch, G. F. Turner & G. M. Hewitt, 1997. Isolation and characterization of microsatellite loci in the cichlid fish *Pseudotropheus zebra*. Molecular Ecology 6: 387–388.

van Oppen, M. J. H., G. F. Turner, C. Rico, R. L. Robinson, J. C. Deutsch, M. J. Genner & G. M. Hewitt, 1998. Assortative mating among rock-dwelling cichlid fishes supports high estimates of species richness from Lake Malawi. Molecular Ecology 7: 991–1001.

Weir, B. S., 1996. Intraspecific differentiation. In Hillis, D. M., C. Moritz & B. Mable (eds), Molecular Systematics, 2nd ed. Sinauer, Sunderland, MA.

Weir, B. S. & C. C. Cockerham, 1984. Estimating F-statistics for the analysis of population structure. Evolution 38: 1358–1370.

Wright, S., 1969. Evolution and the genetics of populations. In The Theory of Gene Frequencies, Vol. 2. University of Chicago Press, Chicago: 511.

Zardoya, R., D. A. Vollmer, C. Craddock, J. T. Streelman, S. Karl & A. Meyer, 1996. Evolutionary conservation of microsatellite flanking regions and their use in resolving the phylogeny of cichlid fishes (Pices: Perciformes). Proceedings of the Royal Society B: Biological Sciences 263: 1589–1598.

Hydrobiologia (2008) 615:81–94
DOI 10.1007/s10750-008-9566-y

SPECIATION IN ANCIENT LAKES

Ecological correlates of species differences in the Lake Tanganyika crab radiation

Saskia A. E. Marijnissen · Ellinor Michel · Marjolein Kamermans · Kim Olaya-Bosch · Moniek Kars · Daniel F. R. Cleary · E. Emiel van Loon · Paola G. Rachello Dolmen · Steph B. J. Menken

Published online: 21 September 2008

Abstract The endemic crabs of Lake Tanganyika include a phenotypically diverse clade that exhibits recent divergence and low phylogenetic species resolution. There are indications that ecological niche segregation has played a prominent role in the divergence of this clade. We used habitat surveys, gut content analyses and stable isotope analyses to test the extent to which morphological species are ecologically different. Our data show some interspecific segregation in depth, substrate type and mean stable isotope signatures. At the same time, a considerable level of ecological niche overlap is evident among species of *Platythelphusa* that coexist in rocky littoral habitats. We consider these results in the framework of adaptive radiation theory, and we discuss general ramifications for the maintenance of species diversity in Lake Tanganyika.

Keywords Decapod · Adaptive radiation · Niche partitioning · Benthic ecology

Guest editors: T. Wilke, R. Väinölä & F. Riedel
Patterns and Processes of Speciation in Ancient Lakes: Proceedings of the Fourth Symposium on Speciation in Ancient Lakes, Berlin, Germany, September 4–8, 2006

S. A. E. Marijnissen · D. F. R. Cleary · P. G. Rachello Dolmen · S. B. J. Menken
Institute for Biodiversity and Ecosystem Dynamics, Section of Evolutionary Biology, University of Amsterdam, Kruislaan 318, 1098 SM Amsterdam, The Netherlands

E. Michel (✉)
Department of Zoology, The Natural History Museum, Cromwell Road, London SW7 5BD, UK
e-mail: e.michel@nhm.ac.uk

M. Kamermans
School of Natural Sciences, Trinity College, Zoology Building, University of Dublin, Dublin 2, Ireland

K. Olaya-Bosch
Department of Geography, Trinity College, University of Dublin, Dublin 2, Ireland

M. Kars
Witmondplein 15, 6706 DV Wageningen, The Netherlands

D. F. R. Cleary
Departamento de Biologia, Centro de Estudos do Ambiente e do Mar (CESAM), Universidade de Aveiro, 3810-193 Aveiro, Portugal

E. E. van Loon
Institute for Biodiversity and Ecosystem Dynamics, Section of Computational Geo-Ecology, University of Amsterdam, Nieuwe Achtergracht 166, 1018 WV Amsterdam, The Netherlands

Introduction

Lake Tanganyika is unique among the African Great Lakes in harbouring an endemic radiation of freshwater crabs (genus *Platythelphusa*) (Cumberlidge et al., 1999; Marijnissen et al., 2004; Reed & Cumberlidge, 2006). In spite of marked levels of phenotypic diversity within the Lake Tanganyika endemic crab clade, phylogenetic resolution based on mitochondrial markers is low (Marijnissen et al., 2006; Marijnissen, 2007). There is evidence that initial divergence of the platythelphusid clade took place less than c. 3 Myr ago (Marijnissen et al., 2006). The relatively young ages of the platythelphusid species could explain the apparent incongruence between their phenotypic diversity and genetic similarity (e.g. Moran & Kornfield, 1993; Albertson et al., 1999). The approximate period of platythelphusid divergence corresponds to inferred periods of major cladogenesis in several Lake Tanganyika cichlid and *Synodontis* catfish lineages (Koblmüller et al., 2004, 2005, 2006, 2007; Duftner et al., 2005; Day & Wilkinson, 2006). It is likely that the contemporaneous divergence of these clades was induced by substantial environmental changes when the lake level dropped considerably during a period of aridification in eastern Africa (Cane & Molnar, 2001). Lake level changes offer ecological opportunities by providing novel resource-rich environments in which ecological diversification and adaptive radiation can take place.

The theory of adaptive speciation predicts that reproductive isolation evolves as a consequence of adaptation to different ecological niches (Schluter, 2000). Adaptive radiations are characterised by elevated levels of divergence among taxa as a response to natural selection acting directionally on ecologically relevant traits. Indeed, there is increasing evidence that the outstanding phenotypic diversity and rapid speciation of the cichlid species flocks in the African Great Lakes has to a large extent been facilitated by ecological niche diversification (Albertson & Kocher, 2006). The majority of studies on adaptive radiations in aquatic ecosystems have focussed on fish (reviewed in Schluter, 2000), although intriguing results are also emerging from studies on Lake Baikal amphipods (Takhteev, 2000), Lake Ohrid limpets (Albrecht et al., 2006) and gastropods from the Malili Lakes on Sulawesi (Von Rintelen et al., 2004). Because Lake Tanganyika harbours a wide range of groups with elevated levels of species richness and endemicity, including not only fish but also ostracods, copepods, gastropods, shrimp, leeches and sponges, it provides an outstanding system to test how ecological opportunities may facilitate adaptive radiation in a diverse array of aquatic taxa. Although there are morphological indications that ecological processes have been important in the evolution of several of these species flocks (Michel, 2000; West et al., 2003; Fryer, 2006), for most Lake Tanganyika endemics, sufficient data to explore adaptive radiation across their entire clade remains lacking.

The platythelphusid crabs are an ideal test case to investigate common effects of shared environmental history and ecological opportunities that have shown to be significant in cichlid speciation. There are some indications that ecological niche partitioning played an important role in facilitating platythelphusid divergence. The majority of the nine platythelphusid species occur sympatrically, and up to five species have been found to coexist in rocky littoral areas. Evert (1970) suggested that some of the unusual phenotypes exhibited by Lake Tanganyika crabs have ecological significance. For instance, the elongated slender legs of the presumed predominantly deep lake dwelling species *Platythelphusa tuberculata* Capart, 1952, and *P. praelongata* Marijnissen et al., 2004, might facilitate movement and tactile predation on mud (Fig. 1; Evert, 1970; Marijnissen et al., 2004). Marked differences also exist in the shape and maximum size of the carapace (Marijnissen et al., 2004: Table 2). The small, rounded body of *P. maculata* Cunnington, 1899, and *P. polita* Capart, 1952, appears to provide advantages for living inside empty gastropod shells. *Platythelphusa conculcata* Cunnington, 1907, *P. immaculata* Marijnissen et al., 2004 and *P. echinata* Capart, 1952, have markedly flattened carapaces (Cumberlidge et al., 1999; Marijnissen et al., 2004), which matches well with the narrow crevices in rocky substrates where they are most commonly found (S. A. E. Marijnissen, personal observations). The largest-bodied platythelphusid species, *P. armata* A. Milne-Edwards, 1887, has greatly enlarged claws that were suggested to have coevolved with Lake Tanganyika's heavily armoured gastropods (West et al., 1991; West & Cohen, 1994). Furthermore, comparative analyses have revealed marked differences in claw functional morphologies between all platythelphusid species, indicating a considerable level of trophic partitioning

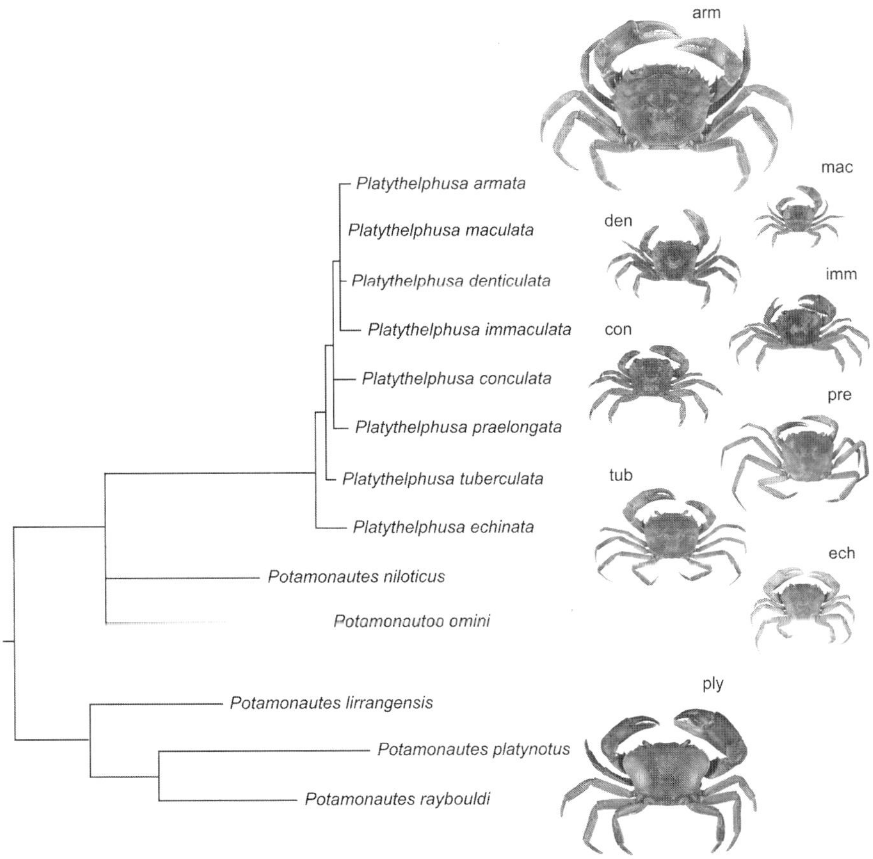

Fig. 1 Hypothesis for the evolution of the endemic Lake Tanganyika crabs (with exception of *P. polita*, for which no genetic sequences are presently available), based on mitochondrial genes (16S rRNA and COII mtDNA, tree adapted from Marijnissen, 2007). arm = *Platythelphusa armata*; con = *P. conculcata*; den = *P. denticulata*; ech = *P. echinata*; imm = *P. immaculata*; mac = *P. maculata*; tub = *P. tuberculata*; pre = *P. praelongata*; ply = *Potamonautes platynotus* (all males, except *P. praelongata*). Scale bar = 50 mm

(Marijnissen, 2007). Although there are several phenotypic clues to differentiation in resource use, it remains to be shown to which extent platythelphusid species occupy distinct ecological niches.

Determining ecological correlates of species differences in Lake Tanganyika endemic crabs is not only relevant from an evolutionary point of view, but also essential for conservation purposes. There is growing concern about the conservation of biodiversity in Lake Tanganyika (McIntyre et al., 2005; Cohen et al., 2005; Darwall et al., 2005). Effective decisions on conservation of biodiversity rely on an understanding of ecological divergence among key taxa. Crabs are common in the lake and they are expected to have a major influence on lacustrine foodweb interactions through their role as prey for fish (Hori, 1983; Coulter, 1991; Hori et al., 1993) and consumers of benthic organisms (e.g. West et al., 1991). If patterns of species divergence and coexistence of Lake Tanganyika's endemic crab species are controlled by ecological differentiation, this could have important implications for conservation decisions as it entails high sensitivity to habitat disturbances. Alternatively, species might be functionally equivalent and diversity could be maintained through versatility (Bellwood et al., 2006).

The aim of this paper is to assess if the platythelphusid species differ in ecological niche realisation. We include *P. denticulata*, although its taxonomic status is uncertain, and it is possibly a hybrid taxon (Marijnissen et al., unpublished data). We also include another endemic Lake Tanganyika species, *Potamonautes platynotus* (Cunnington, 1907), to determine if its ecological niche is distinct from that of the platythelphusid species with which it coexists. Because this is the first study to empirically examine the ecology of the endemic Lake Tanganyika crabs, our approach is exploratory rather than diagnostic. In order to examine the ecological correlates of species differences in these crabs, we use a combination of data from habitat surveys, gut contents and stable isotope analyses. We consider our results in the framework of adaptive radiation theory and we discuss other factors that are likely to play a role in the divergence of the platythelphusid clade.

Methods

Study system and sites

Lake Tanganyika is the oldest (~9–12 my) and deepest (max. 1,470 m) of the East African Rift lakes (Coulter, 1991). The shoreline is steeply sloped in most areas. Different substrate types are interspersed along the shoreline on scales of 10 to 1,000 m, but even on smaller scales substrates can be patchy (Michel et al., 2004). We surveyed an area of approximately 50 km along the Tanzanian shoreline of the lake between 2001 and 2004. Within this area, 16 study sites were selected based on accessibility and substrate composition (Fig. 2). The substrate at 12 sites (MWG to ZGU) was composed of boulders, cobbles, pebbles, and/or aggregations of empty *Pleiodon spekii* (Woodward, 1859) freshwater mussel shells, interspersed with sand. Four sites were situated west of a river delta (Luichi River, see Fig. 2), where the substrate consisted of mud with extensive beds of empty shells from the endemic gastropod *Neothauma tanganyicense* Smith, 1880, and *Coelatura burtoni* (Woodward, 1859) shell hash.

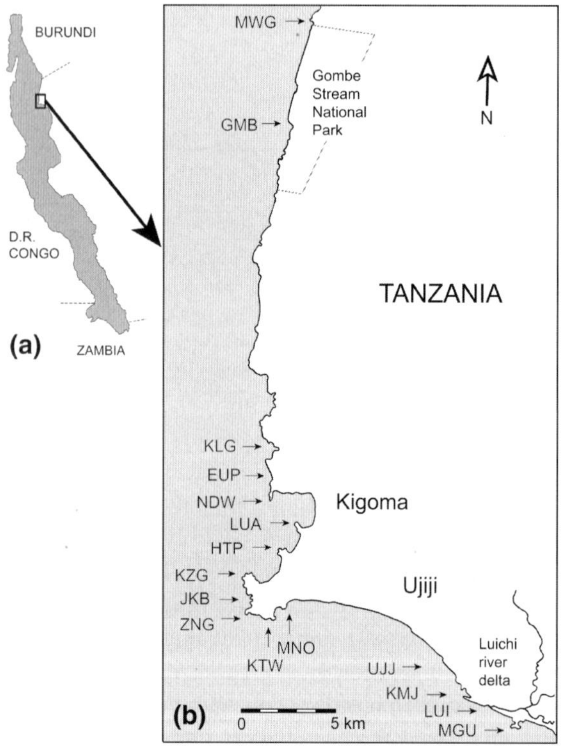

Fig. 2 Lake Tanganyika. (**a**) Map of Lake Tanganyika, (**b**) Study sites that were surveyed for crab habitat specificity. Acronyms (local names in brackets): MWG = Mwamgongo village; GMB = Gombe Stream National Park; KLG = Kalalangabo (Lemba village); EUP = Euphorbia (Kasazi Hill); NDW = Nondwa Point; LUA = Luansa Point (Kigoma Bay); HTP = Hilltop Hotel; KZG = Kazanga/Bangwe Point; JKB = Jakobsen's Beach (Mwamahunga); ZNG = Mzungu; KTW = Kitwe Point; MNO = Meno Hill; UJJ = Ujiji; KMJ = Kangamoja; LUI = Luichi river northernmost outlet; MGU = Mgumile

Habitat specificity

In order to obtain presence/absence data on each crab species, we surveyed 16 study sites at depths ranging between 0 and 30 m. A minimum of five surveys were conducted per site. Each site was systematically searched for crabs using SCUBA or snorkelling. Notes were made on crab species, size (carapace width), life stage (juvenile or adult see Cumberlidge, 1999, p. 324; Marijnissen et al., 2004: Table 2) and sex, as well as on the habitat in which each individual crab was found. Substrate types were categorised according to a modified Udden-Wentworth scale (Wentworth, 1922): fine sediment (0.00025–0.062 mm), sand (0.062–2.00 mm), pebbles (2.00–64.00 mm), cobbles (64.00–256.00 mm), boulders (>256 mm) and aggregations of *P. spekii* or *N. tanganyicense*. Information about the habitat of *P. tuberculata*, which can occur to depths of 190 m (Coulter, 1991), was obtained by combining information on catch localities obtained from fishermen with substrate information from Lewis (2002) and William (2002).

Additional information on habitat specificity for each species was obtained using a stratified random sampling strategy at 10 out of 16 sites (MWG to ZNG, see Fig. 2) in rocky littoral areas. At each site, divers established 20 replicate 50 × 50 cm quadrats at 5-m depth. The slope of the substrate was recorded for each quadrat. We also collected and counted all snails that were visible on rock surfaces within five random quadrats at each site. For each of these quadrats, we quantified chlorophyll *a* concentration as well as the relative amount of organic and inorganic matter, following the method used by McIntyre et al. (2005). By firmly placing a sealed cover over the upward face of a flat lying cobble, a fixed surface area of the underlying sediment and periphyton could be brought ashore for analyses. All material surrounding the cover was scrubbed away and carefully discarded. This provides a sample of attached and unattached material from the rock's surface of a constant and known area. The aufwuchs and loose organic and inorganic material protected underneath the cover was removed, suspended in water and homogenised. Subsamples were collected on precombusted glass–fibre filters (Whatman GF/C, Maidstone, UK). One subsample was extracted in 90% ethanol for 24 h and analysed for chlorophyll *a* after acidification (Moed & Hallegraeff, 1978). Another subsample was dried to a constant mass at 60°C, weighed, combusted at 500°C for 3 h and reweighed to determine the organic content (Sutherland, 1998).

We tested for differences in the habitat specificity among species with a PERMANOVA test on each data set (i.e. surveys of substrate types across depths, and stratified random sampling using quadrats at 5-m depth). After log $(x + 1)$ transformation of the data using PRIMER 6 (Primer-E Ltd, Plymouth UK), a matrix was constructed of pairwise differences in habitat similarity between pairs of individuals based on the Ochiai similarity index (Ochiai, 1957). Similarity indices are often used to assess niche differentiation and overlap between pairs of morphologically or otherwise distinct entities in ecological communities (e.g. Grant et al., 1985; Safran & Omori, 1990; Lehman, 2000). We included the variable 'species' as a fixed factor in a PERMANOVA model, using a type III sum of squares and 9999 permutations under the reduced model using the Ochiai similarity matrix. We also performed a principal coordinates analysis (PCO) on the same matrix to visually examine variation in similarity of habitat specificity among individuals and species. This procedure has the advantage that the ordination can be based on a distance or similarity matrix that is derived from a wide selection of metrics or semimetrics, including presence–absence data (Jackson et al., 1989; Litvak & Hansell, 1990). Because our data includes many overlaying PCO scores that cannot be distinguished using standard biplots, we applied two-dimensional binning to effectively visualise different scores. The scores are displayed by symbols, the size of which is proportional to the number of points in that bin. For all biplots, 50 bins were used on both axes (2,500 squares in the two-dimensional space). The symbol for each score was randomly shifted slightly from the bin centre to prevent superposition of scores for different species.

We also examined whether habitat specificity within the rocky littoral zone is related to crab body size (measured as carapace width). Data were displayed using Box plots as developed by Tukey (Frigge et al., 1989). Kolmogorov–Smirnov tests of normality indicated that the data deviated significantly from normality and would not follow a normal distribution following appropriate transformations ($P < 0.001$). Because of our large sample size ($N = 1401$), parametric methods are expected to have more statistical power than nonparametric methods, and we therefore tested the significance of differences in mean crab carapace width and life stage among substrate types using a general linear model multivariate analysis of variance (GLM MANOVA), followed by Tukey's HSD pairwise comparisons with $\alpha = 0.05$ in SPSS 11.0.4. (SPSS Inc. 2005).

Diet composition

Specimens for dietary analyses were collected between 2002 and 2004 at three sites (HTP, JKB and KMJ, see Fig. 2) along the Tanzanian shoreline of Lake Tanganyika. Crabs were collected by hand at depths between 5 and 10 m using SCUBA, and killed in 98% ethanol within 20 min after collection. Only adult, intermolt specimens were included in the analyses. Foreguts as well as mid- and hindguts were carefully dissected, after which the contents were washed into a Petri dish and viewed under a dissecting microscope. We only used specimens with ≥50% full foreguts. Food items were identified to the lowest possible taxonomic level and also categorized as detritus, vascular plant matter, algae, gastropods, aquatic insects and ostracods. The proportional representation of each category was determined using the percentage occurrence method (Williams, 1981). Percentage frequency of occurrence (PO) was estimated as $(N_1/N_2) \times 100$, where N_1 is the number of individual crabs that consumed food item x, and N_2 represents the total number of individuals.

Stable isotope analyses

The stable isotopic composition of an organism represents an average ratio related to the utilized portion of its diet that is integrated in its tissue over a specific period of time. Stable isotopes of carbon and nitrogen (δ^{13}C and δ^{15}N) accumulate in a predictable way between consumers and their diet, and they can be useful in providing quantitative information on trophic niche segregation between organisms (Newsome et al., 2007 and references therein). We collected samples for stable isotope analyses between August and November 2002 from the same three sites where specimens for gut content analyses were collected, with the exception of *P. tuberculata*. Samples of this species were obtained from specimens collected by fishermen. In total, 83 adult intermolt specimens were collected for stable isotope analyses, including a minimum of six individuals (three males and three females) per species. Each individual specimen was measured and sexed, after which a piece of leg muscle tissue was dissected. All samples were transferred onto ethanol-cleaned aluminium foil, and dried until constant weight at 60°C for $\geq$48 h. Subsamples of dried material were analysed for stable carbon and nitrogen isotopes using a Europa Scientific 20/20 Isotope Ratio Mass Spectrometer (Iso-Analytical Ltd., Cheshire, UK). Isotope ratios are expressed in delta notation (δ) to indicate deviations from standard reference material (Vienna PDB carbon standard and atmospheric nitrogen). Average reproducibility based on replicate measurements was within approximately 0.1‰ for both δ^{13}C and δ^{15}N. In order to examine patterns of trophic segregation between species, a dual-isotope plot δ^{13}C and δ^{15}N values was constructed. We tested for differences in stable isotope composition between sexes and among species with a PERMANOVA test using the programme PERMANOVA+ within PRIMER 6 (Primer-E Ltd, Plymouth UK). After normalizing the data within PRIMER 6, a matrix was constructed based on the Euclidean difference in isotope composition between pairs of species. Sex and species were included as fixed factors in the PERMANOVA model, using a type III sum of squares and 9999 permutations under the reduced model.

Results

Habitat specificity

A total of 1,401 crabs were sampled during random searches at 16 sites along the Tanzanian coastline of Lake Tanganyika, and 404 crabs were sampled during quadrat surveys at a subset of 10 sites. *Platythelphusa polita* and *P. praelongata* were not found during any of our surveys. *Platythelphusa maculata* appears to be restricted to beds of empty *N. tanganyicense* shells near the outlet of the Luichi River (Fig. 2), where the substrate consists of silt and mud. Out of a total of 1,805 crabs that were collected during our surveys, the numbers of individuals per species decreased in the following order: *P. conculcata* (588), *P. echinata* (491), *P. armata* (347), *Po. platynotus* (239), *P. maculata* (67), *P. immaculata* (61), P. *denticulata* (12). *Potamonautes platynotus* is the only species that was also occasionally observed outside of the lake on pebble and cobble beaches. *Platythelphusa tuberculata* was not found during our surveys in the shallow benthic zone between 0 and 30 m; however, this species was regularly caught in the nets of fishermen that reportedly fished on the platform that extends approximately 14 km west from the Luichi River delta (Fig. 2), at depths ranging from approximately 50 to $\geq$100 m. Most of the sediments on this platform are composed of fine silt and mud derived from the Luichi River and/or from pelagic rainout of suspended fine sediments (Lewis, 2002; William, 2002).

The results of PCO based on the surveys of substrate type across depths (0–30 m) are shown in Fig. 3. The first two PCO axes explain 39.9% and 35.1% of the variation in the dataset (Table 1). Substrate types with the highest vector loadings were boulders, cobbles, pebbles and sand. With the exception of *P. tuberculata* and *P. maculata*, all species are associated with these four vectors. Although species distributions along the axes thus show considerable overlap, the frequency of the observations within each substrate type differs for each species. Significant differences were found in the similarity of substrate specificity among species (Pseudo-$F_{5,321} = 10.029$, $P < 0.001$). Significant pairwise differences between species that coexist within the rocky littoral included *P. armata* and *P. conculcata* ($P = 0.002$), *P. armata* and *P. echinata* ($P < 0.001$), *P. armata* and *P. immaculata* ($P = 0.044$), *P. conculcata* and *P. echinata* ($P = 0.010$), *P. conculcata* and *Po. platynotus* ($P < 0.001$) as well as *P. echinata* and *Po. platynotus* ($P < 0.001$). Other pairwise comparisons were not significant ($P > 0.05$).

Additional information on species habitat specificity was obtained from the quadrat surveys at 5-m

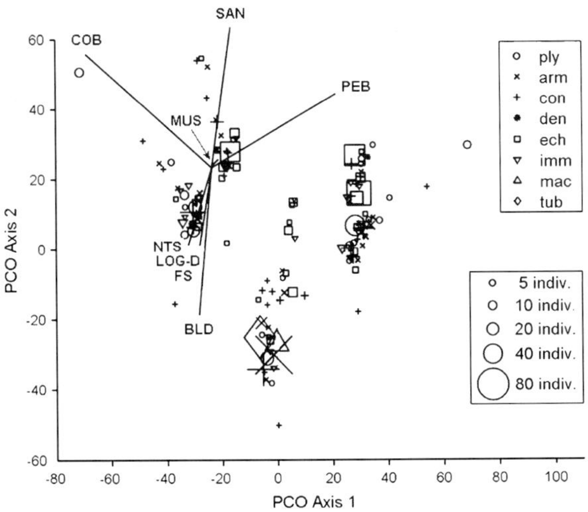

Fig. 3 Principal coordinate (PCO) ordination plot of the two first axes of crab species habitat specificity based on random surveys of different substrate types across depths (0–30 m) at 16 sampling sites along the Tanzanian coastline of Lake Tanganyika. Species abbreviations as in Fig. 1. The relative size of the symbols is proportional for the number of individual observations per species for each habitat type (see text). FS = fine sediment (0.00025–0.062 mm); SAN = sand (0.062–2.00 mm); PEB = pebbles (2.00–64.00 mm); COB = cobbles (64.00–256.00 mm); BLD = boulders (>256 mm); MUS = aggregations of *Pleiodon spekii* freshwater mussel shells; NTS = *Neothauma tanganyicense* shells

Table 1 Percentage of variation explained by individual axes of principal coordinates analysis of the parameters relating to Fig. 2

Axis	Individual (%)	Cumulative (%)
1	39.88	39.88
2	35.07	74.95
3	20.58	95.53
4	15.65	111.18

depth within the rocky littoral zone. The first two axes of the PCO explained 64.2% and 24.6% of the total variance in the species dataset (Fig. 4, Table 2). Influential habitat variables are slope, total organic matter and chlorophyll *a*, as well as relative snail abundance and inorganic matter. *Platythelphusa armata*, *P. conculcata*, *P. echinata* and *Po. platynotus* were associated with all vectors. *Platythelphusa denticulata* was only associated with snail abundance and inorganic matter, while *P. immaculata* was only associated with slope, total organic matter and chlorophyll *a*. Significant differences were

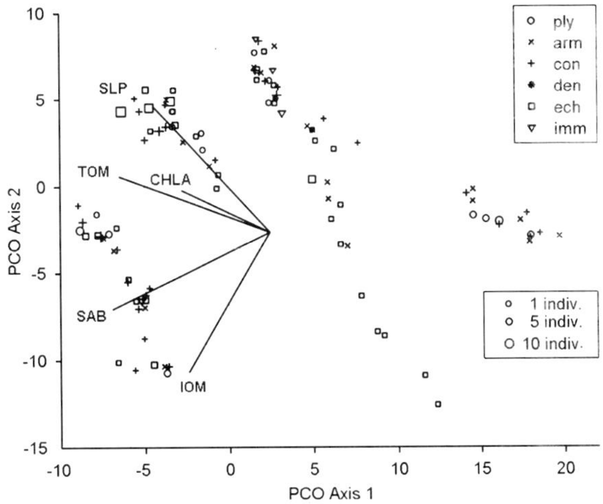

Fig. 4 Principal coordinate (PCO) ordination plot of the two first axes of crab species habitat specificity within the rocky littoral zone at 5-m depth in Lake Tanganyika. Species abbreviations as in Fig. 1. The relative size of the symbols is proportional for the number of individual observations per species for each habitat type (see text). SLP = slope of the substrate, CHLA = chlorophyll *a*; TOM = total organic matter; IOM = total inorganic matter; SAB = relative snail abundance

Table 2 Percentage of variation explained by individual axes of principal coordinates analysis of the parameters relating to Fig. 3

Axis	Individual (%)	Cumulative (%)
1	64.24	64.24
2	24.61	88.84
3	14.79	103.63
4	7.07	110.7

found among species (Pseudo-$F_{7,1463} = 108.680$, $P < 0.001$). Pairwise tests between species showed that all comparisons between species were significant ($P < 0.006$) with the exception of the comparisons between *P. denticulata* and *P. conculcata* ($P = 0.18$), *P. denticulata* and *P. immaculata* ($P = 0.490$) as well as *P. denticulata* and *Po. platynotus* ($P = 0.554$).

Both the mean body size (carapace width) as well as the life stages of crabs had a significant effect on the use of substrate type within the rocky littoral zone (GLM MANOVA size $F = 330.480$, df = 3, $P < 0.001$; life stage $F = 195.716$, df = 3, $P < 0.001$). While relatively broad ranges of size classes are found in all rocky littoral substrate types, cobbles and boulders harbour larger crabs than sand and pebbles (Fig. 5).

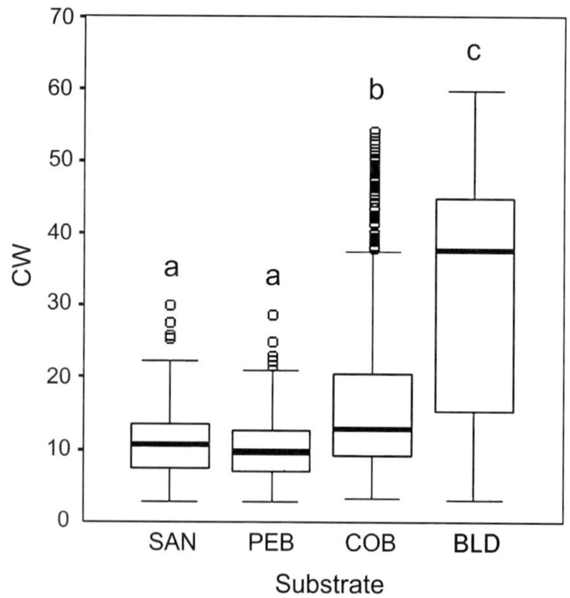

Fig. 5 Distribution of crab carapace widths (CW) among different substrate types in the rocky littoral zone. Box plots represent CW median and interquartile range with whisker ends corresponding to the first and the last decile. Individual observations beyond these limits are plotted as circles. Abbreviations as in Fig. 3. Different letters above box plots indicate significant differences in mean crab CW among substrate types at $\alpha = 0.05$ (Tukey's test: $P < 0.001$)

Diet composition

We collected a total of 574 adult crabs for dietary analyses, of which 386 had ≥50% full foreguts. The diet composition of the six species that were included in the gut content analyses is shown in Fig. 6. The majority of individuals in each species had ingested some amount of detritus (including sand and silt as well as indeterminate organic matter). Vascular plant matter, algae, ostracods, parts of aquatic insects, fragments of gastropod shells and opercula were also found in the guts of individuals from all six species. Vascular plant matter consisted predominantly of bark and plant debris with a terrestrial origin. Algal taxa included typical benthic diatoms (*Encyonema* sp., *Rhopalodia* sp., *Nitzschia* sp. and *Surirella* sp.), green algae (*Oocystis* sp. and *Ulothrix* sp.) as well as cyanobacteria (*Anabaena* sp., *Chroococcus* sp. and *Oscillatoria* sp.). Ostracods included species of Cyprididae, Cyclocypridae (*Mecynocypria* sp.) and Limnocytheridae (*Gomphocythere* sp.). Parts of aquatic insect larvae that could be determined from gut contents were most frequently assigned to caddisflies (including Ecnomidae and Hydropsychidae), mayflies (including Baetidae and Ephemeridae)

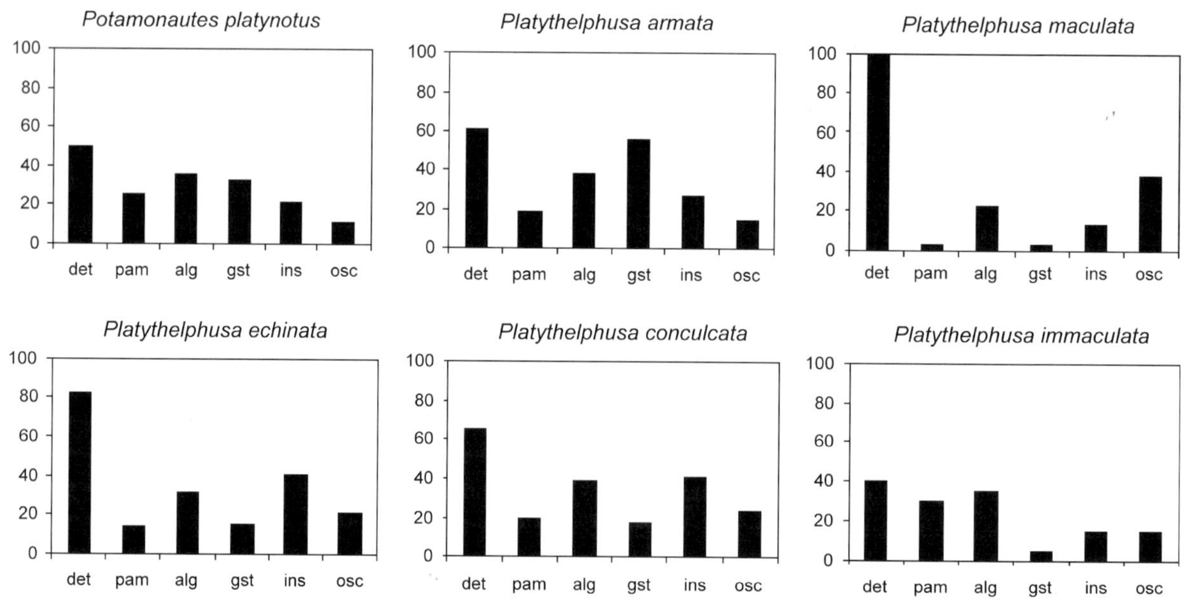

Fig. 6 Percentage occurrence of the various food categories in the six species analysed. Numbers of specimens analysed per species: *Potamonautes platynotus* ($n = 28$); *Platythelphusa armata* ($n = 74$); *P. maculata* ($n = 32$); *P. echinata* ($n = 98$); *P. conculcata* ($n = 128$); *P. immaculata* ($n = 20$). Det = detritus; pam = vascular plant matter; alg = algae; gst = gastropods; ins = aquatic insects; osc = ostracods

as well as chironomids (Chironominae, including *Dicrotendipes* sp.). Gastropod shell fragments were assigned to the endemic Tanganyikan genera *Lavigeria, Stormsia/Reymondia* and *Vinundu.* Gut contents revealed no apparent differences in the types of dietary constituents that were ingested among the six species. However, comparison of percentage-occurrence values in each dietary category revealed several differences. Gut contents of *P. maculata* and *P. echinata* included detritus more frequently than those of other species. Furthermore, *P. armata* appears to consume gastropods more frequently than any of the other species.

Stable isotope analyses

Crab isotopic signatures ranged from −12.4 to −21.2‰ δ^{13}C and from 3.6 to 6.4‰ δ^{15}N. We found a significant difference in isotope composition among species (Pseudo-$F_{6,94} = 23.622$, $P < 0.001$). There was no difference between sexes (Pseudo-$F_{1,94} = 0.008$, $P = 0.920$), nor was there a significant interaction (Pseudo-$F_{6,94} = 1.453$, $P = 0.165$). Pairwise tests between species showed that all comparisons between species were significant ($P < 0.012$) with the exception of the comparison between *P. armata* and *P. maculata*. The distribution of individuals of each species along axes defined by δ^{13}C and δ^{15}N is presented in Fig. 7. The deep, muddy-substrate dwelling *P. tuberculata* is segregated from other platythelphusid species by its high mean δ^{15}N signature. Within the rocky substrate species complex, *P. conculcata* exhibits the lowest δ^{15}N values. Pairwise tests showed significant differences between mean δ^{13}C and/or δ^{15}N between species, with the exception of the comparison between *P. armata* and *P. maculata,* which overlap in both mean δ^{13}C and δ^{15}N values (Table 3).

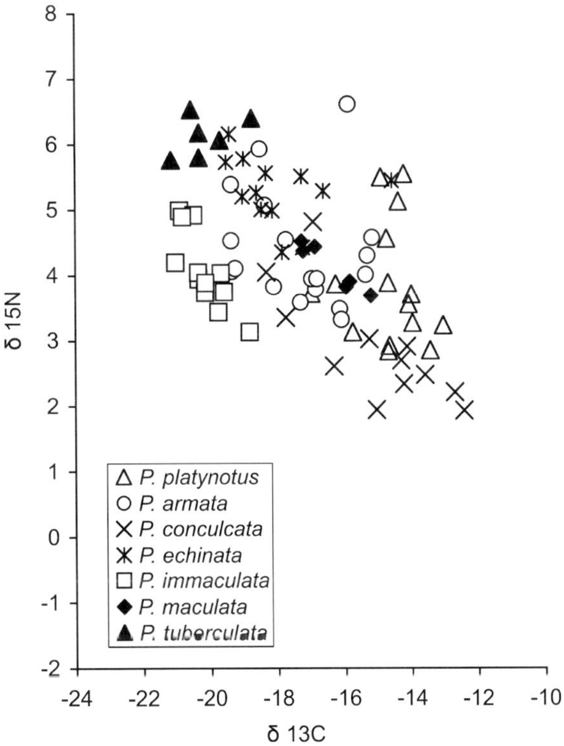

Fig. 7 Stable carbon and nitrogen isotope signatures of Lake Tanganyika endemic crab species. Each data point represents an individual organism

Discussion

The platythelphusid clade meets at least three out of four of Schluter's (2000) criteria for adaptive radiations: (i) common ancestry, (ii) rapid divergence and (iii) differentiation in functional traits (Marijnissen et al., 2006; Marijnissen, 2007). Although the level of genetic resolution between platythelphusid species is low (Marijnissen et al., 2006; Marijnissen, 2007), the addition of ecological data provides more definition and insight into the processes that may have contributed to the divergence of these crabs (Fig. 1). A possible scenario is that Lake Tanganyika was initially invaded by a generalist ancestor, which diverged through adaptation to distinct muddy and rocky habitats. Subsequent divergence occurred predominantly in rocky habitats, where species adapted to fill specific spatial and/or dietary niches.

Our growing understanding of platythelphusid relationships and ecology accommodates this scenario in several major points. The most basal species of the platythelphusid clade is *P. echinata*, a habitat generalist, of which adult specimens occur at a range of depths in rocky habitats, on muddy substrates, as well as in empty *N. tanganyicense* shells. Our results corroborate previous indications that *P. tuberculata,* which is phylogenetically relatively basal and an unambiguously distinct lineage, is found in deep, muddy parts of the lake (Coulter, 1991; Marijnissen et al., 2004), and is trophically highly distinct. The phylogenetic relationships of the species found in

Table 3 Pairwise distances (Tukey's test) of mean $\delta^{13}C$ and $\delta^{14}N$ isotope composition among adult individuals of Lake Tanganyika endemic crab species

	Species	*P. tuberculata*	*P. conculcata*	*P. immaculata*	*P. maculata*	*P. echinata*	*P. armata*
$\delta^{13}C$	*P. conculcata*	5.05*			ns		
	P. echinata	2.12*	2.92*	2.16*	ns		ns
	P. immaculata	ns	5.08*				
	P. maculata	3.73*	ns	3.76*			
	P. armata	2.79*	2.25*	2.83*	ns	ns	
	Po. platynotus	5.48*	ns	5.51*	ns	3.35*	2.68*
$\delta^{14}N$	*P. conculcata*	3.25*					
	P. echinata	ns	2.41*	1.20*	1.16*		
	P. immaculata	2.04*	1.21*				
	P. maculata	1.99*	1.25*	ns			
	P. armata	1.72*	1.52*	ns	ns	0.89*	
	Po. platynotus	2.27*	0.98*	ns	ns	1.43*	ns

* Significant difference of means at $\alpha = 0.05$; ns is non-significant difference at $\alpha = 0.05$

rocky habitats are less resolved, indicating either a more recent divergence or introgression. This is paralleled in ecology, where habitat and trophic preferences overlap to a greater degree than among the species basal in the phylogeny. Three species occur sympatrically in rocky habitats, including *P. conculcata* and *P. immaculata,* as well as *P. armata*. In addition, *P. denticulata* appears to prefer rocky habitats, but the results for this taxon might be biased due to its low sample number. Our surveys furthermore show that the small, round-bodied *P. maculata* is distinct in exhibiting a marked preference for empty *N. tanganyicense* shells. *Potamonautes platynotus* is derived from a lineage that invaded Lake Tanganyika separately. Although the ecological niche of this species exhibits marked overlap with the platythelphusid species that inhabit rocky substrates, *Po. platynotus* is unique in that it is occasionally observed on the lake's margin outside of the water. This contrasts with the platythelphusid crabs, which appear to be fully aquatic.

It is likely that partitioning of ecological niche variables supports species diversity of the Lake Tanganyika crabs. The results of PCO indicate that although individual crabs can be associated with a range of variables, such as depth, chlorophyll *a* concentration and relative amount of organic and inorganic matter, the majority of individuals exhibit species-specific patterns in habitat preferences. Differentiation in habitat specificity within the rocky littoral appears to be mainly a function of substrate size (Fig. 5). Adults of small-bodied species, such as *P. conculcata*, *P. immaculata* and *P. echinata*, are partially relieved from competition by having access to substrates that are too small for adult *P. armata* and *Po. platynotus*. We also found a significant influence of life stage on preferred substrate. This makes it likely that ontogenetic shifts in habitat use play an additional role in facilitating species coexistence within rocky littoral areas. Size-related habitat use is common among crustaceans and is most often attributed to protection from predation (Hudon & Lamarche, 1989; Navarette & Castilla, 1990; Richards, 1992; Platvoet et al., 2008). Predation-mediated habitat use may lead to increased competition within the refuge. Ecological niche partitioning between similar-sized adult individuals of different species that coexist within the same refuge might subsequently take place through specific adaptations allowing exploitation of different food items.

Comparative analyses of claw functional morphologies have revealed marked differences among the endemic Lake Tanganyika crab species, indicating a considerable level of trophic divergence (Marijnissen, 2007). In contrast, our gut content analyses showed no clear evidence of specialization (Fig. 6). In spite of marked divergence in claw functional morphologies, different species show broad overlap in the type of food items that they can handle and ingest. Comparison of percentage occurrence of each dietary category nonetheless revealed several differences. *Platythelphusa armata* consumes gastropods more frequently than any of the other crab species. This is in agreement with expectations based on claw traits and predation experiments, indicating that *P. armata* is a highly derived molluscivore (West et al., 1991; West & Cohen, 1994; Rosales et al., 2002; Marijnissen, 2007; Michel et al., unpublished data). Detritus comprised an important component in the guts of all species that were examined. This can either be the result of non-selective foraging behaviour, or it might reflect partially digested food that could not be identified. Determination of crustacean diets based on gut contents is associated with several well-recognized limitations, due to the effects of mastication and under-representation of readily homogenised food items (e.g. Hill, 1976). Examination of gut content alone might therefore not accurately reflect dietary intake. Furthermore, our gut content analyses are limited because of the fact that the crabs were only collected during daytime. It is likely that foraging patterns are dictated by the risk of being predated upon, and nocturnal activity might thus be different from diurnal activity.

Stable isotopic signatures offer the advantage of spatio-temporal integration of the assimilated food, and are thus useful in providing additional insight into questions of dietary partitioning. Comparison of mean $\delta^{13}C$ and/or $\delta^{15}N$ values between our focal taxa revealed that signatures of individual specimens show considerable overlap. However, there were significant differences in mean $\delta^{13}C$ and/or $\delta^{15}N$ values among species, with the exception of the comparison between *P. armata* and *P. maculata*. These results imply that although there are overall trophic differences among species, individuals within species can exhibit a range of foraging strategies. Work on other phenotypically diverse lacustrine species flocks has shown that closely related endemics often exploit a broader range of resources than what would be expected based on their specialized feeding morphologies (Liem & Osse, 1975; Liem, 1980; Genner et al., 1999). This apparent discrepancy can be explained if specialization is based on non-preferred resources while also allowing exploitation of more preferred common resources (Robinson & Wilson, 1998).

Perhaps the most significant criterion of an adaptive radiation is a correlation between divergent phenotypes and differentiation in ecological niches. Although we have observed segregation in depth, substrate type and mean stable isotope signatures, it is also evident that a considerable level of ecological versatility exists among platythelphusid species that coexist in rocky littoral areas. This may act to relax interspecific competition by allowing opportunistic resource use and could result in some resilience to habitat disturbance. It is also possible that niche partitioning occurs along variables that we have not yet measured. However, since we have addressed the most common ecological niche segregation variables in this study, our data should provide a robust first indication of the level of interspecific differentiation. The possibility that environmental factors are not the sole force driving platythelphusid divergence should also be investigated. Recent field and experimental studies have demonstrated that rapid ecological radiation is often entangled with forces of sexual selection (Van Doorn & Weissing, 2001; Arnegaard & Kondrashov, 2004; Barluenga & Meyer, 2004; Kidd et al., 2006). The presence of marked intraspecific differences in relative claw size and shape between males and females is an indication that sexual selection might have contributed to platythelphusid speciation (Marijnissen, 2007). Furthermore, interspecific hybridization has been pointed out as a process that is likely to be important in adaptive radiations, because it has the potential to elevate rates of response to disruptive or divergent selection (Seehausen, 2004). There is increasing evidence that hybridization has affected speciation in African cichlid radiations (Salzburger et al., 2002; Smith et al., 2003; Schliewen & Klee, 2004; Schelly et al., 2006; Koblmüller et al., 2007). It is readily conceivable that Lake Tanganyika endemic crabs are similarly prone to hybridization, and further investigation of the level of gene flow within the clade is a topic of importance.

It is becoming clear that the platythelphusid radiation exhibits many parallels with species flocks of cichlid fish in the African Great Lakes. Further work on these dynamic invertebrate residents of Lake Tanganyika's benthos promises critical tests of diversification hypotheses that are based so far largely on cichlid systems, and will provide more insight into the ecological functioning of this unusually diverse ancient lake system.

Acknowledgements This work would not have been possible without the invaluable help of George Kazumbe with the fieldwork in Lake Tanganyika. Martin Genner and Pete McIntyre provided comments on the stable isotope analyses; Christine Cocuyt, Hilde Eggermont, Jon Todd and Koen Martens helped with identification of algae and invertebrates for gut content analyses; Jan van Arkel (Institute for Biodiversity and Ecosystem Dynamics) helped with the figures; Keith Ursel, Sven Lange and Fred Schram provided support. We thank the Gombe Stream Research Centre for facilitating surveys along the coastline of Gombe National Park. We are grateful to the Tanzanian Commission for Science and Technology for permission to conduct research at Lake Tanganyika. The Tanzanian Fisheries Research Institute, Kigoma, D. Chitamwebwa and Mr. Muhosa, as well as the Nyanza Project (NSF Paleoclimates ATM 0223920) are thanked for logistical assistance. This research was supported by grants from the Netherlands Foundation for the Advancement of Tropical Research (WOTRO W84-489), the Schure-Beijerinck-Popping Fund of the Royal Netherlands Academy of Arts and Sciences (SBP/JK/2003-27, and SBP/JK/2002-25), Iso-analytical Limited, VSB Fund, Stichting Vrijvrouwe van Renswoude te 's Gravenhage, STUNT Fund, Genootschap ter Bevordering van de Natuur- Genees- en Heelkunde, Stichting AUV Fonds, and Stichting Fonds Dr. Christine Buisman. Finally, we would like to thank the organizers of the SIAL IV symposium in Berlin for the opportunity to present our results in a stimulating forum, with particular thanks to Frank Riedel, Annette Kossler and Thomas Wilke.

References

Albertson, R. C. & T. D. Kocher, 2006. Genetic and developmental basis of cichlid trophic diversity. Heredity 97: 211–221.

Albertson, R. C., J. A. Markert, P. D. Danley & T. D. Kocher, 1999. Phylogeny of a rapidly evolving clade: the cichlid fishes of Lake Malawi, East Africa. Proceedings of the National Academy of Sciences of the USA 96: 5107–5110.

Albrecht, C., S. Trajanovski, K. Kuhn, B. Streit & T. Wilke, 2006. Rapid evolution of an ancient lake species flock: freshwater limpets (Gastropoda: Ancylidae) in the Balkan Lake Ohrid. Organisms, Diversity & Evolution 6: 294–307.

Arnegaard, M. E. & A. S. Kondrashov, 2004. Sympatric speciation by sexual selection alone is unlikely. Evolution 58: 222–237.

Barluenga, M. & A. Meyer, 2004. The Midas cichlid species complex: incipient sympatric speciation of Nicaraguan cichlid fishes? Molecular Ecology 12: 2061–2076.

Bellwood, D. R., P. C. Wainwright, C. J. Fulton & A. S. Hoey, 2006. Functional versatility supports coral reef biodiversity. Proceedings of the Royal Society Biological, Series 273: 101–107.

Cane, M. A. & P. Molnar, 2001. Closing of the Indonesian seaway as a precursor to east African aridification around 3–4 million years ago. Nature 411: 157–162.

Capart, A., 1952. Exploration hydrobiologique du Lac Tanganyika (1946–1947), resultats scientifiques - Crustacés, Décapods, Brachyures. Institut Royal des Sciences Naturelles de Belgique 3: 41–67.

Cohen, A. S., M. R. Palacios-Fest, E. S. Msaky, S. A. Alin, B. McKee, C. M. O'Reilly, D. L. Dettman, H. Nkotagu & K. E. Lezzar, 2005. Paleolimnological investigations of anthropogenic environmental change in Lake Tanganyika: IX. Summary of paleorecords of environmental change and catchment deforestation at Lake Tanganyika and impacts on the Lake Tanganyika ecosystem. Journal of Paleolimnology 43: 125–145.

Coulter, G. W., 1991. Lake Tanganyika and Its Life. Oxford University Press, London, UK.

Cumberlidge, N., 1999. The Freshwater Crabs of West Africa, family Potamonautidae. Faune et Flore Tropicales No. 35. IRD, Paris.

Cumberlidge, N., R. von Sternberg, I. R. Bills & H. Martin, 1999. A revision of the genus *Platythelphusa* A. Milne-Edwards, 1887 from Lake Tanganyika, East Africa (Decapoda: Potamoidea: Platythelphusidae). Journal of Natural History 33: 1487–1512.

Cunnington, W. A., 1899. On a new brachyurous crustacean from Lake Tanganyika. Proceedings of the Royal Society of London 1899: 697–704.

Cunnington, W. A., 1907. Zoological results of the Third Tanganyika Expedition, conducted by Dr. W. A. Cunnington, 1904–1905. Report on the Brachyurous Crustacea. Proceedings of the Royal Society of London 2: 258–276.

Darwall, W., K. Smith, T. Lowe & J.-C. Vié, 2005. The status and distribution of freshwater biodiversity in Eastern Africa. Occasional Paper of the IUCN Species Survival Commission 31. IUCN, Gland, Switzerland: 1–36.

Day, J. J. & M. Wilkinson, 2006. On the origin of the *Synodontis* catfish species flock from Lake Tanganyika. Biology Letters 2: 548–552.

Duftner, N., S. Koblmüller & C. Sturmbauer, 2005. Evolutionary relationships of the Limnochromini, a tribe of benthic deepwater cichlid fish endemic to Lake Tanganyika, East Africa. Journal of Molecular Evolution 60: 277–289.

Evert, M. J., 1970. Le Lac Tanganyika et sa faune. Mémoire de licence. Université de Louvain, Belgium.

Frigge, M., D. C. Hoaglin & B. Iglewicz, 1989. Some implementations of boxplot. American Statistical 43: 50–54.

Fryer, G., 2006. Evolution in ancient lakes: radiation of Tanganyikan atyid prawns and speciation of pelagic cichlid fishes in Lake Malawi. Hydrobiologia 568: 131–142.

Genner, M. J., G. F. Turner, S. Barker & S. J. Hawkins, 1999. Niche segregation among Lake Malawi cichlid fishes? Evidence from stable isotope signatures. Ecology Letters 2: 185–190.

Grant, S. A., D. E. Suckling, H. K. Smith, L. Torvell, T. D. A. Forbes & J. Hodgson, 1985. Comparative studies of diet selection by sheep and cattle: the Hill grasslands. Journal of Ecology 73: 987–1004.

Hill, B. J., 1976. Natural food, foregut clearance rate and activity of the crab *Scylla serrata*. Marine Biology 24: 109–116.

Hori, M., 1983. Feeding ecology of thirteen species of *Lamprologus* (Teleostei: Cichlidae) coexisting at a rocky-shore

of Lake Tanganyika. Physiology and Ecology Japan 20: 129–149.
Hori, M., M. M. Gashagaza, M. Nshombo & H. Kawanabe, 1993. Littoral fish communities in Lake Tanganyika: irreplaceable diversity supported by intricate interactions among species. Conservation Biology 7: 657–666.
Hudon, C. & G. Lamarche, 1989. Niche segregation between American lobster (*Homarus americanus*) and rock crab (*Cancer irroratus*). Marine Ecology Progress Series 52: 155–168.
Jackson, D. A., K. M. Somers & H. H. Harvey, 1989. Similarity coefficients: measures of co-occurrence and association or simply measures of occurrence? The American Naturalist 133: 436–453.
Kidd, M. R., C. E. Kidd & T. D. Kocher, 2006. Axes of differentiation in the bower-building cichlids of Lake Malawi. Molecular Ecology 15: 459–478.
Koblmüller, S., N. Duftner, C. Katongo, H. Phiri & C. Sturmbauer, 2005. Ancient divergence in bathypelagic deepwater cichlids: mitochondrial phylogeny of the tribe Bathybatini. Journal of Molecular Evolution 60: 297–314.
Koblmüller, S., N. Duftner, K. M. Sefc, M. Aibara, M. Stipacek, M. Balnc, B. Egger & C. Sturmbauer, 2007. Reticulate phylogeny of gastropod-shell-breeding cichlids from Lake Tanganyika - the result of repeated introgressive hybridization. BMC Evolutionary Biology 7: 7.
Koblmüller, S., W. Salzburger & C. Sturmbauer, 2004. Evolutionary relationships in the sand dwelling cichlid lineage of Lake Tanganyika suggest multiple colonization of rocky habitats and convergent origin of biparental mouthbrooding. Journal of Molecular Evolution 58: 79–96.
Koblmüller, S., C. Sturmbauer, E. Verheyen, A. Meyer & W. Salzburger, 2006. Mitochondrial phylogeny and phylogeography of East African squeaker catfishes (Siluriformes: Synodontis). BMC Evolutionary Biology 6: 49.
Lehman, S. M., 2000. Primate community structure in Guyana: a biogeographic analysis. International Journal of Primatology 21: 333–351.
Lewis, C., 2002. Elucidating the interplay between tectonic and climatic controls on modern depositional processes in the Luiche delta. Nyanza Project Reports 2002 University of Arizona, Tuscon: 1–7.
Liem, K. F., 1980. Adaptive significance of intra- and interspecific differences in the feeidng repertoires of cichlid fishes. American Zoologist 20: 295–314.
Liem, K. F. & J. W. M. Osse, 1975. Biological versatility, evolution and food resource exploitation in African cichlid fishes. American Zoologist 15: 427–454.
Litvak, M. K. & R. I. C. Hansell, 1990. A community perspective on the multidimensional niche. Journal of Animal Ecology 59: 931–940.
Marijnissen, S. A. E., 2007. Lake Tanganyika Crabs: Evolution, Ecology, and Implications for Conservation. PhD thesis, University of Amsterdam, the Netherlands.
Marijnissen, S. A. E., E. Michel, S. R. Daniels, D. Erpenbeck, S. B. J. Menken & F. R. Schram, 2006. Molecular evidence for recent divergence of Lake Tanganyika endemic crabs (Decapoda: Platythelphusidae). Molecular Phylogenetics and Evolution 40: 628–634.
Marijnissen, S. A. E., F. R. Schram, N. Cumberlidge & E. Michel, 2004. Two new species of *Platythelphusa* A. Milne-Edwards, 1887 (Decapoda, Potamoidea, Platythelphusidae), comments on the taxonomic position of *P. denticulata* Capart, 1952 from Lake Tanganyika, East Africa. Crustaceana 77: 513–532.
McIntyre, P. B., E. Michel, K. France, A. Rivers, P. Hakizimana & A. S. Cohen, 2005. Individual- and assemblage-level effects of anthropogenic sedimentation on snails in Lake Tanganyika. Conservation Biology 19: 171–181.
Michel, E., 2000. Phylogeny of a gastropod species flock: exploring speciation in Lake Tanganyika in a molecular framework. In Rossiter, A. & H. Kawanabe (eds), Advances in Ecological Research. Ancient Lakes: Biodiversity, Ecology and Evolution. Academic Press, San Diego: 275–302.
Michel, E., J. A. Todd, D. F. R. Cleary, I. Kingma, A. S. Cohen & M. J. Genner, 2004. Scales of endemism: challenges for conservation and incentives for evolutionary studies in a gastropod species flock from Lake Tanganyika. Journal of Conchology 3: 155–172.
Milne-Edwards, A., 1887. Observations sur les crabes des eaux douces de l'Afrique. Annales des Sciences Naturelles, Paris 7: 161–191.
Moed, J. R. & G. M. Hallegraeff, 1978. Some problems in the estimation of chlorophyll *a* and phaeopigments from pre- and post-acidification spectrophotometric measurements. Internationale Revue der Gesammten Hydrobiologie 63: 787–800.
Moran, P. & I. Kornfield, 1993. Retention of ancestral polymorphism in the mbuna species flock (Pisces: Cichlidae) of Lake Malawi. Molecular Biology and Evolution 10: 1015–1029.
Navarette, S. A. & J. C. Castilla, 1990. Resource partitioning between intertidal predatory crabs: interference and refuge utilization. Journal of Experimental Marine Biology and Ecology 143: 101–129.
Newsome, S. D., C. Martinez del Rio, S. Bearhop & D. L. Phillips, 2007. A niche for isotopic ecology. Frontiers in Ecology and the Environment 5: 429–436.
Ochiai, A., 1957. Zoogeographic studies on the soleoid fishes found in Japan and its neighbouring regions. Bulletin of the Japanese Society of Scientific Fisheries 22: 526–530.
Platvoet, D., J. T. A. Dick, C. Macneil & G. van der Velde, 2008. Invader-invader interactions in relation to environmental heterogeneity leads to zonation of two invasive amphipods, *Dikerogammarus villosus* (Sowinsky) and *Gammarus tigrinus* Sexton. Biological Invasions (in press).
Reed, S. R. & N. Cumberlidge, 2006. Taxonomy and biogeography of the freshwater crabs of Tanzania, East Africa (Brachyura: Potamoidea: Potamonautidae, Platythelphusidae, Deckeniidae). Zootaxa 1262: 1–139.
Richards, R. A., 1992. Habitat selection and predator avoidance: ontogenetic shifts in habitat use by the Jonah crab *Cancer borealis* (Stimpson). Journal of Experimental Marine Biology and Ecology 156: 187–197.
Robinson, B. W. & D. S. Wilson, 1998. Optimal foraging, specialization, and a solution to Liem's paradox. American Naturalist 151: 223–235.
Rosales, A. B., S. A. E. Marijnissen, E. Michel & P. B. McIntyre, 2002. Snail susceptibility to crab predation: a case study of co-evolution from Lake Tanganyika, East Africa. Integrative Comparative Biology 42: 1303.

Safran, P. & M. Omori, 1990. Some ecological observations on fishes associated with drifting seaweed off Tohoku coast, Japan. Marine Biology 105: 395–402.

Salzburger, W., S. Baric & C. Sturmbauer, 2002. Speciation via introgressiove hybridization in East African cichlids? Molecular Ecology 11: 619–625.

Schelly, R., W. Saltzburger, S. Koblmüller, N. Duftner & S. Sturmbauer, 2006. Phylogenetic relationships of the lamprologine cichlid genus Lepidiolamprologus (Teleastei: Perciformes) based on mitochondrial and nuclear sequences, suggesting introgressive hybridization. Molecular Phylogenetics and Evolution 38: 426–438.

Schliewen, U. K. & B. Klee, 2004. Reticulate sympatric speciation in Cameroonian crater lake cichlids. Frontiers in Zoology 1: 5.

Schluter, D., 2000. The Ecology of Adaptive Radiation. Oxford University Press, Oxford.

Seehausen, O., 2004. Hybridization and adaptive radiation. Trends in Ecology and Evolution 19: 198–207.

Smith, E. A., 1880. On the shells of Lake Tanganyika and of the neighbourhood of Ujiji, Central Africa. Proceedings of the Zoological Society of London 1880: 344–352.

Smith, P. F., A. Konings & I. Kornfield, 2003. Hybrid origin of a cichlid population in Lake Malawi: implications for genetic variation and species diversity. Molecular Ecology 12: 2497–2504.

Sutherland, R. A., 1998. Loss on ignition estimate of organic matter and relationships to organic carbon in fluvial bed sediments. Hydrobiologia 389: 153–167.

Takhteev, V. V., 2000. Trends in the evolution of Baikal amphipods and evolutionary parallels with some marine malacostracan faunas. In Rossiter, A. & H. Kawanabe (eds), Advances in Ecological Research. Ancient Lakes: Biodiversity, Ecology and Evolution. Academic Press, San Diego: 197–220.

Van Doorn, G. S. & F. J. Weissing, 2001. Ecological versus sexual selection models of sympatric speciation: a synthesis. Selection 2: 17–40.

von Rintelen, T., A. B. Wilson, M. Meyer & M. Glaubrecht, 2004. Escalation and trophic specialization drive adaptive radiation of freshwater gastropods in ancient lakes on Sulawesi, Indonesia. Proceedings of the Royal Society London, Series B 271: 2842–2850.

Wentworth, C. K., 1922. A scale of grade and class terms for clastic sediments. Journal of Geology 30: 377–392.

West, K. & A. S. Cohen, 1994. Predator–prey coevolution as a model for the unusual morphologies of the crabs and gastropods of Lake Tanganyika. Archiv für Hydrobiologie/Ergebniss der Limnologie 4: 267–283.

West, K., A. S. Cohen & M. Baron, 1991. Morphology and behaviour of crabs and gastropods from Lake Tanganyika, Africa: implications for lacustrine predator–prey coevolution. Evolution 45: 589–607.

West, K., E. Michel, J. Todd, D. Brown & J. Clabaugh, 2003. The gastropods of Lake Tanganyika: diagnostic key, classification and notes on the fauna. International Association of Theoretical Applied Limnology SIL. Occasional Publication 2: 1–123.

William, E., 2002. Spatial relationship of grain size and coarse sediment mineralogy on the shallow Luiche delta platform and its river streams. Nyanza Project Reports 2002. University of Arizona, Tuscon: 1–7.

Williams, M. J., 1981. Methods for analysis of natural diet in portunid crabs (Crustacea: Decapoda: Portunidae). Journal of Experimental Marine Biology and Ecology 52: 103–113.

Woodward, S. P., 1859. On some new freshwater shells from Central Africa. Proceedings of the Zoological Society of London 1895: 348–351.

Hydrobiologia (2008) 615:95–101
DOI 10.1007/s10750-008-9554-2

SPECIATION IN ANCIENT LAKES

The Great Lakes in East Africa: biological conservation considerations for species flocks

Christian Sturmbauer

Published online: 11 September 2008

Abstract The three largest water bodies of East Africa, Lake Victoria, Tanganyika, and Malawi contain an estimated number of 2,000 endemic cichlid fish species, in addition, to a mostly uncounted wealth of invertebrates. While the terrestrial diversity is reasonably well protected, as economic and touristic interests coincide with biological conservation strategies, this is not the case for most African lakes and rivers. Nonetheless, it must be promoted that these aquatic ecosystems also deserve protection. Conservation strategies for aquatic biota have so far been the same as for terrestrial environments, i.e., by declaring biodiversity hotspots national parks. Such parks also contain rivers and lake shores. Here, I argue that it seems questionable that this strategy will work, given strong micro-geographic structure of the species flocks and the great degree of local endemism. I suggest a novel strategy for protecting African Lake communities that accounts for local endemism, derived from recent molecular phylogenetic and phylogeographic studies on East African cichlid fishes. While connectivity is the major problem for terrestrial and marine national parks, to ensure a large enough effective population size of the protected animals, this is not the case in most taxa of African rivers and lakes, where local endemism prevails. For example, most littoral cichlid species are subdivided into numerous distinct "color morphs" with restricted distribution, and unlike marine fishes with planktonic larvae display brood care with small offspring numbers. It is argued that the establishment of "micro-scale protected areas," a large number of small stretches of strictly protected coast line, each only some hundreds of meters long, is likely to work best to preserve the littoral communities in African lakes. Such protected zones can sustain a reasonably effective population size of littoral species, serve as protected spawning ground or nursery area for pelagic species, and at the same time re-seed neighboring populations that are exploited continuously. As long term stability of littoral fishing grounds is in the immediate interest of village communities, such small protected areas should be managed and controlled by the local communities themselves, and supervised by governmental institutions.

Keywords Cichlid fishes · Adaptive radiation · Explosive speciation · Protection · National parks

Guest editors: T. Wilke, R. Väinölä & F. Riedel
Patterns and Processes of Speciation in Ancient Lakes: Proceedings of the Fourth Symposium on Speciation in Ancient Lakes, Berlin, Germany, September 4–8, 2006

C. Sturmbauer (✉)
Department of Zoology, University of Graz, Universitätsplatz 2, 8010 Graz, Austria
e-mail: Christian.sturmbauer@uni-graz.at

Introduction

East Africa is famous for its outstanding cichlid fauna that—to a great extent—owes its diversity to the formation of the African Rift Valley (Fryer & Iles, 1972). Geological activity shaped the landscape by creating mountain chains with high volcanoes and deep valleys cutting through eastern Africa. The mountains altered the climate by capturing clouds on their way to the east, formed huge water bodies and indirectly shaped the Savannah habitats. These environmental changes induced the evolution of novel species, which adapted to the changing landscapes. Many scientists are convinced that human evolution is also tied to these environmental changes (Bobe & Behrensmeyer, 2004). While the terrestrial diversity is reasonably well protected, as economic interests coincide with conservation biological strategies, this is not the case for the aquatic biodiversity of African rivers and lakes. This has long been seen as an inexhaustible resource of protein, and after the first severe decreases of catch conservation strategies have focused on commercially exploited species only (Nelson & Soul, 1987; Mölsä et al., 2002). Tourist interest was low as was tourism-induced conservation pressure. Considering that the three largest water bodies alone, Lake Victoria, Tanganyika, and Malawi contain an estimated number of 2,000 endemic cichlid fish species (Turner et al., 2001), in addition, to a mostly uncounted wealth of invertebrates, it must become clear that these aquatic ecosystems also deserve protection. Conservation strategies for aquatic biota have so far been the same as for terrestrial environments, i.e., by declaring biodiversity hotspots national parks. Such parks also contain rivers and lake shores. For Lake Tanganyika, these are, for example, Gombe National Park, Mahale National Park (both in Tanzania) and Sumbu National Park in Zambia, Rubondo National Park in the Tanzanian section of Lake Victoria, Ndere Island National Park in Kenya, and Lake Malawi National Park in the south of Lake Malawi. Here, I argue that it seems questionable that this strategy will work, given strong micro-geographic structure of the species flocks and the great degree of local endemism.

Evolutionary biologists discovered those lakes as model systems for the study of explosive speciation, and their results not only foster insights into evolutionary patterns and processes, but have major implications on conservation strategies to preserve these aquatic ecosystems and their diversity (Fryer & Iles, 1972; Coulter, 1991; Barlow, 2000). In contrast to most terrestrial ecosystems, species communities are highly structured on a micro-geographical scale, in that each stretch of coast line comprises its own endemic community that evolved locally. Thus, the traditional management strategy of erecting few, but relatively large national parks cannot effectively preserve the observed diversity.

The goal of this article is to spell out biological conservation implications arising from recent molecular phylogenetic and population genetic studies on the cichlid species flocks of the East African lakes and to suggest strategies to preserve a reasonable fraction of the observed endemic diversity without negating human interests on the exploitation of these rich ecosystems. The insights gained for cichlid fishes about diversity, community structure, and ecological relationships may as well serve as model for other less surveyed groups living in those lakes.

Biological conservation implications for species flocks

What is a species flock and can it be the target of conservation?

Endemism, in situ origin, some taxonomic cohesion and rapid evolution via adaptive radiation are undisputed characteristics of a species flock (Greenwood, 1984). However, there has been considerable discussion about the term, as to whether it implies monophyly of the members and at which level (Mayr, 1942, 1984; Kosswig, 1963; Greenwood, 1984; Sturmbauer, 1998). Those who imply a strict monophyletic origin for the term species flock focus on the notion that one particular group of organisms outruns other seeding lineages in their success to rapidly occupy novel niches (Greenwood, 1984). All members of the species flock descend from a single ancestor that possessed one particular "key-innovation"—a specific set of biological traits—giving it the potential to speciate more rapidly than other species and to colonize the newly available (or species-depleted) habitat. The underlying process was originally described as "schizotypische Artsplitterung," later translated to explosive speciation (Woltereck, 1931). Other evolutionary biologists place stronger weight on the dynamics of the process of explosive speciation or adaptive radiation, in which any successful seeding species will rapidly occupy particular niches by competitive interactions and proliferate. While the outcome is an ecologically coherent, but not necessarily phylogenetically coherent species community, their common ancestor may well be outside the system. However, the seeding species may still share the "key-innovation," which roots back to a common ancestor deeper in time. In this way a species flock may well be polyphyletic, despite the fact that its seeding species share a common ancestor further back in time. The advocates of strict monophyly would call such a polyphyletic assemblage a super-flock. Meanwhile, it turns out that hardly any species flock of cichlid fishes is monophyletic in the strict sense. Lake Victoria was suggested to be a super-flock, in which some descendents of a riverine monophyletic assemblage colonized the lake independently (Meyer, 1993; Verheyen et al., 2003; Salzburger et al., 2005). The same is the case for Lake Malawi, which was also seeded by more than one haplochromine ancestor plus one or more tilapiine species (Albertson et al., 1999; Allender et al., 2003; Genner et al., 2004). The Lake Tanganyika cichlid species flock can be traced back to at least nine cichlid lineages (Koblmüller et al., 2005, 2008). I suggest the less strict definition; monophyly at a deeper level causing a synapomorphic key-innovation, as all players of an evolving group interact with each other and thus co-modulate the self-catalytic process of adaptive radiation, regardless of their immediate evolutionary relatedness.

The next issue about a species flock is whether other more distantly related lineages, which also contributed species to an endemic lake community, should be considered members of a species flock as well. To some extent, they also interact with the other players of the species community and thus also modulate the evolutionary pathways of the flock. Since they have different key-innovations enabling them to speciate into different niches, they might as well be considered members of the species flock. While many evolutionary biologists accept polyphyly, with evolutionary coherence due to monophyly at a deeper phylogenetic level (usually within a family) as criterion for membership, they reject more distant relationships; the term species assemblage applies for the entire species community of a lake. It should be noted that all East African lakes contain smaller (sub-) radiations of other fish families, most of which are endemic as well. Since these lineages were also players in the system, they should be included into evolutionary considerations, even if they may not be seen as members of a species flock. To summarize, a species flock is defined by a reasonably high number of closely related and endemic species that evolved in situ and at a rapid pace to a complex community by occupying new ecological niches or subdividing existing niches. Strict monophyly is no longer an imperative.

Geographic structure

The state of knowledge about origin, phylogeny, and inter-relationships of the cichlid species flocks in East African Lakes is summarized elsewhere (Kornfield & Smith, 2000; Salzburger & Meyer, 2004; Kocher, 2004). For Lake Tanganyika the state of knowledge is summarized in this issue (Koblmüller et al., 2008). I will thus focus on conservation-relevant issues of these results. One of the most intriguing insights about the species flocks in African lakes is their great degree of geographic structure. In Lake Tanganyika, few species occur lake-wide and if so, they are most often subdivided into distinct geographical forms (Kohda et al., 1996). This is the case for Lake Victoria haplochromines (Seehausen, 2006) as well as for Lake Malawi Mbuna (Danley et al., 2000; Danley & Kocher, 2001; Genner et al., 2004). It has been repeatedly shown that several ecological specializations evolved more than once in each system, so that evolutionary parallelism is a widespread phenomenon (Kocher et al., 1993; Rüber et al., 1999; Allender et al., 2003; Sturmbauer et al., 2003; Genner et al., 2004; Sugawara et al., 2005). Moreover, littoral cichlid species often display a fine-scale phylogeographic structure in Lake Tanganyika (Sturmbauer & Meyer, 1992; Verheyen et al., 1996; Rüber et al., 2001; Taylor et al., 2001; Duftner et al., 2006, 2008; Koblmüller et al., 2007; Sefc et al., 2007), as is the case in Lake Malawi cichlids (Arnegard et al., 1999; Markert et al., 1999; Shaw et al., 2000; Rico & Turner, 2002; Pereyra et al., 2004). Perhaps the most impressive examples for locally restricted species are the more than 120 described "geographical races" and sister species of the genus *Tropheus* in Lake Tanganyika (Sturmbauer & Meyer, 1992; Schupke, 2003; Sturmbauer et al., 2005); as well as the numerous sister species of the Mbuna-genera *Metriaclima*, *Labidochromis*, and *Petrotilapia* along the shores of

Lake Malawi (Turner et al., 2001). It follows that each stretch of coast line harbors its own species community that is clearly distinct and unique. In biological conservation terms, fine-scale evolutionary diversity must translate into fine-scale geographic conservation units (Verheyen et al., 1996).

Conflict of interest

The Great East African Lakes are of tremendous significance for feeding the African people and for the economy of the countries. The introduction of the Nile perch is only one example, in which humans have interfered to the detriment of natural fish communities in Africa. Aquaculture activities indirectly caused or even propagated the spread and introduction of *Oreochromis niloticus* in many rivers and lakes, and local governments introduced endemic Tanganyika sardines, *Stolothrissa tanganicae* and *Limnothrissa miodon*, to several other water bodies such as Lake Kariba, Lake Mweru, Lake Kivu etc. While the effect of the introduction of the Nile perch to Lake Victoria is scientifically documented, those of other introductions are not studied to date. Without doubt, introductions have pro's and con's and local economic interests must be balanced with conservation strategies. To my knowledge, only Zambia plans efforts to replace *Oreochromis niloticus* by native *Oreochromis* species in aquaculture settings using cage culture in lakes, due to the enormous risk of escape (P. Ngalande, pers. communication).

Despite the strong interest of exploiting the large African water bodies by local and commercial fisheries, some of the countries have declared and enforced national parks to protect the fish communities of the three Great Lakes. So far, commercial fisheries focused on large or abundant noncichlid fishes such as the endemic Nile perches or sardines in Lake Tanganyika, on introduced Nile perch in Lake Victoria, and on endemic pelagic cichlids in Lake Malawi. Near-shore fish communities were only targeted by artisanal fishermen, which—due to their unprofessional equipment—could not harm these communities (Coulter, 1991). Commercial fishing activities face increasing problems due to overfishing, and governments reacted by enacting catch regulations that are difficult to enforce in large lakes across national borders. Luckily, stocks were somehow self-regulated, in that the high gasoline price made inefficient catches financially impossible, giving the stocks at least some time to recover. However, at least in Lake Tanganyika, the fishing strategy of the large commercial companies changed recently in Mpulungu, toward supporting a franchise system by distributing outboard engines to local fishermen, which are to be paid for in fish returns (H. Phiri, Department of Fisheries Chilanga, pers. communication). This boosted the activities of local fishermen with effects on the fish communities to be seen in the near future. The transport of frozen fish to distant markets was also de-centralized, in that now not only the large trucks of commercial companies deliver fish to larger cities, but also small cool-houses and trucks sustaining several one-man companies.

A second and perhaps more devastating change happened in the exploitation of littoral cichlid fishes by artisanal fishermen. Instead of using gillnets alone and leaving them in the water for several hours, a new strategy can be observed in Lake the southern sections of Tanganyika: snorkeling masks and gill nets are increasingly used by local fishermen, as seen from commercial ornamental fish catchers (author's pers. observation). Snorkeling fishermen now specifically target large littoral fish species on rocky and muddy substrates. Instead of placing the net in the water for long time to catch those fish fitting the net size, divers strategically place the gill nets to actively chase in large individuals that are rapidly depleted in this way. One should also say that some ornamental fishing companies continuously circumvent national borders and must also be limited by conservation politics (T. Veall, Rift Valley Tropicals Ltd., Lusaka, pers. comm.). Given these new developments, action is urgently needed to protect the endemic fish communities.

Strategies to protect local diversity

While it is highly positive that some countries have already enforced larger national parks in and around the lakes, more action is needed (Coulter et al., 2006). In consequence to the finding of the great degree of local endemism, I suggest a novel, socially more acceptable and perhaps easier to enforce conservation strategy: the establishment of several "micro-scale protected areas" instead of few large national parks, in the form of short stretches of shoreline that are fully protected from fishing activities. This concept is analogous to the source and sink metapopulation concept (Hanski & Simberloff, 1997) whereby some subpopulations are being steadily harvested (sinks) and others serving as stabilizing sources (Dias, 1996; Weiss, 2005). While connectivity among geographically separated protected areas is the issue for many other species (Soule, 1980; Franklin, 1980; Ryder, 1986; Moritz, 1994), this is not the case for species with expressed micro-geographic endemism. Unlike many other organisms, cichlid fishes are poor dispersers and display complex brood care in the form of substrate-breeding and guarding the offspring or even female mouthbrooding with small offspring number. Recent genetic data based on nine microsatellite loci for four populations of *Tropheus moorii* resulted in population sizes of up to 10,000 individuals (based on estimated unsampled female population sizes of up to 5,000 individuals under the assumption of equal sex ratios; M. Koch et al., unpublished data). Such small effective population sizes are in line with density estimates of between 6 and 27 individuals per 100 m^2 for larger sized littoral cichlid fish species such as *Petrochromis*, *Lobochilotes*, *Limnotilapia*, *Ophthalmotilapia*, and *Tropheus* (Sturmbauer et al., this volume). Thus, stretches of about 500 m of protected shore would be enough to effectively protect the local fish communities. Such protected zones could sustain a reasonable effective population size of littoral species to serve as reservoirs to re-seed neighboring populations that are exploited continuously (Dias, 1996). A similar conservation strategy was recently suggested for Indo-Pacific grouper species, and the establishment of small no-take-areas increased the population of most grouper species by 30% over 5 years of protected status (Unsworth et al., 2007). The study also showed that a stretch of 500 m of protected area was enough to increase the population of top predatory fish. At the same time micro-scale protected areas would represent undisturbed spawning grounds or nursery areas for pelagic species. It should also be mentioned that such protected zones might be a way to preserve the wealth of unknown invertebrate species, as well as many larger vertebrates such as the water cobra *Boulengerina annulata stormsi* or otters, which are also decreasing rapidly in numbers in Lake Tanganyika (T. Veall, pers. communication). Given the poor means of transport of local fishermen, micro-scale protected areas would not enforce problems to local communities, so that social acceptance may be higher than for large national parks. Concerning management and protection enforcement, small protected zones may be more effectively managed with low human and financial resources. As sustainable management of food resources is in the immediate interest of village communities, such small protected areas should be established, managed, and controlled by local communities. My hope is that each village community will proudly declare its own "micro-scale protected area" to sustain the immediate fish resources and contribute its share to the conservation plan of the entire lake ecosystem.

Resume

The observation of extreme degrees of local endemism in the cichlid fish communities of the three Great East African Lakes renders traditional conservation strategies ineffective. The establishment of micro-scale national parks proposed here would be adequate means to ensure long-term protection of local species communities, both from conservation biological and sustainable exploitation perspectives. Given the relatively small effective population sizes of littoral fish populations, remarkably small stretches of shore line—in the range of several hundreds of meters—would be enough to protect and sustain local faunas. This is not to say that large national parks are not valuable measures of protection, but to stress that they will not be enough to preserve the enormous evolutionary diversity in the East African Lakes.

Acknowledgments This opinion article would not have been possible without the long-term support of and the good relationships to the Department of Fisheries, Ministry of Agriculture and Cooperatives, Republic of Zambia. I would like to particularly thank C. Kapasa, P. Ngalande, H. Phiri, R. Sinyinza, D. Sinyinza, and L. Makasa. Likewise, I wish to thank Prof. L. Mumba, H. G. Mudenda, and C. Katongo from the Department of Biological Sciences at the University of Zambia at Lusaka, and T. Veall from Rift Valley Tropicals Ltd. for his discussion input on conservation issues. I am also obliged to B. Egger, M. Koch, S. Koblmüller, K. Sefc, S. Weiss, and two anonymous reviewers for their insightful comments on the manuscript. Financial support was provided by the Austrian Science Foundation (Grant 17680) and the European Science Foundation (Grant I48).

References

Albertson, R. C., J. A. Markert, P. D. Danley & T. D. Kocher, 1999. Phylogeny of a rapidly evolving clade: the cichlid fishes of Lake Malawi, East Africa. Proceedings of the National Academy of Sciences USA 96: 5107–5110.

Allender, C. J., O. Seehausen, M. E. Knight, G. F. Turner & N. Maclean, 2003. Divergent selection during speciation of Lake Malawi cichlid fishes inferred from parallel radiations in nuptial coloration. Proceedings of the National Academy of Sciences USA 100: 14074–14079.

Arnegard, M. E., J. A. Markert, P. D. Danley, J. R. Stauffer Jr., A. J. Ambali & T. D. Kocher, 1999. Population structure and colour variation of the cichlid fish *Labeotropheus fuelleborni* Ahl along a recently formed archipelago of rocky habitat patches in southern Lake Malawi. Proceedings of the Royal Society, Series B 266: 119–130.

Barlow, G. W., 2000. The Cichlid Fishes: Nature's Grand Experiment in Evolution. Perseus, Cambridge, MA.

Bobe, R. & A. K. Behrensmeyer, 2004. The expansion of grassland ecosystems in Africa in relation to mammalian evolution and the origin of the genus *Homo*. Palaeogeography, Palaeoclimatology, Palaeoecology 207: 399–420.

Coulter, G. W., 1991. Lake Tanganyika and Its Life. Oxford University Press, Oxford.

Coulter, G., V. Langenberg, R. Loewe-McConnell, F. Riedel, F. Roest, J. Sarvala & O. Timoshkin, 2006. The problems confronting survival of biodiversity in ancient lakes. Verhandlungen des Internationalen Vereins für Limnologie 29: 1178–1181.

Danley, P. D. & T. D. Kocher, 2001. Speciation in rapidly diverging systems: lessons from Lake Malawi. Molecular Ecology 10: 1075–1086.

Danley, P. D., J. A. Markert, M. E. Arnegard & T. D. Kocher, 2000. Divergence with gene flow in the rock-dwelling cichlids of Lake Malawi. Evolution 54: 1725–1737.

Dias, P. C., 1996. Sources and sinks in population biology. Trends in Ecology and Evolution 11: 326–330.

Duftner, N., K. M. Sefc, S. Koblmüller, B. Nevado, E. Verheyen, H. Phiri & C. Sturmbauer, 2006. Distinct population structure in a phenotypically homogeneous rock-dwelling cichlid fish from Lake Tanganyika. Molecular Ecology 15: 2381–2396.

Duftner, N., K. M. Sefc, S. Koblmüller, W. Salzburger, M. Taborsky & C. Sturmbauer, 2008. Parallel evolution of facial stripe patterns in the *Neolamprologus brichardi/pulcher* species complex endemic to Lake Tanganyika. Molecular Phylogenetics and Evolution 45: 326–338.

Franklin, I. R., 1980. Evolutionary change in small populations. In Soule, M. E. & B. A. Wilcox (eds), Conservation Biology: An Evolutionary-Ecological Perspective. Sinauer Associates, Sunderland, MA: 135–139.

Fryer, G. & T. D. Iles, 1972. The Cichlid Fishes of the Great Lakes of Africa. T.H.F., Neptune, NJ.

Genner, M. J., O. Seehausen, D. F. R. Cleary, M. E. Knight, E. Michel & G. F. Turner, 2004. How does the taxonomic status of allopatric populations influence species richness within African cichlid species assemblages? Journal of Biogeography 31: 93–102.

Greenwood, P. H., 1984. African cichlid fishes and evolutionary theories. In Echelle, A. A. & I. Kornfield (eds), Evolution of Fish Species Flocks. University of Maine at Orono Press, Orono: 141–154.

Hanski, I. & D. Simberloff, 1997. The metapopulation approach, its history, conceptual domain and application to conservation. In Hanski, I. & M. E. Gilpin (eds), Metapopulation Biology: Ecology, Genetics and Evolution. Academic Press, London: 5–26.

Koblmüller, S., N. Duftner, C. Katongo, H. Phiri & C. Sturmbauer, 2005. Ancient divergence in bathypelagic Lake Tanganyika deepwater cichlids: mitochondrial phylogeny of the tribe Bathybatini. Journal of Molecular Evolution 60: 297–314.

Koblmüller, S., K. M. Sefc, N. Duftner, M. Warum & C. Sturmbauer, 2007. Genetic population structure as indirect measure of dispersal ability in a Lake Tanganyika cichlid. Genetica 130: 121–131.

Koblmüller, S., K. M. Sefc & C. Sturmbauer, 2008. The Lake Tanganyika cichlid species assemblage: recent advances in molecular phylogenetics. Hydrobiologia (this volume). doi:10.1007/s10750-008-9552-4.

Kocher, T. D., 2004. Adaptive evolution and explosive speciation: the cichlid fish model. Nature Reviews Genetics 5: 288–298.

Kocher, T. D., J. A. Conroy, K. R. McKaye & J. R. Stauffer, 1993. Similar morphologies of cichlid fishes in Lakes Tanganyika and Malawi are due to convergence. Molecular Phylogenetics and Evolution 4: 420–432.

Kohda, M., Y. Yanagisawa, T. Sato, K. Nakaya, Y. Rimura, K. Matsumoto & H. Ochi, 1996. Geographical color variation in cichlid fishes at the southern end of Lake Tanganyika. Environmental Biology of Fishes 45: 237–248.

Kornfield, I. & P. F. Smith, 2000. African cichlid fishes: model systems for evolutionary biology. Annual Reviews in Ecology and Systematics 31: 163–196.

Kosswig, C., 1963. Ways of speciation in cichlids. Copeia 1963: 238–244.

Markert, J. A., M. E. Arnegard, P. D. Danley & T. D. Kocher, 1999. Biogeography and population genetics of the Lake Malawi cichlid *Melanochromis auratus*: habitat transience, philopatry and speciation. Molecular Ecology 8: 1013–1026.

Mayr, E., 1942. Systematics and the Origin of Species. Columbia University Press, New York.

Mayr, E., 1984. Evolution of fish species flocks: a commentary. In Echelle, A. A. & I. Kornfield (eds), Evolution of Fish Species Flocks. University of Maine at Orono Press, Orono: 3–11.

Meyer, A., 1993. Phylogenetic relationships and evolutionary processes in East African cichlids. Trends in Ecology and Evolution 8: 279–284.

Mölsä, H., J. Sarvala, S. Badende, D. Chitamwebwa, R. Kanyaru, N. Mulimbwa & L. Mwape, 2002. Ecosystem monitoring in the development of sustainable fisheries in Lake Tanganyika. Aquatic Ecosystem Health and Management 5: 267–281.

Moritz, C., 1994. Defining "evolutionary significant units" for conservation. Trends in Ecology and Evolution 9: 373–375.

Nelson, K. & M. E. Soul, 1987. Genetical conservation of exploited fishes. In Ryman, N. & F. Utter (eds), Population Genetics and Fisheries Management. Washington Sea Grant (Publisher), Seattle: 345–368.

Pereyra, R., M. I. Taylor, G. F. Turner & C. Rico, 2004. Variation in habitat preference and population structure among three species of the Lake Malawi cichlid genus *Protomelas*. Molecular Ecology 13: 2691–2697.

Rico, C. & G. F. Turner, 2002. Extreme microallopatric divergence in a cichlid species from Lake Malawi. Molecular Ecology 11: 1585–1590.

Rüber, L., A. Meyer, C. Sturmbauer & E. Verheyen, 2001. Population structure in two sympatric species of the Lake Tanganyika cichlid tribe Eretmodini: evidence for introgression. Molecular Ecology 10: 1207–1225.

Rüber, L., E. Verheyen & A. Meyer, 1999. Replicated evolution of trophic specializations in an endemic cichlid fish lineage from Lake Tanganyika. Proceedings of the National Academy of Sciences USA 96: 10230–10235.

Ryder, O. A., 1986. Species conservation and systematics: the dilemma of subspecies. Trends in Ecology and Evolution 1: 9–10.

Salzburger, W., T. Mack, E. Verheyen & A. Meyer, 2005. Out of Taganyika: genesis, explosive speciation, key-innovations and phylogeography of the haplochromine cichlid fishes. BMC Evolutionary Biology 5: 17.

Salzburger, W. & A. Meyer, 2004. The species flocks of East African cichlid fishes: recent advances in molecular phylogenetics and population genetics. Naturwissenschaften 91: 277–290.

Schupke, P., 2003. Cichlids of Lake Tanganyika. Part 1. The Species of the Genus *Tropheus*. A.C.S. GmbH Rodgau, Germany: 190 pp.

Seehausen, O., 2006. African cichlid fish: a model system in adaptive radiation research. Proceedings of the Royal Society, Series B 273: 1987–1998.

Sefc, K. M., S. Baric, W. Salzburger & C. Sturmbauer, 2007. Species-specific population structure in rock-specialized sympatric cichlid species in Lake Tanganyika, East Africa. Journal of Molecular Evolution 64: 33–49.

Shaw, P. W., G. F. Turner, M. R. Idid, R. L. Robinson & G. R. Carvalho, 2000. Genetic population structure indicates sympatric speciation of Lake Malawi pelagic cichlids. Proceedings of the Royal Society, Series B 267: 2273–2280.

Soule, M. E., 1980. Thresholds for survival: maintaining fitness and evolutionary potential. In Soule, M. E. & B. A. Wilcox (eds), Conservation Biology: An Evolutionary-Ecological Perspective. Sinauer Associates, Sunderland, MA: 151–169.

Sturmbauer, C., 1998. Explosive speciation in cichlid fishes of the African Great Lakes: a dynamic model of adaptive radiation. Journal of Fish Biology 53(Supplement A): 18–36.

Sturmbauer, C., U. Hainz, S. Baric, E. Verheyen & W. Salzburger, 2003. Evolution of the tribe Tropheini from Lake Tanganyika: synchronized explosive speciation producing multiple evolutionary parallelism. Hydrobiologia 500: 51–64.

Sturmbauer, C., S. Koblmüller, K. M. Sefc & N. Duftner, 2005. Phylogeographic history of the genus *Tropheus*, a lineage of rock-dwelling cichlid fishes endemic to Lake Tanganyika. Hydrobiologia 542: 335–366.

Sturmbauer, C. & A. Meyer, 1992. Genetic divergence, speciation and morphological stasis in a lineage of African cichlid fishes. Nature 358: 578–581.

Sugawara, T., Y. Terai, H. Imai, G. F. Turner, S. Koblmüller, C. Sturmbauer, Y. Shichida & N. Okada, 2005. Parallelism of amino acid changes at the RH1 affecting spectral sensitivity among deep-water cichlids from Lakes Tanganyika and Malawi. Proceedings of the National Academy of Sciences USA 102: 5448–5453.

Taylor, M. I., L. Rüber & E. Verheyen, 2001. Microsatellites reveal high levels of population substructuring in the species-poor eretmodine cichlid lineage from Lake Tanganyika. Proceedings of the Royal Society London, Series B 268: 803–808.

Turner, G. F., O. Seehausen, M. E. Knight, C. J. Allender & R. L. Robinson, 2001. How many species of cichlid fishes are there in African lakes? Molecular Ecology 10: 793–806.

Unsworth, R. K. F., A. Powell, F. Hukom & D. J. Smith, 2007. The ecology of Indo-Pacific grouper (Serranidae) species and the effects of a small-scale no take area on grouper assemblage, abundance and size frequency distribution. Marine Biology 152: 243–254.

Verheyen, E., L. Rüber, J. Snoeks & A. Meyer, 1996. Mitochondrial phylogeography of rock dwelling cichlid fishes reveals evolutionary influence of historic lake level fluctuations in Lake Tanganyika, Africa. Philosophical Transactions of the Royal Society London, Series B 351: 797–805.

Verheyen, E., W. Salzburger, J. Snoeks & A. Meyer, 2003. Origin of the superflock of cichlid fishes from Lake Victoria, East Africa. Science 300: 325–329

Weiss, S., 2005. Keynote address. conservation genetics of freshwater organisms. Bulletin Français de la Pêche et de la Pisciculture 376: 571–583.

Woltereck, R., 1931. Beobachtungen und Versuche zum Fragenkomplex der Artbildung I. Biologisches Centralblatt 51: 231–233.

Hydrobiologia (2008) 615:103–140
DOI 10.1007/s10750-008-9558-y

SPECIATION IN ANCIENT LAKES

Ancient Lake Ohrid: biodiversity and evolution

Christian Albrecht · Thomas Wilke

Published online: 22 September 2008

Abstract Worldwide ancient lakes have been a major focal point of geological, biological, and ecological research, and key concepts in, for example, evolutionary biology are partly based on ancient lake studies. Ancient lakes can be found on most continents and climate zones with most actual or putative ancient lakes in Europe being restricted to the Balkan Region. The arguably most outstanding of them is the oligotrophic and karstic Lake Ohrid, a steep-sided graben of rift formation origin situated in the central Balkans. Here, an attempt is made to summarize current knowledge of the geological, limnological, and faunal history of Lake Ohrid. Additionally, existing data on biodiversity and endemism in Lake Ohrid are updated and evaluated, and patterns and processes of speciation are reviewed in the context of the Ohrid watershed, including its sister lake, Lake Prespa. Whereas the geological history of the Ohrid Graben is relatively well studied, there is little knowledge about the limnological and biotic history of the actual lake (e.g., the age of the extant lake or from where the lake first received its water, along with its first biota). Most workers agree on a time frame of origin for Lake Ohrid of 2–5 million years ago (Mya). However, until now, the exact limnological origin and the origin of faunal or floral elements of Lake Ohrid remain uncertain. Two largely contrasting opinions either favour the theory of de novo formation of Lake Ohrid in a dry polje with a spring or river hydrography or a palaeogeographical connection of Lake Ohrid to brackish waters on the Balkan Peninsula. Whereas neither theory can be rejected at this point, the data summarized in the current review support the de novo hypothesis. An assessment of the fauna and flora of Lake Ohrid confirms that the lake harbours an incredible endemic biodiversity. Despite the fact that some biotic groups are poorly studied or not studied at all, approximately 1,200 native species are known from the lake, including 586 animals, and at least 212 species are endemic, including 182 animals. The adjusted rate of endemicity is estimated at 36% for all taxa and 34% for Animalia. In terms of endemic biodiversity, Lake Ohrid is with these 212 known endemic species and a surface area of 358 km^2 probably the most diverse lake in the world, taking surface area into account. Preliminary phylogeographical analyses of endemic Lake Ohrid taxa indicate that the vast majority of respective sister taxa occurs in the Balkans and that therefore the most recent common ancestors of Ohrid- and non-Ohrid species likely resided in the region when Lake Ohrid

Guest editors: T. Wilke, R. Väinölä & F. Riedel
Patterns and Processes of Speciation in Ancient Lakes: Proceedings of the Fourth Symposium on Speciation in Ancient Lakes, Berlin, Germany, September 4–8, 2006

C. Albrecht · T. Wilke (✉)
Department of Animal Ecology and Systematics, Justus Liebig University Giessen, Heinrich-Buff-Ring 26-32 (IFZ), 35392 Giessen, Germany
e-mail: tom.wilke@allzool.bio.uni-giessen.de

came into existence. These data also indicate that there is relatively little faunal exchange and overlap between Lake Ohrid and its sister lake, Lake Prespa, despite the fact that the latter lake is a major water supplier for Lake Ohrid. Studies on selected species flocks and scatters, mostly in molluscs, point towards the assumption that only few lineages originally colonized Lake Ohrid from the Balkans and that the majority of endemic species seen today probably started to evolve within the lake during the early Pleistocene. Within the Ohrid watershed, endemism occurs at different spatial and taxonomic scales, ranging from species endemic to certain parts of Lake Ohrid to species endemic to the whole watershed and from subspecies to genus level and possibly beyond. Modes of speciation in the Ohrid watershed are largely affected by its degree of isolation. Observational evidence points towards both allopatric (peripatric) and parapatric speciation. Though sympatric speciation within a habitat is conceivable, so far there are no known examples. Today, the lake suffers from increasing anthropogenic pressure and a "creeping biodiversity crisis". Some endemic species presumably have already gone extinct, and there are also indications of invasive species penetrating Lake Ohrid. The comparatively small size of Lake Ohrid and the extremely small range of many endemic species, together with increasing human pressure make its fauna particularly vulnerable. It is thus hoped that this review will encourage future research on the ecology and evolutionary biology of the lake's taxa, the knowledge of which would ultimately help protecting this unique European biodiversity hot spot.

Keywords Ancient lake · Lake Ohrid · Lake Prespa · Balkans · Sister lakes · Geology · Limnology · Biodiversity · Endemism · Speciation

Introduction

Evolutionary patterns in ancient lakes is a topic that has fascinated biologists for decades, and a number of landmark publications have resulted from studies of ancient lake taxa, significantly increasing our knowledge about evolutionary processes in these unique ecosystems and beyond. Basic concepts of sympatric speciation (e.g., Genner et al., 2007), sexual selection (e.g., Seehausen, 2006), adaptive radiation (e.g., von Rintelen et al., 2004; Herder et al., 2006), hybridization (e.g., Seehausen, 2004), and punctuated equilibrium (e.g., Williamson, 1981; Gould, 1992) are, at least partly, based on studies of ancient lake taxa.

Moreover, ancient lakes, i.e., extant lakes that have continuously existed for hundred thousand or even million years (see Glossary), have long been recognized as centres of biodiversity and endemism (Brooks, 1950; Martens et al., 1994; Martens, 1997; Rossiter & Kawanabe, 2000) and are a major focal point for systematic, ecological, and conservation research (Coulter et al., 2006).

A principal problem in ancient lake studies is that due to their prominence, some workers narrowly focus on these systems, whereas the closer or wider surroundings often receive less attention, potentially leading to an overestimation of degree of endemism in ancient lakes. Other problems include lack of knowledge and/or controversy about the limnological history of many ancient lakes and the unknown origin of respective faunal and floral elements.

Although ancient lakes undisputedly differ in their origins and characteristics, it is their diversity and endemism that typically distinguish them from short-lived, post-glacial lakes. In fact, these parameters often serve as proxies for the recognition of ancient lakes (Martens, 1997), and it has recently been demonstrated that species diversity in modern and fossil records of lake faunas correlates with lake longevity (Gierlowski-Kordesch & Park, 2004). Most workers, however, consider longevity to be the only objective criterion for ancient lakes as there is a considerable number of worldwide lakes that are presumably old but lack a high degree of diversity and endemism. Reasons for that are, for example, recent extinction events as in Lake Kivu (Degens et al., 1973; Martens, 1997) or harsh climate conditions as in Lake Tahoe (Gardner et al., 2000). On the other hand, there are lakes with incredible biotic diversity that are considered to be relatively young (e.g., Lake Victoria, Seehausen, 2006). An unclear limnological history and often poor faunal knowledge also partly account for the uncertain status of a number of putative ancient lakes in the world that are outshone by their more famous counterparts (Albrecht et al., 2006a; Wilke et al., 2007). Though studies on ancient lake biota are ongoing for a number of taxa and lakes, no unifying theory on patterns and

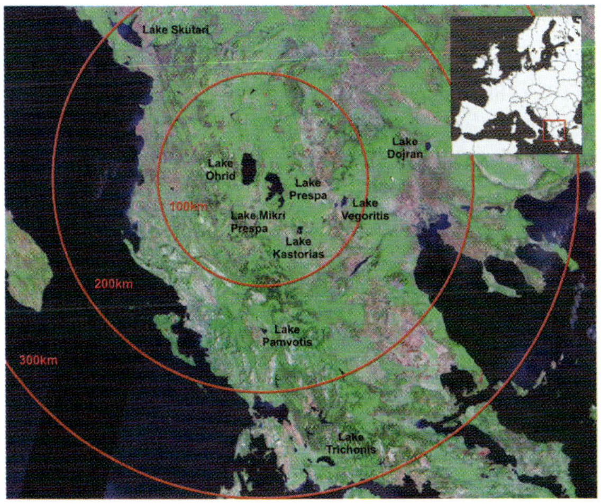

Fig. 1 Position of ancient Lake Ohrid and its sister lake, Lake Prespa, in the central Balkans relative to other putatively old lakes in Europe harbouring endemic species, which are all located within a radius of 300 km around the former lakes with the exception of the transcontinental Caspian Lake

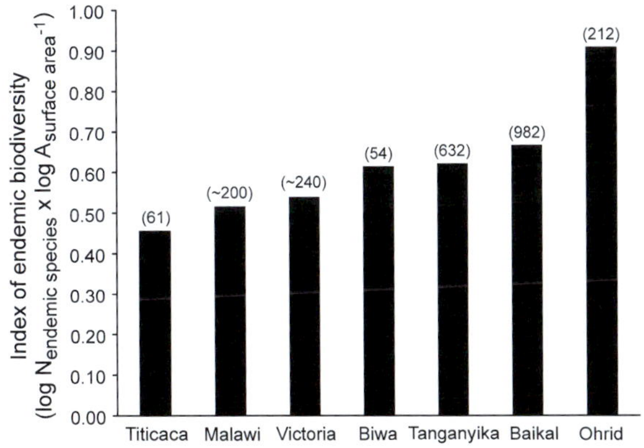

Fig. 2 Index of endemic biodiversity for selected worldwide ancient lakes and Lake Victoria taking surface area into account. Numbers in parentheses above the lakes indicate the total number of respective endemic species. The number for Lake Ohrid is a result of the present study. The numbers for the other lakes are taken from Matzinger et al. (2007)

processes of evolution in ancient lakes exists (Rossiter & Kawanabe, 2000).

Ancient lakes can be found on most continents and climate zones but appear to be more common in seismic active zones where tectonic activities can compensate for ongoing sedimentation (e.g., Hinderer & Einsele, 2001). In Europe, most actual or putative ancient lakes are restricted to the Balkan Region (see Fig. 1).

The arguably most outstanding of them is the oligotrophic and karstic Lake Ohrid, a steep-sided graben with rift formation origin situated in the South Adriatic–Ionian biogeographic region (Banarescu, 1991). With a limnological age of likely 2–5 million years, it is one of the oldest lake in Europe harbouring an outstanding degree of biodiversity in several groups of organisms (e.g., Stankovic, 1960). It also is home to a disproportional number of endemic species. Though the total number of endemic species is highest in Lake Baikal with 982 endemic species described, followed by Lake Tanganyika with 632 endemic species, and lakes Victoria and Malawi with approximately 200–240 endemic species each, Lake Ohrid is with 212 known endemic species probably by far the most diverse lake in the world taking surface area into account (Fig. 2).

In this review, we summarize current knowledge of the limnological and faunal history of Lake Ohrid, with notes on its physiography, hydrology, and geology. Additionally, we update and evaluate existing data on biodiversity and endemism and review patterns and processes of speciation in Lake Ohrid. By paying special attention to results that emerged during the past decade, including first genetic data from endemic taxa, we show that Lake Ohrid is a key study site for biodiversity and speciation research.

It is also hoped that the current review serves as a first basis for ultimately answering four vexing questions related to the evolution of Lake Ohrid:

(A) What is the actual degree of biodiversity and endemism in Lake Ohrid?
(B) What is the phylogeographical origin of biotas within the Ohrid watershed, and what are the biogeographical relationships of the lake?
(C) When and how often did outside species invade the basin and did they diversify within Lake Ohrid?
(D) What are the modes of speciation in Lake Ohrid and its watershed?

History of Lake Ohrid geological and biological studies

The recognition of Lake Ohrid as a peculiar lake with an outstanding biodiversity began in 1891, when two zoologists from Vienna, Franz Steindachner and Rudolf Sturany, visited the lake, later reporting their discoveries and describing new fish and mollusc species (e.g., Steindachner, 1892; Sturany, 1894). Brusina (1896) first drew attention to the biogeographical implications of these early findings. Major geological, geomorphological, and limnological explorations were carried out by the Serbian geographer Jovan Cvijic (1865–1927), and much of the current geological knowledge still rests on these early explorations.

The uniqueness of Lake Ohrid was only appreciated after World War I, and systematic research started to be devoted to many groups of organisms (e.g., Doflein, 1921; Augener, 1926; Polinski, 1929; Hrabe, 1931; Stankovic, 1932; Arndt, 1938; An der Lan, 1939; Schneider, 1943).

Basic faunistic and floristic inventories were mainly carried out prior to World War II. In 1932, the Serbian scientist Sinisa Stankovic (1892–1974) published a landmark article discussing the origin of the relic fauna of Lake Ohrid. In 1935, Stankovic founded the Hydrobiological Station Ohrid (HBI) with its own periodicals, thus triggering a wealth of organized scientific studies (also see the extensive bibliography on the lake compiled by Noveska et al., 1985).

After World War II, numerous faunistic contributions were made particularly to macrozoobenthic groups (e.g., Hadzisce, 1953, 1956; Sapkarev, 1966). All these studies are summarized in the outstanding monograph on Lake Ohrid by Stankovic (1960).

Starting with the early days of scientific exploration of Lake Ohrid, research has been characterized by international cooperation, first among individual workers and later between the Hydrobiological Institute in Ohrid and counterparts in other parts of the world. Unfortunately, the institute had to cope with increasing isolation in the 1980s and particularly 1990s, when only few international research programmes could be conducted at the lake due to the political instabilities on the Balkans in general and at the border zone of former Yugoslavia, Albania, and Greece in particular. Thus, the latest review of the characteristics of Lake Ohrid given by Salemaa (1994) mainly had to be based on older data.

A revival of scientific support and collaboration in the last decade has led to significant progress in many fields of Lake Ohrid research. Limnology and hydrology (e.g., Matzinger et al., 2006a, b, 2007), tectonics (Dumurdzanov et al., 2004, 2005), and particularly environmental issues (Watzin et al., 2002) have attracted much scientific and political attention. Preliminary drilling campaigns and seismic studies have highlighted the potential of Lake Ohrid as archive for palaeoenvironmental conditions (Wagner et al., 2008a, b). Last but not least, workers studying questions related to the evolution of the lake's fauna and flora have begun to use molecular techniques and to consider a wider geographic scope as well as a comparative context involving other lake systems in the Balkans and surrounding regions.

Physiography and hydrology

Physiography

Lake Ohrid (*Ohridsko Ezero* [Macedonian], *Liqeni i Ohrit* [Albanian]) is part of the European lake group called Dessaretes located on the Balkan Peninsula. The Dessaretes consist of Lake Ohrid (Macedonia, Albania), Lake Prespa (Macedonia, Albania, Greece), Lake Mikri Prespa (Greece, Albania), and Lake Maliq (Albania). The latter, however, was drained after World War II. Lake Ohrid is not only the largest and oldest lake of the Dessaretes but also of the whole Balkan Peninsula. In fact, many workers even consider Lake Ohrid as the only European ancient lake (e.g., Salemaa, 1994), when disregarding the transcontinental Caspian Lake. Within a radius of approximately 300 km around Lake Ohrid, there is, however, a series of relatively large, often relatively deep, originally mostly oligotrophic, and potentially old lakes harbouring endemic species (Fig. 1).

Lake Ohrid itself (Figs. 3, 5A) is located on the central Balkan with approximately two-third of its surface area belonging to the Former Yugoslav Republic of Macedonia (from hereon called Macedonia) and about one-third belonging to the Republic of Albania (from hereon called Albania). The maximum length and width of the lake are 30.3 and 15.6 km, respectively, with a surface area of 358 km^2 and a maximum depth of 288.7 m (Fig. 4, Table 1). The local climate is influenced by the proximity to the Adriatic Sea, by the surrounding mountains, and by the thermal capacity of Lake Ohrid itself (Watzin et al., 2002). Average monthly air temperatures range from 26°C during summer to −1°C during winter. Precipitation averages around 750 mm year^{-1} and is

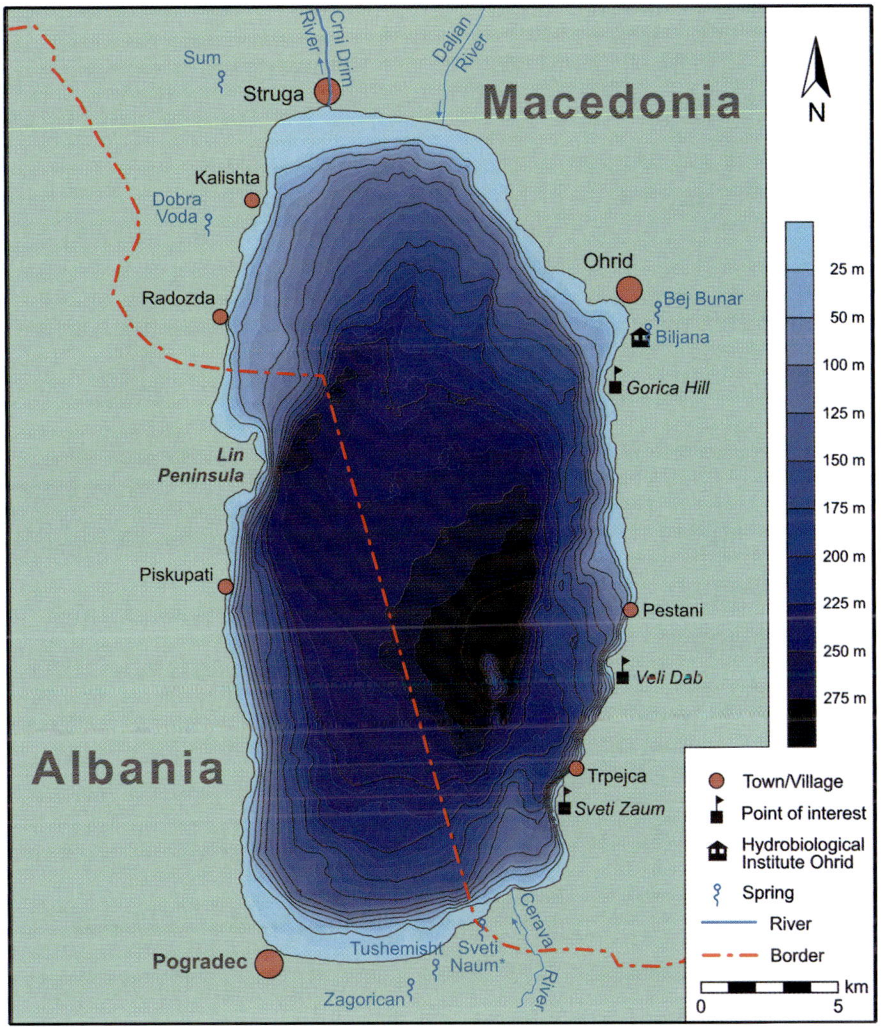

Fig. 3 Map of Lake Ohrid. Bathymetry with 25 m contour intervals modified from Denner (2006). *Note*: The Sveti Naum spring complex is also the source of the Crni Drim River, the northern effluent part of which constitutes the only surface outflow of the lake

at a minimum during summer (Watzin et al., 2002). Prevailing wind directions are northerly or southerly, being topographically affected by the shape of the Ohrid Valley and the surrounding mountains. The temperature of Lake Ohrid's pelagic water ranges from 6°C (below 150 m depth, year-round) to 24–27°C (at the surface during summer) (Naumoski et al., 2007; Matzinger et al., 2006b).

Lake Prespa (*Prespansko Ezero* [Macedonian], *Liqeni i Prespës* [Albanian], *Limni Megali Prespa* [Greek]) is a transboundary lake located 9.0 km (shortest strait line distance) east of Lake Ohrid, across the Galicica Mountains (Figs. 4, 5B). In contrast to the deep and relatively regular topography of Lake Ohrid, Lake Prespa is shallow, with few deeper holes. With 58 m, the deepest spot appears to be Kazan ("caldron") just off the coast between Stenje and Konjsko on the Macedonian side (Albrecht and Wilke, unpublished data).

With a maximum length and width of 27.3 and 17.0 km, respectively, Lake Prespa is smaller than Lake Ohrid and located at an altitude more than 150 m higher than the latter lake (Fig. 4, Table 1).

The smallest of the extant Dessaretes, Lake Mikri Prespa (*Limni Mikri Prespa* [Greek], *Liqeni i Prespës se vogel* [Albanian]), is close to Lake Prespa. These two lakes once formed a single water body before sedimentation caused an isthmus separating them (Papoutsi-Psychoudaki & Psychoudakis, 2000). Today, Lake Mikri Prespa drains into Lake Prespa via a channel (Fig. 5C). Since 1969, the channel is used to control the water level of Lake Mikri Prespa

Fig. 4 WNW–ESE transect through lakes Ohrid and Prespa showing the underground karstic connection between the lakes

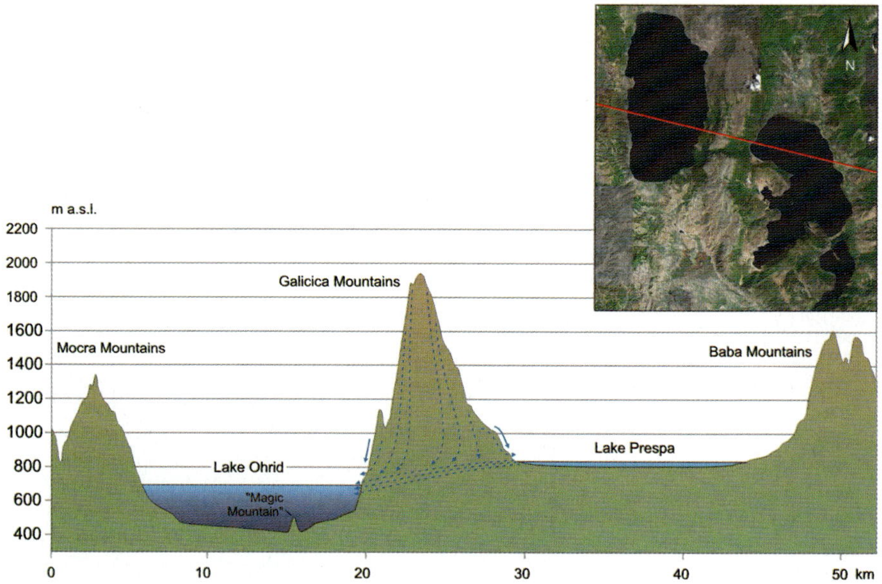

Table 1 Summarized geographical, physiographical, and hydrological characteristics of lakes Ohrid, Prespa, and Mikri Prespa compiled from LandSat 7 GIS data

	Lake Ohrid	Lake Prespa	Lake Mikri Prespa
Location	40.900°–41.174° N; 20.628°–20.810° E	40.763°–41.007° N; 20.898°–21.119° E	40.691°–40.809° N; 21.029°–21.125° E
Surface area, km^2	358[a]	254[a]	53[c]
Altitude, m a.s.l.	683[a]	849[a]	853[c]
Length (maximum), km	30.3	27.3	14.3
Width (maximum), km	15.6	17.0	6.5
Depth (mean), m	155[a]	14[a]	4.1[c]
Depth (maximum), m	288.7[a]	58	8.4[c]
Volume, km^3	55[a]	3.6[a]	0.22[c]
Watershed, km^2	1,002[b]	~1,000[a]	260[c]
Residence time, a (year)	70[a]	11[a]	3.4[c]
Surface water temperature (maximum), °C	27[a]	25	28[c]

[a] Matzinger et al. (2006b)

[b] Popovska & Bonacci (2007)

[c] Zacharias et al. (2002)

and to keep it relatively constant throughout the year. This distinguishes Lake Mikri Prespa from Lake Prespa, which not only shows lake-level fluctuations of up to several metres within a period of few years but also significant changes throughout the year (Hollis & Stevenson, 1997).

Whereas there is a direct connection between Lake Prespa and Lake Mikri Prespa (Fig. 5C), the former is connected to Lake Ohrid via underground karstic channels (Figs. 4, 5D). Lakes Ohrid and Prespa are bounded to the east and west by high mountain chains, reaching to about 1,500 m a.s.l. to the west of Lake Ohrid (Mokra Mountains), 2,250 m a.s.l. between the two lakes (Galicica Mountains and Mali i Thate Mountains), and 2,600 m a.s.l. to the east of Lake Prespa (Baba Mountains). Whereas the mountain ranges to the west of Lake Ohrid and to the east of Lake Prespa are mostly crystalline with limited water permeability, the bedrock separating the two lakes consists of karstified carbonates forming aquifers with high hydraulic conductivity (Cvijic, 1908; Eftimi et al., 2001; Wagner et al., 2008b; Fig. 4). Lake Ohrid and Lake Prespa are graben-type lakes, which possibly resulted from collapsed karstic fields, so called "poljes" (Stankovic, 1960; also see below).

Fig. 5 Images of the Dessarete Lakes and their hydrological connections. (**A**) View of Lake Ohrid from the Galicica Mountains to the northwest. (**B**) Greek part of Lake Prespa with the Albanian side in the background. (**C**) Channel connecting lakes Prespa and Mikri Prespa with the latter lake seen in the background. (**D**) Swallow holes (ponors) on the Albanian side of Lake Prespa near Zavir (where the water boils) during low water. These ponors are portals of the karstic underground connection between lakes Prespa and Ohrid. (**E**) Part of the Sveti Naum feeder spring complex south of Lake Ohrid. These feeder springs are effluent seepages of the karstic underground connection with Lake Prespa and a major water supplier for Lake Ohrid. (**F**) Short channel (affluent part of the Crni Drim River) connecting the feeder spring complex in Sveti Naum with Lake Ohrid

Preliminary seismic investigations of Lake Ohrid indicate a rather uniform and bioturbated sedimentation in the central part of the basin with sedimentation rates of approximately 0.5–1.0 mm year^{-1} (Wagner et al., 2008b). The succession of undisturbed sediments reaches a thickness of >700 m (Krastel, personal communication).

Both a presumed large original depth and the low sedimentation rate might be reasons why Lake Ohrid may have formed a deep, permanent water body for time spans beyond the lifetime of short-lived lakes. In addition, the area is still tectonically active and might thus compensate sedimentation by subduction (Stankovic, 1960). The tectonic activity of the region is underlined by frequent earthquakes, with the last prominent earthquake having taken place in February 1911 close to the Lake Ohrid basin ($M = 6.7$, corresponding to European Macroseismic Scale X; Wagner et al., 2008b).

Hydrology

The water column of Lake Ohrid above 150 m depth is thermally stratified from March to November and mixed in winter, thus following the typical seasonality of temperate lakes. Below 150 m depth, the water is stratified by salinity. This salinity gradient is only overcome by complete mixing roughly once per decade (Matzinger et al., 2006b). Transparency varies between 9 and 17 m (Naumoski et al., 2007). The water balance of the lake is characterized by average in- and output rates of approximately 37.9 $m^3 s^{-1}$ (Matzinger et al., 2006b). According to these workers (also see Watzin et al., 2002), about two-third of the output occurs via the River Crni Drim and one-third through evaporation. Today's water input of Lake Ohrid is characterized by inflows from karstic aquifers ($\sim$53%), direct precipitation on the lake surface ($\sim$23%), and river inflow ($\sim$23%). The river discharge was even lower before River Sateska was artificially diverted into the lake in 1962 (Matzinger et al., 2006a).

A characteristic of Lake Ohrid's water balance is the high inflow from karstic springs with sublacustrine ($\sim$49%) and surface springs ($\sim$51%) contributing almost equally to the inflow balance (Matzinger et al., 2006a). The most important surface spring complexes are the one near the monastery Sveti Naum at the southern edge of the lake, which forms a spring lake connecting to the actual lake via a short channel (affluent Crni Drim), and the sister spring complex Zagorican/Tushemisht on the Albanian side (Fig. 3). Smaller spring complexes and springs include those in the northwestern part of the lake (e.g., Dobra Voda and Sum) as well as the Biljana Spring complex (at the Hydrobiological Institute Ohrid) and the nearby spring Bej Bunar in the northeastern part (Fig. 3). Many sublacustrine and surface springs, particularly on the southeastern and southern side of Lake Ohrid, are charged by neighbouring Lake Prespa as well as by mountain range precipitation seeping through the karstic rocks and mixing with the waters originating in Lake Prespa (Anovski et al., 1980; Eftimi & Zoto, 1997; Matzinger et al., 2006a; Amataj et al., 2007; Popovska & Bonacci, 2007; Fig. 4).

The catchment area of Lake Ohrid is relatively small, approximately 2,600 km^2 (including Lake Prespa; Matzinger et al., 2007). Excluding Lake Prespa, it is even smaller with 1,002 km^2 (Popovska & Bonacci, 2007; Table 1).

Though Lake Ohrid can still be classified as oligotrophic, progressing eutrophication has recently been identified through a phosphorus increase in the lake sediment over the past century (Matzinger et al., 2007), shifts in phytoplankton species close to polluted inflows (Watzin et al., 2002), as well as decreasing Secchi depths by about 30 cm $year^{-1}$ since the 1920s (Matzinger et al., 2006b). The observed eutrophication can be explained by a sharp increase in tourists (Ernst Basler and Partners, 1995) and resident population (+100,000 inhabitants) since the late 1940s (Matzinger, personal communication). Moreover, under a global warming scenario, amplification of detrimental effects of increased nutrient inputs has been demonstrated (Matzinger et al., 2007).

Geological and limnological history of Lake Ohrid

Most workers agree that the Ohrid Basin is of tectonic origin (e.g., Cvijic, 1911; Stankovic, 1960; Dumurdzanov et al., 2004; Gams, 2005; Wagner et al., 2008b). According to Aliaj et al. (2001), the Ohrid graben zone was formed by uplift of surrounding mountains and simultaneous subsidence of the Ohrid Plain. Dumurdzanov et al. (2004, p. 16) gave as the time frame of the formation of the Ohrid Graben (as well as the Prespa Graben) the "Uppermost Meotian-Pontian", i.e., a geological age of some 5–8 million years (My). Radoman (1985, p. 209) pointed out that the Ohrid Basin first underwent a "prelimnic phase"—a phase of a polje—before being flooded and becoming a lake. The actual limnological age of the lake, however, remains uncertain. Part of the problem is that in the literature, it is not always clear whether workers refer with their age estimations to the geological age of the Ohrid formation or the limnological age of the actual lake. In the literature, most age-related information is directly or indirectly based on the work of Cvijic (1911, p. 716), who stated that "the Ohrid valley probably originated during the Pliocene", i.e., some 2–5 Mya. Stankovic (1960, p. 209) referred to Cvijic (1911) and stated "According to Cvijic, the Ohrid Basin was formed at the end of the Pliocene in the upper reach of an ancient pregrabenian river valley ...", i.e., some 2–3 Mya. Salemaa (1994) gave, without reference, a lake age of 2–3 Mya. Spirkovski et al. (2001) and Susnik et al. (2006) noted a time frame for the onset of lake formation of 4–10 Mya. The latter workers cited Banarescu (1991) as their source. Unfortunately, we were unable to find this information in Banarescu's text book.

Even more controversial than the age of Lake Ohrid is the question of the origin of its water. This question is of particular interest for biologists as these waters may have provided its first species and been the basis for its present high degree of endemism. As Stankovic (1960) pointed out in his text book on the "living world" of Lake Ohrid, the little serious limnological data on the lake are largely based on the works of Cvijic (1906, 1908, 1911). These and subsequent studies (e.g., Goebel, 1919; Bourcart, 1922; Kossmat, 1924; Nowack, 1929; Laskarev, 1935; Lukovic, 1935) are far from being sufficient to explain the limnological history of Lake Ohrid. Even today, this situation has not changed significantly. It should also be noted that an older molluscan or other fossil record (i.e., pre-Pleistocene) that could be used to infer the limnological history of the lake so far is lacking in Lake Ohrid.

Despite these problems and in the context of newer lithological–palaeogeographical data for the region of the former Paratethys (e.g., Popov et al., 2004), four hypotheses for the limnological origin of extant Lake Ohrid can be advanced (Fig. 6):

Hypothesis 1 Mesohellenic Trough derivate hypothesis

During the late Eocene and the middle Miocene, the Mesohellenic Trough was a narrow basin extending from northern Greece in northwest direction to the Korca Basin (which passes north into the Ohrid Basin) (van Hinsbergen et al., 2005), comprising a series of marine, brackish-water and freshwater beds (Stankovic, 1960; also see Popov et al., 2004). Haug (1922) even suggested an open marine strait ("sillon transégéen") from the Skutari Basin southeast to today's Gulf of Volos in Greece. According to Radoman (1985, p. 212) "This marine furrow, running closely to the south margin of Lake Ohrid, is important for the palaeogeographical history of the region of Ohrid". Some workers (reviewed in Stankovic, 1960) suggested that this marine or brackish-water connection is the source of marine-like faunal elements in Lake Ohrid.

(1A) Pontian River hypothesis

Mio-Pliocene deposits are attributed to a Pontian palaeoriver that ran through the Mesohellenic Trough, originating near the Aegean Palaeolake, running along the Albanian side of what is today Lake Ohrid, leaving characteristic terraces, stretching north along the Skumbi River system, and finally ending near Mati in the Adriatic Sea. Such a river could have filled a sunken Graben, creating extant Lake Ohrid (Cvijic, 1911; Radoman, 1985; but also see Haug, 1922).

(1B) Aegean Lake derivate hypothesis

According to Bourcart (1922) in Radoman (1985), the individual lakes of the Dessaretes (Ohrid, Prespa, Mikri Prespa, and Maliq) once were connected via the Korca Depression in Albania and possibly communicated with the great "Aegean Lake" (which filled many depressions in which today some 19 individual lakes in southern Macedonia and northern Greece exist). A connection between today's Ohrid Basin and the Aegean Lake presumably existed twice: in the middle to late Miocene and in the middle to late Pliocene (Fig. 6, but also see Fig. 7).

Hypothesis 2 "Tethys" derivate hypothesis

This hypothesis of Radoman (1985), which is based on Haug (1922), stated that Lake Ohrid was once part of a complex system of lacustrine basins that had previously broken away from the "Tethys" (i.e., today's Adriatic Sea) during the lower and middle Pliocene by the inundation of the sea into lowland depressions. The Dessaretes—originally lowland lakes—later rose with the Hellenid

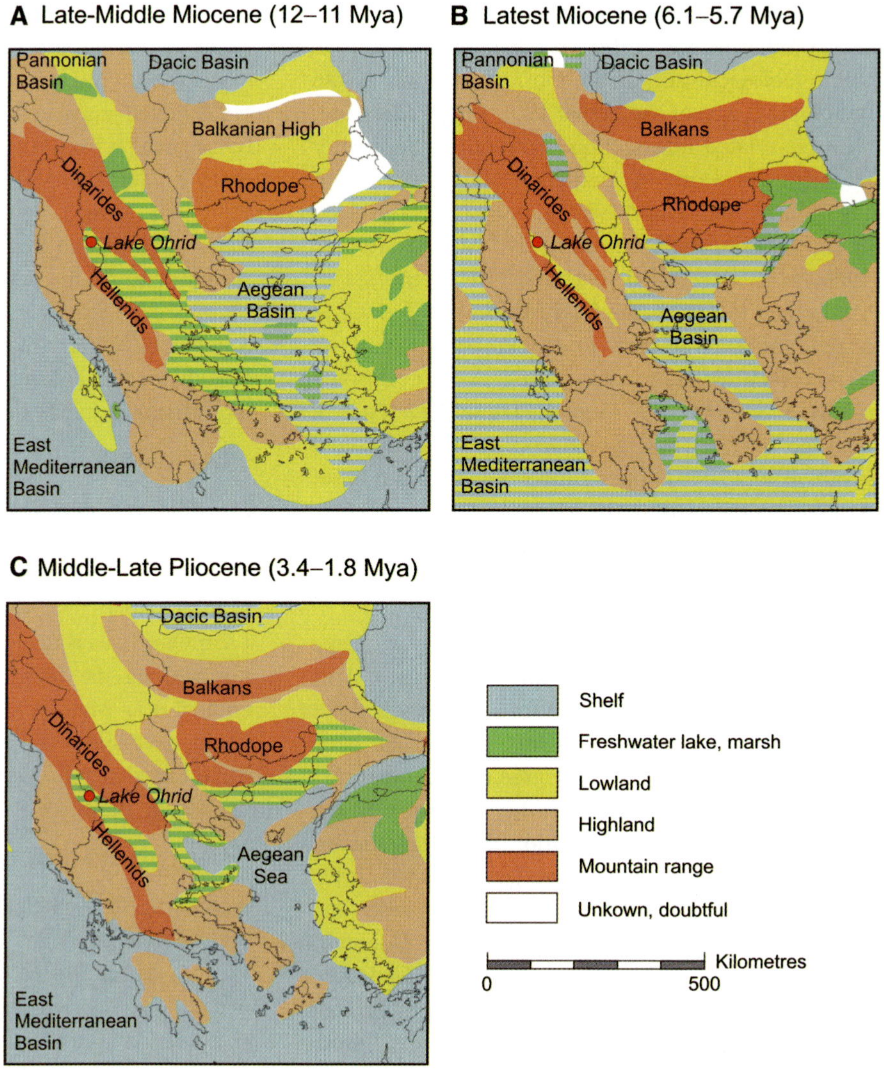

Fig. 6 Palaeogeographical maps of the central and southern Balkans for the late middle Miocene (**A**), the latest Miocene (**B**), and the middle-late Pliocene (**C**). The position of extant Lake Ohrid is indicated by a red dot. Note that according to these maps, during the Miocene the area of today's Ohrid basin was connected both to the Paratethys and to the Aegean Basin. These direct connections ceased at the end of the Miocene. Maps modified from Popov et al. (2004) with written permission of the publisher

Mountains. According to Radoman (1985, p. 88), "these lakes could be considered the 'ultimate scions' of those freshwater lacustrine systems that originated by 'freshening' of the preceding brackish lacustrine phases directly separated from the Tethys owing to the upheaval of the Alpides in these areas".

Hypothesis 3 Lake Pannon derivate hypothesis

This hypothesis suggests that during the middle to late Miocene, today's Ohrid Basin was temporarily connected with the Pannonian Basin via a series of freshwater lakes (e.g., the Serbian Lake and the Macedonian Lake), or at least received some faunal elements from there (Stankovic, 1960; Müller et al., 1999). A stepping stone function of Lake Pannon for descendent lake systems, such as Lake Dacia, is nowadays widely accepted (Harzhauser & Mandic, 2008).

Hypothesis 4 De novo hypothesis

Cvijic (1911) suggested that the individual lakes of the Dessarete Lake group (including lakes Ohrid and Prespa) have formed de novo in a dry polje from springs (or rivers). This hypothesis is supported by new hydrological data (Matzinger et al., 2006b) showing that today's water balance of Lake Ohrid is dominated by inflow from karst aquifers. Stankovic (1960) argued that the valley of Cerava (south affluent) and the valley of Crni Drim (north effluent) are remnants of a river valley that existed prior to the

Fig. 7 Palaeolimnological reconstruction of the maximum extent (red lines) of the Dessarete Lakes (lakes Ohrid, Prespa, Mikri Prespa, and Maliq) and the remnants of the Aegean Palaeolake. Figure modified from Cvijic (1911)

formation of the graben and that could have supported the formation of Lake Ohrid. He also suggested that the Crni Drim River is older than the lake itself. Interestingly, the feeder spring complex at Sveti Naum south of Lake Ohrid is not only a major supplier of water for Lake Ohrid, it also is the source of the Crni Drim (Fig. 3). Thus, Lake Ohrid is actually fed by a short, channel-like, affluent part of this river. This, in turn, raises the possibility that the evolution of Lake Ohrid was more affected by the feeder springs and the affluent part of the Crni Drim than by the affluent Cerava River.

It should be noted that the above hypotheses are not mutually exclusive, and the possibility that Lake Ohrid went through several stages of existence and disappearance, each with its own limnological history (i.e., palaeolakes), cannot be excluded. However, until now, the exact limnological origin and the origin of faunal or floral elements of Lake Ohrid remain uncertain. The above scenarios reflect two major opinions reviewed in Radoman (1985): Cvijic's (1911) theory of de novo formation of Lake Ohrid in a dry polje with a spring or river hydrography versus a palaeogeographical connection of Lake Ohrid and other lakes to brackish waters on the Balkan Peninsula.

Whereas Radoman (1985) and Banarescu (2004) reject the de novo hypothesis mainly based on biogeographical data of gastropods and other faunal elements, Stankovic (1960, pp. 229, 230) noted that "the fauna and flora of Lake Ohrid ... belong as whole to a freshwater world" and that "There is no much point in assuming the presence of the marine and brackish immigrants in Lake Ohrid". Moreover, Cvijic (1911) showed in his palaeolimnological reconstruction that Lake Ohrid never had a direct connection neither to Lake Prespa or Lake Maliq nor to remnants of the ancient Aegean Lake or any other brackish water body (Fig. 7; also see Stankovic, 1960, Fig. 91).

The de novo hypothesis is also supported by new limnological data (see above) as well as by modern phylogeographical data on endemic radiations in Lake Ohrid (see section "Patterns and processes of speciation in Lake Ohrid").

Acknowledging the uncertain role of the Meso-hellenic Trough (see the hatched freshwater/lowland area northwest and southeast of Lake Ohrid in Fig. 6C), we regard the de novo scenario as the most probable. Nonetheless, the biogeographical history of Lake Ohrid has to be considered as poorly studied, and Stankovic (1960) pointed out that for a better understanding of the biogeographical history of the lake and the evolution of its inhabitants, it is crucial to (A) establish when the limnological phase of the Ohrid Basin begun, (B) study the origin and character of waters that filled it, (C) unravel the history of connection with neighbouring waters, (D) establish the time frame of continuity of existence of the lake, and (E) understand the dynamics of hydrological, ecological, and climatic changes during the lake's history. Almost 50 years after Stankovic, all these questions are still open.

Whereas geological and hydrological studies remain the key for understanding the evolution of Lake Ohrid, great advances in molecular systematics not only provide an opportunity to test previous geological hypotheses but also help us unravelling the limnological history of a lake based on evolutionary data from key endemic taxa.

Biodiversity and endemism

The extraordinarily high degree of biodiversity and endemism in many groups of Lake Ohrid organisms has long been recognized, with most of the relevant studies conducted in the first half of the twentieth century (summarized in Stankovic, 1960). No attempt, however, has been made thus far to review the emerging new results on systematics, biogeography, and evolutionary relationships of most groups that had been studied since the monograph of Stankovic (1960), with the possible exception of microgastropods (Radoman, 1983, 1985). Though an extensive bibliography on the lake had been compiled (Noveska et al., 1985), accessibility of the lake and the (mostly local) literature on it became increasingly difficult due to political reasons. Only the recent years have seen emerging English literature in international journals and new approaches towards an understanding of the evolution of the lake's biota. These studies resulted in new insights into patterns of biodiversity and endemism.

In contrast to African and southeast Asian ancient lakes where vertebrate species play a major role in ancient lake species flocks, no such extensive species flocks have been reported for vertebrate species in temperate ancient lakes, with the exception of Lake Baikal. However, unlike vertebrate species, endemism in European ancient lake invertebrates is high, with several species flocks being reported (see section "Phylogenetical patterns"). Among these invertebrate groups, benthic and semi-benthic taxa appear to be particularly diverse.

We here list the information available on the diversity and degree of endemism of major taxa in Lake Ohrid. Special emphasis is put on species-rich assemblages and radiations. This review is based on thorough literature search including such databases as the *Zoological Record* (Thomson Scientific), the *Web of Science* (Thomson Scientific), and *Fauna Europaea* (2004). It has to be noted, however, that there is a considerable amount of literature published in local journals and in local grey literature. We have attempted to include this information, though coverage may not be complete. If possible, additional expert interviews have been conducted for some groups.

It should also be noted that relatively little information is available for the biodiversity of higher plant and protist species in Lake Ohrid, a problem already noted by Stankovic (1960). Thus, our review is biased towards animal groups with only selected "macrophyte" and "microphyte" (e.g., blue green algae, protists) groups being mentioned. It should also be noted that the higher systematics of many groups of organisms remain controversial. Therefore, the system used here may not reflect the latest findings in systematics but simply serves the purpose of an informal classification.

Bacteria

Cyanobacteria

The known diversity of Cyanobacteria is moderate, and 55 taxa belonging to 18 genera have recently been identified for Lake Ohrid. Most of these inhabit the littoral (Patceva, 2005). Few taxa have been regarded as being endemic, though the actual rate of endemism remains open (see Cado, 1978).

Chlorobionta

"Macrophytes"

A recent census indicated a relatively low diversity of macrophytes in Lake Ohrid (Talevska & Trajanovska, 2005). Four major zones of macrophyte vegetation can be found in Lake Ohrid (Trajanovska & Talevska, 2007; Talevska, 2007). These zones are the *Cladophora* zone, the *Phragmites australis* zone, the *Potamogeton* zone, and the *Chara* zone, with the latter having been studied in detail. The *Chara* belt extents into the lower littoral, and a total of at least 15 *Chara* species have been recognized. Lake Ohrid thus is seen as a hot spot of *Chara* diversity (Blazencic, 2004; Trajanovska, 2005). There are species that are restricted to the lake and only few other localities, for example, *Chara ohridana* Kostic, presumably only occurring in lakes Ohrid and Dojran (Blazencic et al., 2006). A recent census of *Potamogeton* spp. revealed the occurrence of five species in Lake Ohrid (Talevska, 2007). Other common species belong to the genera *Myriophyllum*, *Ceratophyllum*, *Vallisneria*, and *Zanichellia*. A total of 22 macrophytic species were found during a census along the Macedonian shoreline of Lake Ohrid (Watzin et al., 2002). On the Albanian side, an additional floating leave plant zone may occur (Watzin et al., 2002). Due to the still insufficient knowledge on the macrophytic vegetation of Lake Ohrid, we here use a conservative biodiversity estimate of ~25 species. Regarding invasive species, there is great concern about the further spread of *Elodea canadensis* Michx. (Watzin et al., 2002).

Protista

"Diatoms" (Heterokontobionta)

Diatoms can be considered to be the best studied group of Lake Ohrid microalgae. The enormous diversity has been subject to a monograph by Levkov et al. (2007). According to these authors, a total of 500 nominal taxa are present in lakes Ohrid and Prespa. "Relictary" taxa and many new species have been described, making Lake Ohrid a hotspot of diatom diversity. There is, however, controversy about the actual number of species-level taxa and particularly about the degree of endemism (e.g., Rakaj et al., 2000). Miho & Lange-Bertalot (2003) found more than 350 taxa in the Albanian part of the Lake alone. About 25% of them are considered to be rare or endemic. The particular impact of nomenclature and species concepts on the perception of diversity assessment in Balkan lake diatoms has been exemplified by Reed et al. (2004). Therefore, a reliable assessment of endemism in Lake Ohrid is not yet possible.

"Rhizopoda"

This group of organisms has been considered to be "certainly the least studied group of Lake Ohrid" (Stankovic, 1960, p. 170). A recent study on testate amoebas (Rhizopoda, Testacea) revealed the occurrence of 14 species in Lake Ohrid (Golemansky, 1994).

Ciliophora

A total of 30 endemic species of Ciliophora are listed in Ministry of Environment and Physical Planning (2003). This corresponds to the list given by Stankovic (1960).

Many taxa are parasitic and presumably co-evolved with their native host, mainly oligochaetes and tricladids (Stankovic, 1960). For molluscs, Raabe (1965, 1966), however, found only few endemic elements of parasitic ciliates. Molloy et al. (2008) provided more detailed information on parasites of the mussel *Dreissena presbensis* (=*Dreissena stankovici*, see section on bivalves).

Animalia (=Metazoa)

Porifera

The origin and evolutionary history of freshwater sponges is generally poorly understood (Itskovich et al., 2006), and their biodiversity is probably underestimated (Manconi & Pronzato, 2008). There are few contributions to the sponge fauna of the Balkans including Lake Ohrid (e.g., Arndt, 1923, 1937, 1938; Hadzisce, 1953). The lake is inhabited by five species of freshwater sponges, four of them (80%) endemic. The endemic genus *Ochridaspongia* is represented with two species in Lake Ohrid. The rounded sponge, *Ochridaspongia rotunda* Arndt, 1937, lives in deeper zones, lacking a gemmule stadium in its life cycle. This led workers to propose its relictary status with a tertiary origin (Gilbert & Hadzisce, 1975, 1977). Another representative of this genus, *O. interlithonis* was described by Gilbert & Hadzisce (1984). These workers also described the endemic genus *Ohridospongilla* with its only representative, *O. stankovici* Gilbert & Hadzisce, 1984. The fourth endemic species, *S. stankovici* Arndt, 1938, belongs to the genus *Spongilla*. The only non-endemic species of Lake Ohrid is the widely distributed *Eunapius fragilis* (Leidy, 1851), which is also found in other Balkan lakes.

The phylogenetic position of the genera *Ohridospongilla* and *Ochridaspongia* remains unclear; possible convergence to Lake Baikal and Lake Tiberias forms had been noted (Gilbert & Hadzisce, 1975, 1977). In a taxonomic review, the former theory of close phylogenetic relationships among ancient lake sponges (Por, 1964) has been re-erected, and a new family, Malawispongiidae, encompassing five genera occurring in lakes Kinneret, Malawi, Poso, and Ohrid has been described (Manconi & Pronzato, 2002). A recent molecular study (Meixner et al., 2007), however, showed that the family Malawispongiidae is paraphyletic, and in their latest article, Manconi & Pronzato (2008) regard the genus *Ohridospongilla* as *incertae sedis*.

Tricladida (Plathelminthes)

The Tricladida is a diverse group that has been studied in some detail (e.g., Kenk, 1978). At least 21 species had been recognized from Lake Ohrid alone, with a high degree of endemicity (Sapkarev et al., 1998).

According to the latest investigations of Krstanovski (1994), the tricladid fauna of the Ohrid watershed is comprised of three families (Dugesiidae, Planariidae, and Dendrocoelidae), with a total of five genera (*Dugesia, Planaria, Phagocata, Crenobia,* and *Dendrocoelum*). The same author provided a list of 29 taxa of planarians for Lake Ohrid and adjacent springs, with 23 species (79%) being endemic and 21 species (72%) being restricted to the actual lake.

Within the Tricladida, the genus *Dendrocoelum* is characterized by a particular high degree of endemism. Sixteen endemic species belong to this genus and presumably form a species flock. According to Stankovic (1960), the genus and its species represent an obvious example of autochthonous radiation. Representatives of the Tricladida inhabit all three major depth zones of Lake Ohrid: the littoral, the sublittoral, and the profundal up to 120 m depth and show the highest densities during winter and spring seasons (Krstanovski, 1994). A preliminary analysis of selected endemic and widespread species from Lake Ohrid utilizing enzyme electrophoresis suggested that diversification among groups of endemics happened during different time frames (Sywula et al., 2003). Recently, populations of a presumed widespread species occurring in Lake Ohrid turned out to be phylogenetically distinct (Sywula et al., 2006).

Rhabdocoela (Plathelminthes)

In contrast to the tricladids, other plathelminth groups are poorly studied. The data on Rhabdocoela almost entirely go back on the work of An der Lan (1939), who listed 25 species, of which 10 (40%) are now considered to be endemic to Lake Ohrid (Ministry of Environment and Physical Planning, 2003). Sapkarev et al. (1998) listed 25 species, with 11 of them being endemic. No reference is given as to the origin of the latter data.

Trematoda (Plathelminthes)

The knowledge on the trematod fauna of Lake Ohrid has lately increased. Parasitological research has revealed the presence of at least 20 species of monogenean trematods (Stojanovski et al., 2004). This diversity is considered to be high with a pronounced host specificity. According to Stojanovski et al. (2004), two of the species are new to science. However, the author does not provide estimations of endemicity for the remaining taxa, and one year later Stojanovski et al. (2005) noted the presence of only 15 species of monogeneans in Lake Ohrid.

Rotatoria ("Nemathelminthes")

The fauna of rotifers is monitored on an annual basis by the zooplankton research group at the Hydrobiological Institute Ohrid. During a comprehensive survey in 2001–2002, 11 pelagic and 38 littoral species of Rotatoria have been registered (Kostoski et al., 2004). A total of 14 pelagic species were mentioned in Guseska et al. (2007). No data on endemicity are available for this planktonic group.

Nematoda ("Nemathelminthes")

Decraemer & Coomans (1994) reviewed the fauna of free-living nematodes in ancient lakes, including Lake Ohrid. They followed the line of research on Lake Ohrid's nematode fauna, which started with the work of Schneider (1943), who already discussed the presumed ancientness of Lake Ohrid nematodes. After taxonomic adjustments, 24 species are known from Lake Ohrid (Decraemer & Coomans, 1994). Sapkarev et al. (1998), however, only listed 23 species, with four of them being endemic. Decraemer & Coomans (1994) only acknowledged the presence of three endemic species, *Punctodora ohridensis* W. Schneider, 1943, *Daptonema subsetosa* (W. Schneider, 1943), and *Domorganus bathybius* (W. Schneider, 1943). This view is widely accepted and the degree of endemism therefore would be 13%.

Oligochaeta (Annelida)

The oligochaete fauna of Lake Ohrid has been studied by Sapkarev (1964, 1966), who mentioned 36 taxa (species and subspecies) from the lake, including 17 endemic forms. In a review on oligochaetes in worldwide ancient lakes, Martin (1994) listed a total of 22 species for Lake Ohrid, with five species being endemic. Recently, the Macedonian biodiversity report (Ministry of Environment and Physical Planning, 2003) listed 27 Ohrid species.

Nine of them (plus eight subspecies) are considered to be endemic to Ohrid (33%). Regardless the actual value of endemism, the great number of forms ("subspecies") makes the oligochaete fauna a very interesting study object. Unusual life history strategies, such as the aquatic lifestyle of otherwise strictly terrestrial lumbricids, make the group interesting for evolutionary studies.

Hirudinea (Annelida)

The Hirudinea from Lake Ohrid comprise eurybathic forms that can be found in all three depth zones. Similar to other components of the macrozoobenthos, the leech fauna of Lake Ohrid is diverse and highly endemic. The investigation of Lake Ohrid leeches has a long tradition (Doflein, 1921; Augener, 1925, 1926), with further studies carried out by Remy (1934, 1937), Pawlowski (1936), and Augener (1937). Extended research on ecology and systematics of Hirudinea were made by Sapkarev (1964, 1966, 1975, 1990) and Sket (1968, 1981). According to Sket & Sapkarev (1992), the leech fauna of Lake Ohrid comprises at least 24 species belonging to five families: Piscicolidae (2 species), Glossiphonidae (7 species), Hirudinidae (1 species), Haemopiidae (1 species), and Erpobdellidae (13 species). Twelve species of them are considered to be endemic. The authors assume the possibility of new, still undescribed endemic erpobdellid species occurring in the lake. An interesting group is the species flock of the nominal genus *Dina*. It includes at least eight valid species according to Sket (1989). Trajanovski et al. (2006) mentioned the existence of at least two more undescribed species, which raises the total endemism rate of Hirudinea to about 54%. One additional species, *Dina latestriata* Neubert & Nesemann, 1995, was recently described from Lake Mikri Prespa (Greece). In a systematic treatment of Erpobdellidae, Siddall (2002), however, reduced this family to a monogeneric taxon, with *Erpobdella* being the sole genus.

A recent molecular study of the "*Dina*" species flock suggests that all taxa in Lake Ohrid are, indeed, monophyletic and that they either evolved intralacustrine or that few ancestral species living in prelake habitats, such as springs or rivers, gave rise to the diversity seen today (Trajanovski et al., unpublished data). Interestingly, this study also shows that nominal *D. latestriata* from Lake Mikri Prespa is not a member of the Lake Ohrid *Dina* species flock.

Acari (Arthropoda: Arachnida)

There are relatively few studies on the water mite fauna of Lake Ohrid (Viets, 1936; Georgiev, 1957; Pesic, 2003; Baker et al., 2007). The total number of known species is currently estimated at 43 (Baker et al., 2007; Pesic, personal communication); 41 of them belong to the Hydrachnidia, none of them considered to be endemic. In addition, there are two representatives of Halacaroidea, with one of them, *Copidognathus profundus* (Viets, 1936), being endemic and restricted to depths of 80–100 m. The total water mite endemism thus is 2%.

Cladocera (Arthropoda: Crustacea)

During a census conducted in 2001/2002, a total of 31 species of Cladocera were found in Lake Ohrid. Of these, 26 represent benthic and 5 pelagic species (Kostoski et al., 2004). The only endemic species seems to be *Alona smirnovi* Petkovski & Flössner, 1972, which lives in the lake and its feeder springs at Sveti Naum (Petkovski & Flössner, 1972). The rate of endemicity thus is 3%.

Recent studies revealed changes in species compositions with two non-indigenous cladocerans, *Leptodora kindtii* (Focke, 1844) and *Diaphanosoma birgei lacustris* Korinek, 1981, having invaded the lake (Kostoski et al., 2005).

Ostracoda (Arthropoda: Crustacea)

The Balkan ostracod fauna has long been recognized as highly diverse and rich in endemic species (e.g., Stankovic, 1960; Petkovski, 1969). This is particularly true for Lake Ohrid (Griffiths & Frogley, 2004). Endemicity is most pronounced in two characteristic ancient lake groups, the Candoninae and the Cytheroidea (Forrester, 1991). In a recent review of the Balkan ostracod fauna, 52 ostracod species were mentioned from Lake Ohrid, of which 33 (63%) are endemic (Griffiths & Frogley, 2004).

According to Petkovski et al. (2002), there is a large number of endemics belonging to the complex *Candona neglecta* G.O. Sars, 1887. Interestingly, these endemic taxa are characterized by an extraordinary range of shell shapes. Altogether, 22 *Candona* spp. are known to be endemic to Lake Ohrid, whereas one species, *Candona marginatoides* Petkovski, 1960, also occurs in Lake Prespa. There is another interesting genus present in Lake Ohrid, *Leptocythere*, the species of which normally inhabit salt- and brackish-water habitats.

Copepoda (Arthropoda: Crustacea)

The latest taxonomic review of the Copepoda listed 36 known species for Lake Ohrid of which six species (17%) appear to be endemic (Petkovski & Karanovic, 1997).

The compositional change of pelagic and littoral forms is subject to an ongoing monitoring programme carried out by the Hydrobiological Institute Ohrid. The census for 2001–2002, for example, recorded four pelagic and 10 species restricted to the littoral (Kostoski et al., 2004).

Recent studies on copepods of Lake Ohrid mainly focus on life history and ecology of selected species (Shumka, 2002). Some of them show interesting life history patterns, such as *Ochridacyclops arndti* Kiefer, 1937, which lives inside endemic sponges (Stankovic, 1960).

Preliminary mitochondrial DNA data suggest a distinct but rather unexpected phylogenetic position of the dominant endemic species, *Cyclops ochridanus* Kiefer, 1932 (Kostoski et al., 2007). The biogeographic affinities of *Arctodiaptomus steindachneri* (Richard, 1897) within the Balkans were studied by Michaloudi et al. (1997).

Decapoda (Arthropoda: Crustacea)

Two decapod species are known from Lake Ohrid, the widespread European crayfish *Astacus astacus* (Linnaeus, 1758) and the freshwater crab *Potamon fluviatilis* (Herbst, 1785). The latter is common in the western Balkans. Both species are rare and locally restricted within Lake Ohrid (Albrecht & Wilke, unpublished data).

Isopoda (Arthropoda: Crustacea)

Four species of Isopoda occur in Lake Ohrid, three of which are endemic (75%). These three endemic species belong to the genus *Proassellus* Dudich, 1925. Several subspecies or morphological forms have been described from different bathymetric zones of the lake (Karaman, 1953). Based on allozyme data, Kilikowska et al. (2006) suggested that intralacustrine speciation led to the current diversity and that ecological segregation is involved in these processes.

Recently, Wysocka et al. (2008) investigated the phylogenetic relationships and divergence times of the endemic *Proasellus* species from Lake Ohrid based on mitochondrial DNA sequencing data. They found two divergent, well-supported groups, which were largely corresponding to depth zones. They, too, suggested intralacustrine speciation to have taken place within these groups. The tree, however, did not reflect the present morphology-based classification.

Amphipoda (Arthropoda: Crustacea)

Karaman (1985) recognized 11 amphipod species for the Ohrid watershed, whereas Sapkarev et al. (1998) listed nine endemic species and one non-endemic for Lake Ohrid (rate of endemicity = 90%). The dominant "*Ochridagammarus*" endemic species flock is not only considered to be ecologically and morphologically highly heterogeneous, it also is karyologically highly diverse, with chromosome numbers ranging from $n = 12–32$ (Salemaa & Kamaltynov, 1994). These authors suggested that diverse habitats, such as littoral springs and access to underground karstic waters, increase the degree of partial isolation. They also suggested several ancient immigration waves of *G. salemaai* ($n = 12$) as a "palaeoendemic relict" surviving in the area. Tentatively, they accept polyploidy as primary source for the exceptional chromosomal range observed in Lake Ohrid gammarids. Despite the works on Lake Ohrid species (Karaman, 1977, 1985), no detailed account on Balkan Gammaridae is yet available. Preliminary mitochondrial DNA data on the endemic *Ochridagammarus* complex suggest high levels of genetic differentiation, possibly corresponding to depth zones (Sell et al., 2007).

Insecta (Arthropoda)

The entomological exploration of Lake Ohrid is still in its infancy. With the exception of the Chironomidae (Angelovski & Sapkarev, 1991; Smiljkov & Sapkarev, 1999), no other major insect taxon has been studied in detail. The countrywide review of biodiversity lists no endemic insect species for Lake Ohrid (Ministry of Environment and Physical Planning, 2003). According to Smiljkov (2002), three subfamilies of Chironomidae are present in Lake Ohrid: the Tanypodinae, Orthocladiinae, and Chironominae. The total species number is 48, with 32 species belonging to the Chironominae alone. This recent study yielded 15 species new for the fauna of Lake Ohrid, with the highest diversity being found in the littoral zone. To our knowledge, no other analysis as to endemism or biogeographical affiliation of any insect group inhabiting Lake Ohrid exists at present. For reasons of completeness, we list in Table 2 the data published by Stankovic (1960) for Ephemeropta, Trichoptera, and Odonata.

Gastropoda (Mollusca)

Studies on the outstanding gastropod fauna of Lake Ohrid already started at the end of the nineteenth and the beginning of the twentieth century (e.g., Sturany, 1894; Brusina, 1896; Polinski, 1929, 1932), and among all groups of organisms in Lake Ohrid, they

Table 2 Biodiversity and endemism of Lake Ohrid

Taxon	Total number of native species	Number of endemic species	Rate of endemism
Bacteria			
Cyanobacteria	55	n.a.	n.a.
"Macrophytes"	~25	0	0%
Protista			
"Diatoms" (Heterokontobionta)	~500	n.a.	n.a.
"Rhizopoda"	14	n.a.	n.a.
Ciliophora	34	30	91%
Animalia			
Porifera	5	4	80%
Tricladida (Plathelminthes)	29	23	79%
Rhabdocoela (Plathelminthes)	25	10	40%
Trematoda (Plathelminthes)	15	2	13%
Rotatoria ("Nemathelminthes")	52	n.a.	n.a.
Nematoda ("Nemathelminthes")	24	3	13%
Oligochaeta (Annelida)	27	9	33%
Hirudinea (Annelida)	26	14	54%
Acari (Arachnida)	43	1	2%
Cladocera (Crustacea)	31	1	3%
Ostracoda (Crustacea)	52	33	63%
Copepoda (Crustacea)	36	6	17%
Decapoda (Crustacea)	2	0	0%
Isopoda (Crustacea)	4	3	75%
Amphipoda (Crustacea)	10	9	90%
Chironomidae (Insecta)	49	0	0%
Ephemeroptera (Insecta)	6	0	0%
Odonata (Insecta)	26	0	0%
Trichoptera (Insecta)	22	0	0%
Gastropoda (Mollusca)	72	56	78%
Bivalvia (Mollusca)	13	1	8%
"Osteichthyes" (Chordata)	17	7	41%
Total number	~1,200	212	36%[a]
(Animalia)	(586)	(182)	(34%[a])

For references and comments see text

Total values for native animals (Metazoa) are given in parentheses

n.a., Information not available

[a] Adjusted rate based on only those groups for which the number of endemic species is known

are probably the best studied (e.g., Hadzisce, 1956; Radoman, 1983). Stankovic (1960, p. 185) stated: "This is certainly the most important group of Lake Ohrid's animal population, especially because of its extraordinary endemism". Watzin et al. (2002) noted a degree of 86% endemism (but no reference or actual data were given). Thus, we here refer to the latest comprehensive treatment of the gastropod fauna of Lake Ohrid (Radoman, 1985). The author listed 72 gastropod species, of which 56 (78%) are endemic. This is most likely a conservative estimate, since preliminary molecular data show that some of the assumed widespread taxa occurring in Lake Ohrid represent distinct lineages or species (Albrecht & Wilke, unpublished data).

A characteristic of Lake Ohrid's gastropod fauna is the presence of several species flocks in such major gastropod taxa as pulmonates (e.g., *Carinogyraulus* spp.) and hydrobiids (e.g., Pyrgulinae). The micro-gastropod family Hydrobiidae accounts for the majority of species known from the lake, partly including point endemics (e.g., Radoman, 1983). It is also remarkable that several gastropod groups include representatives with peculiar shell characters, such as ribs, scalariform shell shapes, and screw-shaped opercula (Radoman, 1983; Albrecht et al., 2006b; Wilke et al., 2007), and these conspicuous species have received considerable attention. It should, however, be noted that all known species flocks also appear to include inconspicuous taxa with few informative shell characters (Fig. 8).

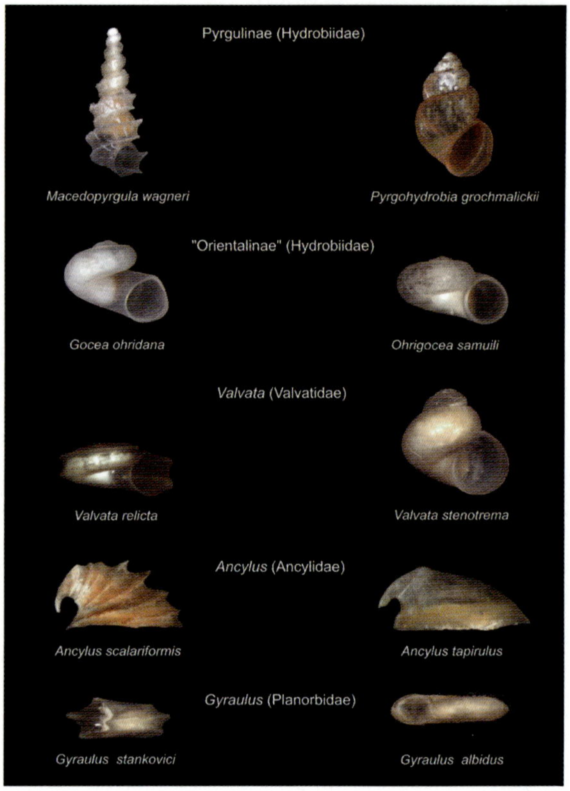

Fig. 8 Contrasting patterns of shell shape and sculpturing within selected Lake Ohrid gastropod groups. For each taxon, conspicuous endemic representatives are shown on the left and inconspicuous endemic counterparts on the right. Individuals are not to scale

Bivalvia (Mollusca)

Recently, a review of the bivalve fauna of the southern Balkans was published by Korniushin (2004). Accordingly, Lake Ohrid is home to at least 13 bivalve taxa of three families (Unionidae, Sphaeriidae, and Dreissenidae), whereas nine species of bivalves occur in Lake Prespa. Based on this study, lakes Ohrid and Prespa harbour two and one endemic species, respectively. Whether the other Balkan lakes harbour endemics remain unclear. A number of nominal endemic taxa (mostly at the forma level) have been described, the validity of which, however, has not been confirmed.

Although the bivalve fauna of Lake Ohrid is poor in endemic species compared to the Gastropoda, some workers used the faunal relationships of endemic bivalve species in Lake Ohrid with their widespread European congeners to discuss the origin of the lake's fauna. According to Starobogatov (1994), the occurrence of the endemic *Dreissena stankovici* L'vova & Starobogatov, 1982, in Lake Ohrid indicates a strong affinity to the Paratethys (also see Korniushin, 2004). However, recently, it has been shown that *D. stankovici* is a junior synonym of *D. presbensis* Kobelt, 1915 and that the taxon is not endemic but widespread throughout the western-central Balkans (Albrecht et al., 2007). Moreover, it has been demonstrated that a relatively young sister species of *D. presbensis*, *D. blanci* Westerlund, 1890 populates the southern Balkans. Thus, a direct faunal connection between the Paratethys and Lake Ohrid becomes less convincing (also see section "Patterns and processes of speciation in Lake Ohrid").

For the cryptic pea clam genus *Pisidium* (Sphaeriidae), Schultheiß et al. (2008) suggested the presence of nine species, including one undescribed taxon, for Lake Ohrid and seven species, including one undescribed taxon, for Lake Prespa. Among them, there is one endemic species each in lakes Ohrid and Prespa.

The three unionid species of Lake Ohrid are not yet studied in detail, but they include the rare and disjunctly distributed *Microcondylaea bonelli* (A. Férussac, 1827).

Combining the information presented by Korniushin et al. (2000), Korniushin (2004), Albrecht et al. (2007), and Schultheiß et al. (2008), the total number of bivalve species in Lake Ohrid appears to be 13 with one of them (= 8%) being endemic.

"Osteichthyes" (Chordata)

An exact estimation of the actual fish diversity and degree of endemism in Lake Ohrid is difficult, partly due to the unresolved taxonomic status of many of the native fishes. Most often a total number of 17 native fish species (2 species of Salmonidae, 12 species of Cyprinidae, 2 species of Cobitidae, and 1 species of Anguillidae) are given, with 10 of them being endemic (e.g., Watzin et al., 2002). This number, however, is likely to change when more molecular data become available. A recent example is the demonstrated taxonomic non-validity of *Gobio gobia ohridanus*, Karaman, 1924 (Sanda et al., 2005). Acknowledging the uncertainties, especially within the *Salmo letnica* complex, we here follow the latest listings of the fish fauna for Lake Ohrid that includes 17 native species, with seven of them being endemic (Ministry of Environment and Physical Planning, 2003). Accordingly, rate of fish endemism in Lake Ohrid would be 41%.

The fish fauna of Lake Ohrid has long attracted intensive research, mainly triggered by the famous Ohrid trout, *Salmo letnica* (Karaman, 1924), which has a considerable economic importance. There are still uncertainties as to the taxonomic status of at least five endemic taxa of *Salmo*. A distinct endemic species is the belvica, *Salmo ohridana* (Steindachner, 1892), formerly known as *Acantholingua ohridana* (Phillips et al., 2000; Susnik et al., 2006). The monophyly of *S. letnica* has been demonstrated using mtDNA sequences (Susnik et al., 2007).

Cypriniformes make up the majority of fish species if one considers lakes Ohrid and Prespa together. Cyprinid endemism seems to be mostly on subspecific level, though *Rutilus ohridanus* (Karaman, 1924) and *Poxinellus epiroticus* (Steindachner, 1896) are listed as endemic species for Lake Ohrid (Ministry of Environment and Physical Planning, 2003). Again, information is ambivalent, as a recent study (Sanda, 2007) suggested that none of the cypriniform fishes would be unequivocally endemic solely to Lake Ohrid, but that most of them occur in the Ohrid–Drim–Skutari system. Similarities between the endemic cyprinid fauna of lakes Ohrid and Skutari were also noted by Maric et al. (2007).

Interestingly, effective fish endemism is considered to be higher in Lake Prespa than in Lake Ohrid. Crivelli et al. (1997) list 12 indigenous species for the former lake, of which seven are endemic. Future studies have to show whether the presumed higher endemism rate in Lake Prespa is due to a more effective isolation of Lake Prespa (e.g., Karaman, 1971; Sanda, 2007) or simply due to a taxonomic and/or biogeographical bias (e.g., that species assumed to be endemic for Lake Prespa are actually more widely distributed).

In both lakes, there is concern about decreasing population sizes of endemic fish species, especially in salmonids. At the same time, at least seven non-native fish species have been introduced into Lake Ohrid (Watzin et al., 2002) and 11 into Lake Prespa (Crivelli et al., 1997).

Summarizing biodiversity research in Lake Ohrid (Table 2), considerable progress has been made in the past decades, particularly for such taxa as the Gastropoda, Bivalvia, Isopoda, and salmonid fishes. Other taxa, however, remain poorly or even unstudied, particularly within the Protista, Coelenterata, Trematoda, Cestoda, Nemertina, Tardigrades, and Insecta. Moreover, much more work is necessary in order to obtain a precise estimate of overall endemism, both at the species and at the supraspecific level. From the well-studied groups, it is clear that Lake Ohrid harbours high endemic biodiversity, making it an exceptional lake (Fig. 2).

Interestingly, the total number of endemic species summarized here has not changed significantly compared to older works. Even the new molecular data have not led to significant shifts in estimated endemism rates: so far, most endemic species identified based on morphological data have been confirmed based on molecular data, and the few species that have been inferred to be not endemic were compensated for by a similar number of newly identified species. However, we expect a significant increase in the total number of endemic species when data for other major groups harbouring endemic species (e.g., diatoms) become available.

Patterns and processes of speciation in Lake Ohrid

Faunal relationships

The controversial origin and relationships of Lake Ohrid faunas were discussed above. Problems involve the complex biogeographic situation on the Balkans (e.g., Reed et al., 2004) as one of the major European refugial areas and the unresolved interrelationship of Lake Ohrid to both the surrounding watershed and the Neogene lake systems (e.g., Harzhauser & Mandic, 2008).

Several relationships in space and time have been proposed for different taxa, with most work having been done in molluscs (e.g., Radoman, 1985; Korniushin, 2004) and fishes (e.g., Banarescu, 2004). They include (A) a Paratethys connection, (B) an Aegean–Anatolian connection, and (C) a more regional Adriatic–Ionian connection.

A possible example for a Paratethys (Lake Pannon) relationship (Fig. 6; section on bivalve diversity) is the mussel genus *Dreissena*. Most workers agree that the ecological and climatic history of the Paratethys region produced the pattern seen today in *Dreissena*. Considering the time frame for the split of the Balkan taxa suggested by Stepien et al. (2003) and the duration of Lake Pannon (11.5–4.0 Mya; Harzhauser & Mandic, 2004), an invasion of the Balkans from the Pannonian Basin (or a remnant of it) appears to be possible. However, Lake Ohrid probably is too young for a direct invasion from the Pannonian Basin and may therefore have been colonized via, what Radoman (1985) called, secondary or tertiary basins on the Balkans. Moreover, Albrecht et al. (2007) pointed out that other biogeographical scenarios for the colonization of the Balkans by the most recent common ancestor of the subgenus *Carinodreissena* cannot be dismissed (e.g., colonization from Aegean or Anatolian areas during the Messinian Salinity Crisis).

A possible Aegean–Anatolian faunal relationship of Lake Ohrid has recently been tested for endemic species of the microgastropod subfamily Pyrgulinae (Wilke et al., 2007). The study could neither confirm a direct Aegean–Anatolian nor a Paratethys link. Instead, the closest relatives of endemic Ohrid taxa are to be found in the western Balkans.

Such a close faunal relationship to the western Balkans, particularly to the South Adriatic–Ionian Region, was discussed by Radoman (1985) and Banarescu (2004). Recent molecular studies indicate that a number of Ohrid taxa have a close regional, i.e., Balkan or southeast Mediterranean relationship. Examples include not only the already mentioned mussel *Dreissena presbensis* and the gastropod subfamily Pyrgulinae, but also the freshwater limpet genus *Ancylus* (Albrecht et al., 2006b), the freshwater gastropod genus *Radix* (Albrecht et al., 2008), the pea clam genus *Pisidium* (Schultheiß et al., 2008), and the endemic Ohrid trout *Salmo letnica* (Susnik et al., 2007).

These studies add to the increasing body of evidence that the most recent common ancestor of many endemic Ohrid faunal elements resided in the area when Lake Ohrid came into existence. This, in turn, would support the de novo hypothesis of Lake Ohrid.

Phylogenetical patterns

Two hypotheses are often involved in ancient lake speciation studies: ancient lakes themselves can act as *evolutionary reservoirs* that assure the survival of glacial, marine, or other relic species; at the same time, extant species may evolve through *intralacustrine speciation* (e.g., Rossiter & Kawanabe, 2000; Wilson et al., 2004; Albrecht et al., 2006b). Though, the general idea of these concepts may be compelling, many aspects (particularly of the relic concept) remain poorly understood and are often difficult to apply.

Until the advance of molecular tools about two decades ago, relicts in ancient lakes were most often proposed based on such morphological characteristics as "thalassoid" shell or armature forms, presumably indicating their marine origin (reviewed in Martens, 1997). Today, it is commonly accepted that most of these morphological features are subject to homoplasy. As pointed out by Martens (1997, p. 177), "Ancient lakes … are not old seas, but their faunas can show various convergent aspects". Moreover, for a long time, it was assumed that ancient lake relic species, as remnants of a previous geological period (see Glossary), show a slow rate of evolution (the stasigenesis concept; see Rensch, 1947; Huxley, 1958; Dobzhansky et al., 1977; but also see Swetlitz, 1995), assuring their survival in isolated parts of an ancient area without particular change (reviewed in Stankovic, 1960). Many evolutionary biologists, however, believe that relic species in ancient lakes change over time, if not in morphology and ecology, then at least in gene frequencies due to, for example, spontaneous mutations. This "gradual evolution" is generally referred to as anagenesis (i.e., a phyletic change in an entire population without splitting in two or more species; see Glossary) and means that a relic species gradually evolves over time into a new ("chrono-") species, with the ancestral population becoming extinct. Acknowledging that there is controversy as to the amount of change and the time necessary for this process to take place, it is conceivable that, in general, the long existence of an ancient lake may support anagenesis in relic species. This, in turn, would mean that relic species may also be subject to intralacustrine speciation. On the other hand, taking the literal meaning of the word relict as "remnants of ancient faunae and florae, whose close relatives have disappeared" (Stankovic, 1960, p. 231), many taxa in ancient lakes may qualify as relicts even though they are subject to intralacustrine speciation.

Given these conceptual difficulties, we here refrain from further discussions of relict versus intralacustrine speciation concepts. Rather, we will focus on two of the fundamental underlying questions: (1) How often and when did a particular taxon invade the lake? and (2) Did the invading species split into sister species (cladogenesis) within the lake, thus increasing the lakes' endemic biodiversity?

The two concepts that could help answering these questions are the "species flock" and "species scatter" concepts. Originally, the definition of species flock referred to a group of species that are characterized by the three requirements speciosity, monophyly, and endemicity (Greenwood, 1984). Today, species flocks in ancient lakes are seen as a monophyletic group of at least three species that are endemic to the lake but which may include taxa that today occur outside the lake (Schön & Martens, 2004). It is generally assumed that an ancient lake species flock evolved within the respective lake (i.e., intralacustrine). But in theory, the concept may also apply to flocks that originated under pre-lake conditions (e.g., in water bodies that later gave rise to an ancient lake).

Given the criterion monophyly, a species flock, by definition, coalesces to a single ancestor, i.e., a colonization event of a single species if the flock evolved intralacustrine. In contrast, different species of a particular taxon (e.g., genus) may have invaded a lake independently (multiple colonization), resulting in a group of closely related taxa that are characterized by speciosity and endemicity but which usually do not represent a monophyletic entity. This pattern has been called ancient lake species scatter (Hauswald et al., 2008).

Acknowledging that species scatter and species flock are mutually not exclusive (a species scatter may, at least in theory, include a species flock), we feel that the application of these two concepts may help unravelling the evolutionary history and endemic biodiversity of ancient lake taxa.

Major species flocks in worldwide ancient lakes, for example, the Baikal gammarid amphipod flock (Sherbakov, et al., 1998) or the cichlid radiations in the African Rift Valley (Seehausen, 2006), have been subject to intense studies and many basic evolutionary concepts built, at least partly, upon these flocks (see section "Introduction").

For Lake Ohrid, a number of species flocks have been proposed, but only few of them were tested based on molecular data. A principal problem in the study of species flocks in Lake Ohrid is the close relationship of the lake to its sister lake, Lake Prespa (Albrecht et al., 2008). There are species flocks (e.g., *Ancylus* spp.) that are restricted to Lake Ohrid. Other species flocks (e.g., the Pyrgulinae, Wilke et al., unpublished data) have representatives occurring in both lakes. However, given the wider definition of a species flock discussed above, we here consider a monophyletic group of endemic taxa occurring in lakes Ohrid and Prespa to represent a species flock. It should be noted that there is no evidence for the existence of species flocks exclusively occurring in Lake Prespa.

Recent molecular studies of selected animal groups in Lake Ohrid confirmed the existence of several (mostly small) species flocks. Albrecht et al. (2006b) showed that four endemic species of the gastropod *Ancylus* (Fig. 8) form such an ancient lake species flock. Interestingly, the flock contains an undescribed taxon exclusively occurring in the southern feeder springs of the lake that is the sister species to the three taxa occurring in the actual lake. The phylogenetic age of the whole flock (including the endemic taxon from the southern feeder springs) was estimated as 1.4 ± 0.6 My (Albrecht et al., 2006b), whereas the age of the three lake taxa is roughly 0.9 ± 0.4 My (Albrecht et al., unpublished data). In addition, the springs northeast and northwest of Lake Ohrid are exclusively inhabited by an undescribed widespread southern European *Ancylus* taxon, which does not occur in the lake proper. These patterns indicate that Lake Ohrid and its southern feeder springs probably were successfully colonized only once, that the basal split involves cladogenesis into spring and lake taxa, and that later only the species in the actual lake further diversified. The data also indicate that there are barriers between the actual lake and the surrounding springs, today effectively preventing a colonization of the lake by spring taxa or vice versa.

A species flock was also confirmed for the microgastropod subfamily Pyrgulinae (Wilke et al., 2007). This taxon (Fig. 8) is represented with 13 nominal species in Lake Ohrid and one species in Lake Prespa. Preliminary molecular clock analyses indicate that this flock, like *Ancylus*, is relative young. Based on the data presented in Wilke et al. (2007), the minimum age of the Pyrgulinae flock can be estimated as 1.7 ± 0.2 My. (Only four representatives of this flock were included in this study, and this figure thus has to be considered a minimum age due to a potential bias introduced by missing taxa.) The preliminary age of the split of the Ohrid flock from its sister group outside the lake is 2.8 ± 0.3 My.

Some of the individual representatives of the Pyrgulinae flock show interesting distribution patterns. Few of these endemic taxa occur in the springs adjacent to Lake Ohrid (both in the feeder springs and in springs in the northeast and northwest; Radoman, 1983); a single taxon of the flock colonized Lake Prespa only few hundred thousand years ago (Wilke et al., unpublished data); and today, a single representative of the flock only occurs remotely outside the lake. This taxon, *Pyrgula annulata* (Linnaeus, 1767), has a scattered distribution ranging from Lake Skutari (Fig. 1), which is connected to Lake Ohrid via the Crni Drim River, to some northern Italian lakes. It is hypothesized that the common ancestor of this taxon occurred in Lake Ohrid and that a part of the population left the lake via the Crni Drim River, giving rise to populations of extant *P. annulata* (Wilke et al., unpublished data).

Another small flock was found by Wysocka et al., (2008) within the isopod genus *Proasellus*, comprising three species with a total of four subspecies. The workers suggested an age of the most recent common ancestor of 2.3–2.8 My.

If we accept that the *Proasellus*, *Ancylus*, and Pyrgulinae species flocks evolved within Lake Ohrid, then already these few flocks point towards a high degree of cladogenesis within the lake and a comparable low number of species that colonized the lake from outside, at least in gastropods. Moreover, if the molecular clock data presented above are roughly correct, then the limnological age of Lake Ohrid must be at least 2.3 My. In contrast to these species flocks, Hauswald et al. (2008) found endemic species of the gastropod genus *Valvata* in Lake Ohrid to be not monophyletic. They suggested at least three independent colonisations of the lake.

The above studies on ancient lake species flocks and scatters in Lake Ohrid point to complex biogeographical patterns within the Ohrid watershed. Most endemic taxa are restricted to the lake proper. Few endemic taxa can be found in the southern feeder springs and even fewer in other springs to the northeast and northwest of the lake. The latter springs also harbour a number of wide-spread species that do not occur in the feeder springs or in the actual lake. Interestingly, the southern feeder springs (and presumably the northwestern and northeastern springs as well) harbour an own set of endemic species (Radoman, 1983; Albrecht et al., 2006b, 2008) that are absent from the lake proper. This is astonishing as, for example, some of the southern feeder springs are only few metres away from the actual lake and as a minor lake level increase of approximately 0.5 m would lead to an inundation of these springs. This likely has happened during evolutionary times (Fig. 7), even though archaeological data suggest that the water level was several metres lower than today for extended time periods during the past 1,000 years (Milevski et al., 1997; Matzinger et al., 2006a).

As for the faunal relationships between the sister lakes Ohrid and Prespa, the endemic biodiversity in Lake Prespa is much lower than in Lake Ohrid. From the data presented above, as well as from additional data on the gastropod genus *Radix* (Albrecht et al., 2008) and the bivalve genus *Dreissena* (Albrecht et al., 2007), there is an increasing amount of evidence that

(A) there has been generally very little faunal exchange between these sister lakes since evolutionary times;
(B) a considerable number of endemic species in Lake Prespa, however, have their most recent common ancestor in Lake Ohrid and not vice versa;
(C) the species in Lake Prespa are often relatively young; and
(D) there is little evidence for cladogenesis in endemic Lake Prespa species.

This all points to the conclusion that the mostly young endemic fauna in Lake Prespa is affected by either Lake Ohrid or other Balkan lakes and that this, in turn, renders the interesting hypothesis of Radoman (1985) that the biodiversity of Lake Ohrid is shaped by organisms of the potentially older Lake Prespa unlikely.

Nonetheless, the peculiar relationship of endemic elements within Lake Ohrid on one side and between lakes Ohrid and Prespa on the other side warrants a closer look at patterns of endemicity.

Patterns of endemicity

Spatial level

Three general patterns of endemism can be distinguished for lakes Ohrid and Prespa: taxa that are endemic to either lake with no close relatives in the respective sister lake (e.g., *Ochridaspongia rotunda* of Lake Ohrid or *Gyraulus stankovici* from Lake Prespa; Fig. 8), closely related but distinct endemic taxa (sister species) in both lakes (e.g., *Radix relicta* and *R. pinteri* in lakes Ohrid and Prespa, respectively), and shared endemic taxa occurring in both lakes (e.g., *Candona marginatoides* Petkovski, 1960).

Within the Ohrid watershed, endemism occurs at different spatial scales (also see section "Phylogenetical patterns"):

(A) species endemic to Lake Ohrid and its adjacent springs,
(B) species endemic to surrounding springs,
(C) species endemic to the lake proper, and
(D) species endemic to certain parts of the lake separated by horizontal and vertical barriers.

Particularly the latter scale is of special importance in Lake Ohrid and involves a number of so-called key endemic areas (Wilke & Albrecht, 2007). During ongoing surveys of the invertebrate faunas of the Ohrid Basin, several of these areas were identified. The arguably most important one, Veli Dab (Figs. 2, 11C), is situated along the karstic eastern shore of the lake. Other key endemic areas are the littoral near Sveti Zaum, the feeder spring complex at Sveti Naum, and its sister complex Tushemisht/Zagorican in Albania (Figs. 2, 5E).

Unfortunately, the lake suffers from increasing anthropogenic pressure and a "creeping biodiversity crisis" (Wilke & Albrecht, 2007, p. 44). The comparatively small size of Lake Ohrid and the extremely small range of many endemic species together with ever increasing human pressure make its fauna particularly vulnerable, and some endemic species presumably have already gone extinct. Whereas anthropogenic effects are noticeable all around the lake, adverse human impact on key endemic areas could have grave consequences. The destruction of the habitats around Veli Dab could lead to a permanent loss of >10% of the lake's biodiversity (Wilke & Albrecht, 2007).

Taxonomic level

Endemic taxa are described at various taxonomic levels, including the subspecies ("forma") level (e.g., formae of Charophyta), the species level (e.g., the copepod *Cyclops ochridanus*), the genus level (e.g., the sponge taxon *Ohridospongilla*), the subfamily level (e.g., the gastropod taxon Chilopyrgulinae), and the family level (e.g., the gastropod taxon Micropyrgulidae).

Whereas there are numerous good examples for endemism at species and genus levels, endemism below and above these levels remains controversial.

Part of the problem is that the subspecies concept is not universally accepted. Whereas the biological species concept explicitly acknowledges the existence of subspecies, ordinarily under the condition of geographical separation (Mayr, 1940; O'Brien & Mayr, 1991), they are not recognized by the phylogenetic species concept. Where the subspecies concept is accepted, it is best reserved for populations that are reproductively isolated (usually by geographical barriers) and that exhibit recognizable phylogenetic partitioning due to the time-dependent accumulation of genetic differences (e.g., O'Brien & Mayr, 1991; Wilke & Pfenninger, 2002). These geographical barriers are conceivable for subspecies living within the lake versus their sister taxa outside. For subspecies within the lake, the situation might be more complicated (also see the section "Speciation"). There are several cases of nominal subspecies within the lake where molecular studies could not find genetic subdivisions between the taxa. An example includes the two sympatric *Salmo* forms, *S. letnica typicus* and *S. letnica aestivalis*, for which Susnik et al. (2007) could not confirm genetic structuring.

As for endemism above the genus level, molecular studies mainly in molluscs could show that some of the endemic subfamily- and family-level taxa are probably not valid. Accordingly, Radoman's (1973) endemic subfamily Chilopyrgulinae and his endemic family Micropyrgulidae might not be valid and rather at best represent genus-level endemic taxa (e.g., Wilke et al., 2007).

Whereas mollusc diversity and endemism are generally high in Lake Ohrid, there is also a distinct pattern of absence of certain taxonomic groups in Lake Ohrid. One example of such faunal absence in the Ohrid Basin is the snail family Bithyniidae, a group widely distributed in the Balkans (Glöer et al., 2007). It is unknown whether bithyniids never occurred in Lake Ohrid or whether ancestral populations got wiped out when other major gastropod radiations started. It is a well-recognized pattern in caenogastropod evolution that in ancient lakes (but also in other freshwater systems) either cerithioid or rissooid groups radiate (Boss, 1978; Michel, 1994; but also see Haase & Bouchet, 2006). This "exclusion rule" could also apply to the Hydrobiidae and Bithyniidae in the case of Lake Ohrid.

As Radoman (1985) pointed out, so far no author could offer conclusive explanations for the low degree of faunal exchange (immigration and emigration) between Lake Ohrid and neighbouring water systems. He believed in an inherent structure of the biocenosis that prevents such events. The highly adapted fauna of Lake Ohrid may outcompete most invading species but is probably inferior outside native Lake Ohrid—a pattern called eco-insularity (Boss, 1978).

Special features of endemism

Besides distinct spatial and taxonomic patterns seen in Lake Ohrid endemics, they are also characterized by some features that are typical to ancient lakes or even unique to Lake Ohrid.

One of these features, which has already been discussed, are "thalassoid" shell or armature forms in many benthic Lake Ohrid organisms. Such patterns are known from several ancient lake groups and have been extensively studied in, for example, Lake Tanganyika (West et al., 2003; Glaubrecht, 2008).

It is noteworthy that not only the morphospace within many benthic groups in Lake Ohrid is much larger than in comparable non-ancient lake groups (Gorthner, 1992), at the same time, the total size range within ancient lake groups can be comparably higher as well. The Pyrgulinae species flock in Lake Ohrid might again serve as a useful example of this pattern. Non-Ohrid species of the subfamily Pyrgulinae range in size from 5.7 to 9.9 mm. In contrast, the endemic Lake Ohrid species range from 2.0 and 15.4 mm (Radoman, 1983), with the species *Chilopyrgula sturanyi* Brusina, 1896, possibly being the largest representative of the species-rich family Hydrobiidae *s.s.* worldwide. Future studies would be needed to show whether this character displacement is one of the evolutionary strategies in these ancient lake taxa enabling the coexistence of a multitude of species in a comparable small area, and possibly pertaining to trophic specialization.

Finally, some endemic taxa in Lake Ohrid show distinct adaptations to the lake's unique ecological and hydrological settings. They involve, among others, strategies to adapt to living in greater depths.

One example includes the rounded sponge, *Ochridaspongia rotunda*. It lives in deeper zones under relative constant conditions and lost the gemmule stadium in its life cycle. Gemmules are normally typical for freshwater sponges and allow them to survive unfavourable periods such as winter or droughts.

Other examples for adaptations to greater depths can be found in the air-breathing water snail genus *Radix*. *Radix relicta* Polinski, 1929, occurs down to 30 m and has a special physiological adaptation with a completely closed pneumostome, less vascularised tissues, and body surface respiration (Kaiser, 1959), making it the only *Radix* taxon that can stay in greater depths during its whole life cycle.

To summarize the state of the art of endemicity in Lake Ohrid, we point out that many biogeographical and evolutionary aspects of Lake Ohrid endemism are still poorly understood. Problems involve the often restricted geographic focus of relevant studies and a large taxonomic bias, with molluscs and fishes being well studied and other groups largely neglected. Nonetheless, the data available suggest a hierarchical spatial system of endemicity in the Ohrid watershed with distinct taxonomic patterns and special adaptations. So far, preliminary molecular data also confirm the high degree of endemism previously proposed for Lake Ohrid.

Speciation

The field of speciation in ancient lakes has received considerable attention (e.g., Martens, 1997; Rossiter & Kawanabe, 2000) and concepts like sympatric versus allopatric speciation (Martens et al., 1994; Schliewen et al., 1994; Bunje et al., 2007) or adaptive versus non-adaptive radiation (e.g., Schön & Martens, 2004) have been on the mind of evolutionary biologists for decades.

Speciation processes in worldwide ancient lakes are particularly well studied in the African Rift Valley lakes (Wilson et al., 2004; Seehausen, 2006), in Lake Baikal (Sherbakov, 1999; Macdonald et al., 2005), and increasingly in ancient lakes on Sulawesi (von Rintelen et al., 2004, 2007; Herder et al., 2006; Glaubrecht & von Rintelen, 2008). In contrast, there is very little and mostly older explicit information on speciation processes in Lake Ohrid (Hadzisce, 1956; Radoman, 1985).

Given the lack of relevant information, we here intend to summarize some observational evidence for patterns and processes of speciation in Lake Ohrid, mainly from the best studied taxonomic group in Lake Ohrid, the Gastropoda. As most previous discussions of speciation in Lake Ohrid involved a spatial component (i.e., the role of horizontal and vertical barriers within the Ohrid watershed and within the lake proper; e.g., Hadzisce, 1956; Hubendick, 1960; Radoman, 1985), we here follow the concept of Martens et al. (1994) and Schön & Martens (2004), who used geographical patterns in respect to the degree of isolation to discriminate speciation processes in ancient lakes.

Allopatric speciation

This type of speciation involves a strict geographical barrier that prevents or considerably reduces gene flow among subpopulations. According to Schön & Martens (2004), such barriers could be isolated sub-basins during low lake stands or water bodies associated to the main lake (e.g., wetlands, riverine deltas, lagoons, or a string of separate lakes). Whereas isolated sub-basins might play an important role for speciation processes in, for example, Lake Tanganyika (Coulter, 1994) or Lake Baikal (Sitnikova, 2006), the bathymetrical characteristic of Lake Ohrid may not support isolated basins under basically any lake level change (Fig. 2). Thus, allopatric speciation in Lake Ohrid might be best considered in the context of its watershed. Water bodies with a certain degree of isolation in the Ohrid watershed are Lake Ohrid itself, its southern feeder springs, the springs to the northeast and northwest, the effluent Crni Drim River, and Lake Prespa (Fig. 9A).

As outlined in the section "Patterns of endemicity", at least three of these five water bodies support their own endemic species—Lake Ohrid, Lake Prespa, and the southern feeder springs. For the time being, it is not clear whether the Crni Drim River or the northeastern or northwestern springs, indeed, harbour their own endemic species as previously suggested (e.g., Radoman, 1985). The microgastropod species *Pyrgohydrobia jablanicensis* Radoman, 1955, for example, was originally described as endemic species for an artificial lake near the Sum Spring northeast of Lake Ohrid. Genetic analyses (Wilke et al., unpublished data), however, strongly

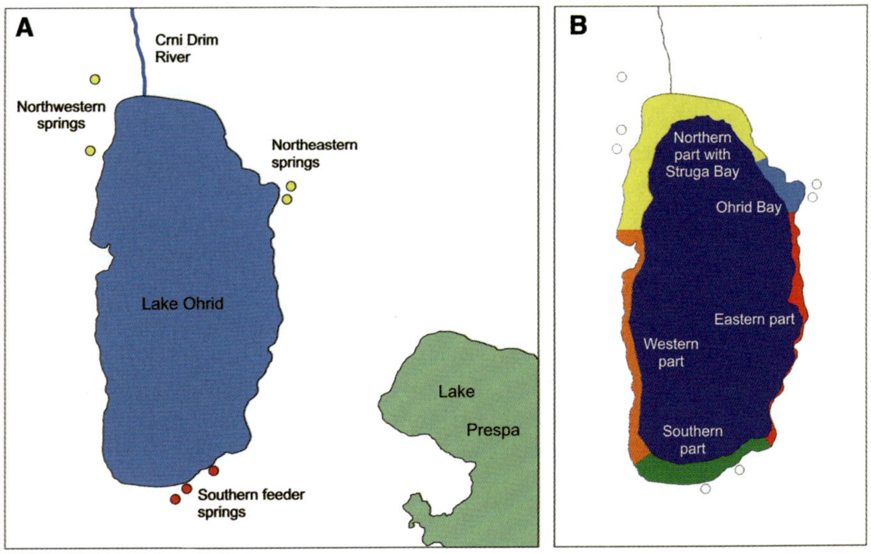

Fig. 9 Horizontal zonation of the Ohrid watershed (**A**) and of the littoral/sublittoral of the lake proper (**B**). The zonation of the lake proper (**B**) shows the approximate spatial extent of five major habitat zones in the littoral/sublittoral of Lake Ohrid (<50 m water depth) that are defined by geological (e.g., type of rocks), hydrological (e.g., amount of sublacustrine springs), and ecological features (e.g., substratum and vegetation). The profundal (>50 m) is shown in dark blue

suggest that this species is conspecific with the lake taxon *P. grochmalickii* (Polinski, 1929).

Whereas allopatric speciation for taxa in Lake Ohrid versus taxa in Lake Prespa might be plausible, the situation of Lake Ohrid versus its springs, in general, is less clear. As mentioned above, the southern feeder springs, though probably rather old, may temporarily have been inundated, thus potentially enabling gene flow between lake and springs. The frequency of inundation is unknown and thus parapatric speciation processes (see below) cannot be excluded. Likewise, some (or even all) populations in the southern feeder springs may have formed small peripheral populations subject to founder effects and subsequent strong genetic drift (i.e., peripatric speciation). On the other hand, even if we assume allopatric speciation for taxa in the southern feeder springs, it remains unclear whether vicariance (i.e., that these springs were originally sublacustrine and later emerged to become geographically isolated from the lake proper) or dispersal from littoral species into the spring (or vice versa) are the historical processes involved (discussed by Radoman, 1985).

If it is already difficult to assess modes of speciation in these relatively distinct water bodies, it is even more difficult to distinguish geographical modes of speciation within the lake. Though many workers suggest "microgeographical" speciation within the lake due to the existence of intralacustrine barriers in both bathymetrical (vertical) and horizontal dimension to be an important mode of speciation in Lake Ohrid (e.g., Hadzisce, 1956; Hubendick, 1960; Radoman, 1985), it remains open whether these processes are allopatric (peripatric) or parapatric. As "microgeographical" speciation (sensu Radoman, 1985) is, like allopatric speciation, often discussed within the conceptional framework of the ecological niche, speciation processes associated with horizontal and vertical barriers within the lake are therefore discussed under parapatric speciation below.

Parapatric speciation

In parapatric speciation, there is divergence of neighbouring populations that are separated but do overlap. Due to insufficient geographical barriers, there is modest gene flow that subsequently leads to reproductive isolation (Coyne & Orr, 2004; also see Glossary).

Parapatric speciation in ancient lakes has received much attention, and Schön & Martens (2004) suggested the following instances for parapatric speciation specifically in ancient lakes:

(A) speciation along a geographical gradient,
(B) speciation along an ecological gradient, and
(C) speciation based on mosaic patterns of distribution (e.g., in the fragmented littoral).

All these instances may also apply to Lake Ohrid with its horizontal (Fig. 9B) and vertical (Fig. 10)

Fig. 10 Transect showing the vertical (bathymetrical) zonation of major habitat types in Lake Ohrid (based on Radoman, 1985). The depth ranges of these zones may vary throughout the lake, and the numbers given here are typical for Lake Ohrids' karstic eastern part. The *G. munda* subspecies pair figured to the left constitutes an example of vertically separated sister taxa

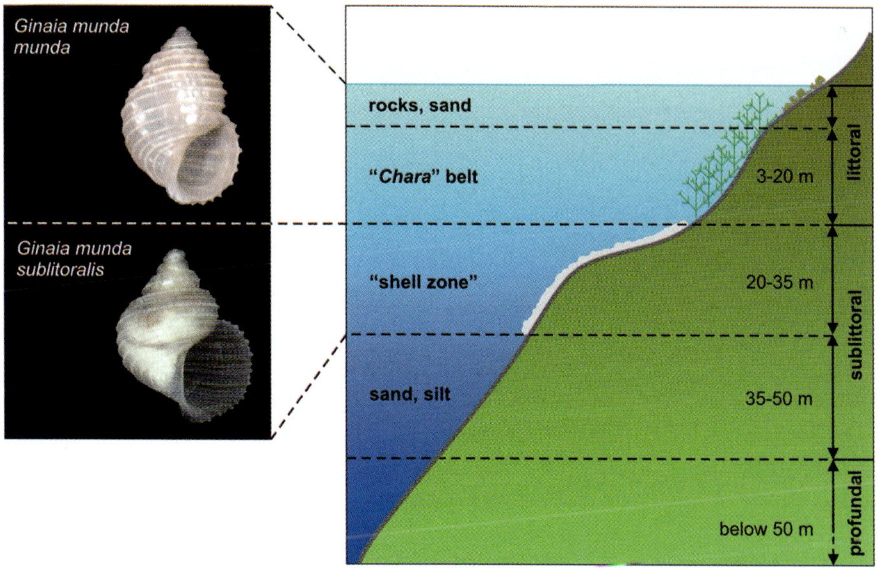

zonation, as well as with its partially highly fragmented habitats, though geographical gradients may be hard to envision for this relatively small lake.

Given the physiographical setting and the specific limnological situation of Lake Ohrid, there are, indeed, different major habitat zones recognizable in the littoral and sublittoral of Lake Ohrid (Fig. 9B; also see Radoman, 1985). These zones include, among others, the steep eastern part of the lake with its numerous karstic springs (Fig. 11D) and the shallow, sandy northern part of the lake. Though raising the possibility that this zonation may lead to "horizontal speciation", Radoman (1985, p. 129) also pointed out that "Far more detailed study … is needed for the reliable judgment of the 'firmness' of the horizontal spatial barriers …".

We do not know of any study that statistically compared the distribution of endemic species among those zones, but there is some empirical evidence that the degree of endemism is particularly high in the relatively species-rich eastern part of the lake, whereas the highest number of widespread species can be found in the shallow northern part of the lake, which today is also the part that, together with the Ohrid Bay, is beginning to show signs of eutrophication (Matzinger et al., 2006b).

If parapatric speciation among these horizontal zones does indeed take place, it remains open whether it is more related to ecological or geographical clines (see above).

The same might apply to the vertical differentiation within the lake. Radoman (1985) described the distinct vertical zonation of Lake Ohrid (Fig. 10), which includes, from top to bottom, a rocky or sandy upper littoral, the *Chara*-belt (Fig. 11F), the shell zone (with both dead and live *Dreissena* specimens), a sandy and silty part without shells, and the profundal. These vertical zones not only differ in substratum, vegetation, and degree of water movement, but also by the degree of sunlight penetration and the range of water temperature variation. Radoman (1985) and Hubendick (1960) particularly stressed the implication of the *Chara*-belt both as habitat for some taxa and as barrier against the expansion of others, whereas Hadzisce (1956) considered bathymetrical barriers, in general, to be important factors for speciation.

Indeed, there are several pairs of sister species that are subject to bathymetrical segregation. One example (Fig. 10) involves the two nominal gastropod subspecies *Ginaia munda munda* (Sturany, 1894) and *G. munda sublitoralis* Radoman, 1978 occurring in the *Chara*-belt and shell zone, respectively.

Other examples for vertical separation of closely related taxa involve *Ancylus* spp. (Albrecht et al., 2006b) in the Ohrid specific "littoral interlithon" (0 to <15 m deep; Fig. 11E) versus the shell zone or *Acroloxus macedonicus* Hadzisce, 1959 from the upper littoral versus *Acroloxus improvisus* Polinski, 1929 in the shell zone (Hubendick, 1960). Moreover, Kilikowska et al. (2006) and Wysocka et al. (2008)

Fig. 11 Horizontal and vertical zonation of Lake Ohrid. (**A**) Ohrid Bay, which today shows signs of beginning eutrophication. (**B**) Shallow southern part of the lake, which is strongly affected by the adjacent feeder springs. (**C**) The site Veli Dab at the karstic eastern part of the lake—one of the lake's major hotspots of endemic biodiversity. (**D**) Littoral part off Veli Dab with sublacustrine springs (indicated by a red arrow). (**E**) Karstic rocks in the littoral zone of Veli Dab ("littoral interlithon"). These rocks are home to a high number of endemic benthic organisms. (**F**) *Chara*-belt off Veli Dab. The belt supposedly constitutes an effective barrier for some benthic taxa

provided data for isopods showing the effectiveness of such barriers in Lake Ohrid, and Sell et al. (2007) found high levels of genetic differentiation in the *Ochridagammarus* complex, possibly corresponding to depth zones.

Several characteristics of the *Chara*-belt may cause vertical isolation of benthic invertebrates. Migration within and through the belt is impeded since these macrophytic algae form a dense net, which would enlarge distance to be taken by individual snails (Fig. 11F). Moreover, interstitial water becomes eventually anoxic or even toxic at the base of the *Chara*-cover (Berger & Schagerl, 2003).

The example of *G. munda* ssp. shown in Fig. 10, however, raises questions about the effectiveness of the vertical geographical barriers in Lake Ohrid. Occurring *within* the *Chara*-belt versus *below* the belt may hint towards parapatric speciation along an ecological cline rather than a geographical cline. The situation may be different for taxa occurring *below* and *above* the *Chara*-belt or the shell zone. Thus, future studies have to show if and for which species the horizontal zonation in Lake Ohrid constitutes a strong geographical barrier supporting allopatric speciation and for which species a geographical/ecological gradient supporting parapatric speciation.

The third possible instance of parapatric speciation in Lake Ohrid involves speciation based on mosaic patterns of distribution, mainly in the fragmented littoral. A prime example would be the "littoral interlithon" along the steep eastern shore of the lake. The area is characterized by sandy bottoms with sublacustrine spring fields and large rocks that harbour a high number of scattered and often very small endemic species. Some of these taxa, such as the gastropod *Gocea ohridana* Hadzisce, 1956 (Fig. 8) and closely related species, have a very patchy distribution with a total range of likely less than 1 km^2. For these small lithophilous taxa, the sandy parts among the rocks may prevent barriers for migration.

Sympatric speciation

In sympatric speciation, evolutionary barriers evolve within a single, initially randomly mating population, which is occupying the same geographic location (see Glossary). Often cited examples of sympatric speciation are found in insects that become dependent on different host plants in the same area (reviewed in Coyne & Orr, 2004). The existence of sympatric speciation as an evolutionary mechanism, however, is still debated. Some workers have argued that many evidences for sympatric speciation are, in fact, examples of micro-allopatric, or parapatric speciation. Indeed, much of the often critical discussion pertaining to sympatric speciation in Lake Ohrid relates to micro-geographical differences (e.g., species on the lower side of a stone versus the upper side; Radoman, 1985).

Today, sympatric speciation is mostly seen in a strictly non-geographical context invoking, for example, disruptive selection, polyploidization, or speciation via hybrid formation (Bolnick & Fitzpatrick, 2007; Coyne, 2007).

In contrast to some other ancient lakes, sympatric speciation is by far the least understood and least studied mode of speciation in Lake Ohrid. Hybridization in Lake Ohrid species has been investigated by Susnik et al. (2006), who found evidence of rare hybridization between *Salmo ohridanus* and the Lake Ohrid brown trout (*Salmo "letnica"*). However, interspecific hybridization per se is a natural process that normally leads to sterile offsprings (postzygotic barrier). Therefore, any discussion about the potential of hybridization for sympatric speciation in Lake Ohrid might be premature in the absence of more detailed information.

Summarizing the knowledge on speciation in Lake Ohrid, we have to emphasize that relevant studies are rare and that all modes suggested are largely based on observational evidence. Nonetheless, given the horizontal and vertical zonation of the Ohrid watershed and the lake proper, the following speciation modes based on degree of isolation are suggested:

(A) Lake Ohrid versus Lake Prespa: allopatric speciation;
(B) Lake Ohrid versus its northeastern and northwestern springs: allopatric or peripatric speciation (but so far it is unclear whether these springs, indeed, harbour their own endemic species);
(C) Lake Ohrid versus its southern feeder springs: allopatric or peripatric speciation;
(D) Horizontal littoral/sublittoral zones within Lake Ohrid: probably mostly parapatric speciation, but allopatric speciation cannot be excluded;
(E) Vertical (bathymetrical) differentiation within Lake Ohrid: parapatric or allopatric speciation;
(F) Fragmented habitats within Lake Ohrid: probably parapatric speciation;
(G) Within habitats of Lake Ohrid: potentially sympatric speciation, but so far no known examples.

Conclusions

Lake Ohrid, with 212 known endemic species and a surface area of 358 km^2, appears to be the most diverse lake in the world, taking surface area into account. The total number of endemic species summarized here has not changed significantly compared to older works, though a significant increase is expected as soon as additional data for other major groups harbouring endemic species (e.g., for diatoms) becomes available.

Preliminary phylogeographical analyses of endemic Lake Ohrid taxa indicate that their respective sister taxa and therefore also the most recent common ancestor likely resided in the area when extant Lake Ohrid came into existence. This, in turn, supports the de novo hypothesis of Lake Ohrid origin, i.e., that Lake Ohrid newly developed in a polje from springs or rivers, probably during the Pliocene. Phylogeographical data also indicate that there is relatively little faunal exchange and overlap between Lake Ohrid and its sister lake, Lake Prespa. However, it seems that a number of endemic Lake Prespa taxa have their most recent common ancestor in Lake Ohrid and not vice versa. This indicates that biodiversity and faunal evolution in Lake Ohrid are little affected by Lake Prespa, despite the fact that Lake Prespa is a major water supplier for Lake Ohrid.

Studies on selected species flocks and scatters (mostly in molluscs) point towards the assumption that only few lineages originally colonized Lake Ohrid from the Balkans and that the majority of endemic species seen today started to evolve within the lake probably during the early Pleistocene.

Within the Ohrid watershed, endemism occurs at different spatial (ranging from species endemic to certain parts of Lake Ohrid to species endemic to the whole watershed) and taxonomic scales (from subspecies to genus level and possibly beyond). Some endemic invertebrate species in Lake Ohrid also show such special features as "thalassoid" forms and enlarged morphospace, as well as special adaptations to life in greater depths.

Modes of speciation in the Ohrid watershed are largely affected by the respective degree of isolation. Evidence points towards both allopatric (peripatric) and parapatric speciation. "Microgeographical" speciation often mentioned for Lake Ohrid might mostly be best explained with parapatric speciation along an ecological cline. Though sympatric speciation within a habitat is conceivable, so far there are no examples.

The importance of the rich endemic Lake Ohrid biota for maintaining biodiversity of the area and for understanding patterns and processes of evolution is unquestioned. Unfortunately, the lake suffers from increasing anthropogenic pressure and a "creeping biodiversity crisis". Some endemic species presumably have already gone extinct, making evolutionary studies increasingly difficult. The comparatively small size of Lake Ohrid and the extremely small range of many endemic species, together with ever increasing human pressure, make its fauna particularly vulnerable. Whereas anthropogenic effects are noticeable all around the lake, adverse human impact on key endemic areas could have grave consequences. There are also indications of invasive species penetrating Lake Ohrid.

Though considerable progress has been made in both Macedonia and Albania to reduce pollution and eutrophication and to protect flagship species, the threats to the biodiversity of Lake Ohrid remain. Part of the problem is a lack of scientific knowledge on the ecology and evolutionary biology of many of the lake's taxa, and we hope that the current review could stimulate further research, which would ultimately help protecting this unique European hot spot of biodiversity.

Acknowledgements We are extremely grateful to our colleagues from the Hydrobiological Institute Ohrid (HBI) for their immense support and hospitality and for encouraging our work. Without the continuous help from our dear friends, particularly from the institutes' director, Goce Kostoski, and the benthos expert, Sasho Trajanovski, our work would have been impossible. Sonja Trajanovska, Biljana Budzakoska, and Bruno Streit took part in some field trips and/or provided valuable logistic support. Dimce Georgiev was always a very experienced scout and Zoran Brdarovski a safe skipper of the HBI research vessel. Both also helped a lot with sampling. Over the years, many graduate students took part in field trips, contributed to sampling efforts, and generated valuable data. They are gratefully acknowledged here. Kirstin Schreiber digitized the maps shown in Fig. 6; Anne-Kathrin Hauswald provided the photographs of *Valvata* spp. shown in Fig. 8; and Andreas Kubicek took the two underwater photographs shown in Fig. 11. We thank Mathias Harzhauser, Oleg Mandic, Andreas Matzinger, Sebastian Krastel, Jerzy Sell, and Bernd Wagner for valuable comments on an earlier version of the manuscript and for the fruitful discussions on Lake Ohrid issues. We are also very grateful to Thomas von Rintelen and Matthias Glaubrecht for reviewing the manuscript. Our work has been greatly inspired by the outstanding scientific contributions of the world authority on the evolution of Lake Ohrid—Sinisa Stankovic (1892–1974).

Glossary

Allopatric speciation	Evolution of reproductive barriers in populations that are prevented from exchanging genes at more than a negligible rate by a geographic barrier (Futyuma, 2005).
Anagenesis	Directional evolutionary change of various characteristics within a lineage (Futyuma, 2005).
Ancient lake	Extant lake that has existed since before the last glacial maximum. There is controversy as to the minimum age of an ancient lake. Whereas Gorthner (1994) proposed an age of at least 0.1 My, other workers suggest a considerably older minimum age.
Ancient sister lake	Ancient sister lakes are ancient lakes lying in close geographic proximity, sharing a related origin and significant time of co-existence, usually having hydrological connection as well as a balanced degree of faunal overlap and distinctness (Albrecht et al., 2008).
Aquifer	A formation, group of formations, or part of a formation that contains sufficient saturated permeable material yielding water to wells and springs (Field, 2002).
Cladogenesis	Branching of a lineage into two or more descendant lines (Futyuma, 2005).
Convergence	Similarities which have arisen independently in two or more organisms that do not share a common ancestry.
Endemic	Pertaining to a taxon that is restricted to the geographic area specified (Lomolino et al., 2006).
Intralacustrine	Within a lake.
Karst	A terrane, generally underlain by limestone or dolomite, in which the topography is chiefly formed by the dissolving of rock and which may be characterized by sinkholes, sinking streams, closed depressions, subterranean drainages, and caves (Field, 2002).
Lake proper	Actual continuous lake body excluding peripheral water bodies or effluents/affluents.
Long-lived lake	A lake of long existence that can be extant (ancient lake) or fossil (palaeolake) (Gorthner, 1994; Martens, 1997).
Palaeolake	A fossil ancient lake that existed in the past when hydrological conditions were different (Cohen, 2003).
Parapatric speciation	Divergence of neighbouring populations, between which there is modest gene flow, with subsequent reproductive isolation (Futyuma, 2005).
Peripatric speciation	Divergence of a small population from a widely distributed ancestral form (Futyuma, 2005).
Polje	A large, flat floored depression in karst limestone (Slavic *polje*: field) whose long axis is developed parallel to major structural trends (Field, 2002).
Ponor	Hole or opening (Slavic *ponor*: swallow hole) in the bottom or side of a depression where a surface stream or lake flows either partially or completely underground into the karst groundwater system (Field, 2002).
Proto-lake	The term "proto-lake" is sometimes used synonymously with the term "palaeolake". Here, the term is used in association with pre-lake settings.
Radiation	Event of rapid cladogeneses.
Relict	Surviving taxon of a previous geological period or of ancient faunas and floras, whose close relatives have disappeared (e.g., Stankovic, 1960).
Short lived lake	Mostly post-glacial lakes that go through a normal cycle of lake succession (see also ancient or long-lived lake).
Speciation	Evolutionary process leading to new species.
Species flock	In ancient lakes, monophyletic group of at least three species that are endemic to the lake but which may include taxa that today occur outside the lake (Greenwood, 1984; Schön & Martens, 2004).

Species scatter	A group of closely related taxa that are characterized by the criteria speciosity and endemicity but which do not represent a monophyletic entity (Hauswald et al., 2008).
Sympatric speciation	Evolution of reproductive barriers within a single, initially randomly mating population (Futyuma, 2005).
Thalassoid	Marine-like (Greek *thalassa*: sea). It refers to taxa that resemble marine taxa without necessarily having a marine origin. In ancient lake research, the term "thalassoid" is most often used to describe morphological or structural features of shells and other external body parts that are usually known from marine taxa (i.e., ribs, spines).
Watershed	Catchment area of a drainage basin.

References

Albrecht, C., R. Schultheiß & D. Lohfink, 2006a. Dramatic decline and loss of mollusc diversity in long-lived lakes in Greece. Tentacle 14: 11–13.

Albrecht, C., R. Schultheiß, T. Kevrekidis, B. Streit & T. Wilke, 2007. Invaders or endemics? Molecular phylogenetics, biogeography and systematics of *Dreissena* in the Balkans. Freshwater Biology 52: 1525–1536.

Albrecht, C., S. Trajanovski, K. Kuhn, B. Streit & T. Wilke, 2006b. Rapid evolution of an ancient lake species flock: Freshwater limpets (Gastropoda: Ancylidae) in the Balkan lake Ohrid. Organisms Diversity & Evolution 6: 294–307.

Albrecht, C., C. Wolff, P. Glöer & T. Wilke, 2008. Concurrent evolution of ancient sister lakes and sister species. The freshwater gastropod genus *Radix* in lakes Ohrid and Prespa. Hydrobiologia (this volume).

Aliaj, S., G. Baldassarre & D. Shkupi, 2001. Quaternary subsidence zones in Albania: Some case studies. Bulletin of Engineering Geology and the Environment 59: 313–318.

Amataj, S., T. Anovski, R. Benischke, R. Eftimi, L. Gourcy, L. Kola, I. Leontiadis, E. Micevski, A. Stamos & J. Zoto, 2007. Tracer methods used to verify the hypothesis of Cvijic about the underground connection between Prespa and Ohrid Lake. Environmental Geology 51: 749–753.

An der Lan, H., 1939. Zur rhabdocoelen Turbellarienfauna des Ochrida-Sees (Balkan). Aus den Sitzungsberichten der Akademie der Wissenschaften in Wien: Mathematisch Naturwissenschaftliche Klasse. Abteilung I 148: 195–254.

Angelovski, P. & J. Sapkarev, 1991. Contribution to the knowledge of taxonomy and ecology of the larvae of chironomid fauna (Diptera: Chironomidae) of Lake Ohrid. Godishen Zbornik Biologija Prirodno-Matematichki Fakultet na Univerzitetot Kiril i Metodij Skopje 43(44): 35–49.

Anovski, T., J. Naumovski, D. Kacurkov & P. Kirkov, 1980. A study of the origin of waters of St. Naum Springs, Lake Ohrid (in Macedonian). Fisika 12: 76–86.

Arndt, W., 1923. Balkanspongilliden. Mit einer Bemerkung über ungarische und chinesische Kolonien von Spongilla carteri Carter. Zoologischer Anzeiger 56: 74–81.

Arndt, W., 1937. *Ochridaspongia rotunda* n.g. n.sp., ein neuer Süßwasserschwamm aus dem Ochridasee. Archiv für Hydrobiologie 31: 636–677.

Arndt, W., 1938. Spongiologische Untersuchungen am Ochridasee. Archiv für Hydrobiologie 34: 48–80.

Augener, H., 1925. Blutegel von der Balkanhalbinsel. Zoologischer Anzeiger 62: 161–211.

Augener, H., 1926. Nachtrag zur Blutegelfauna der Balkanhalbinsel nebst Bemerkungen über Egel aus anderen Gebieten. Zoologischer Anzeiger 68: 229–247.

Augener, H., 1937. Hirudineen aus jugoslavischen Seen. Festschrift zum 60. Geburtstag vom Embrik Strand 2: 403–413.

Baker, R. A., R. Gerecke, V. M. Pesic, S. Stojanovski & N. Hristovski, 2007. A comparative analysis of the water mite fauna of three transboundary lakes in the Balkans. 1st International Symposium for Protection of the Natural Lakes in Republic of Macedonia: 176–182.

Banarescu, P. M., 1991. Zoogeography of Fresh Waters, Vol. 2: Distribution and Dispersal of Freshwater Animals in North America and Eurasia. Aula Verlag, Wiesbaden.

Banarescu, P. M., 2004. Distribution patterns of the aquatic fauna of the Balkan Peninsula. In Griffith, H. I., et al. (eds), Balkan biodiversity. Kluwer Academic Publishers, Dordrecht, Boston, London: 243–260.

Berger, J. & M. Schagerl, 2003. Allelopathic activity of *Chara aspera*. Hydrobiologia 501: 109–115.

Blazencic, J., 2004. The Ohrid Lake—An important centre of Charophyta diversity. 2nd Congress of Ecologist of the Republic of Macedonia with International Participation, Ohrid: 378–383.

Blazencic, J., B. Stevanovic, Z. Blazencic & V. Stevanovic, 2006. Red data list of charophytes in the Balkans. Biodiversity and Conservation 15: 3445–3457.

Bolnick, D. H. & B. M. Fitzpatrick, 2007. Sympatric speciation: Models and empirical evidence. Annual Review of Ecology, Evolution and Systematics 38: 459–487.

Boss, K. J., 1978. On the evolution of gastropods in ancient lakes. In Fretter, V. & J. Peake (eds), Pulmonates, Vol. 2a: Systematics, Evolution and Ecology. Academic Press, London, New York, San Francisco: 385–428.

Bourcart, J., 1922. Les confins albanais administrés par la France. Revue de Geographie 10: 1–307.

Brooks, J. L., 1950. Speciation in ancient lakes. Quarterly Review of Biology 25(30–60): 131–176.

Brusina, S., 1896. Bemerkungen über macedonische Süsswasser-Mollusken. Annalen des kaiserlich-königlichen Naturhistorischen Hofmuseums Wien 3: 365–370.

Bunje, P. M. E., M. Barluenga & A. Meyer, 2007. Sampling genetic diversity in the sympatrically and allopatrically speciating Midas cichlid species complex over a 16 year time series. BMC Evolutionary Biology 7: 25.

Cado, I., 1978. Littoral zone of Lake Ohrid and its microphytic vegetation. Verhandlungen der Internationalen Vereinigung für Limnologie 20: 1085–1089.

Cohen, A. S., 2003. Paleolimnology. History and Evolution of Lake Systems. Oxford University Press, Oxford.

Coulter, G. W., 1994. Lake Tanganyika. In Martens, K., B. Goddeeris & G. W. Coulter (eds), Speciation in Ancient Lakes, Vol. 44. Archiv für Hydrobiologie Special Issues Advances in Limnology: 13–18.

Coulter, G., V. Langenberg, R. Lowe-McConnell, F. Riedel, F. Roest, J. Sarvala & O. Timoshkin, 2006. Survival of Ancient Lake Biodiversity. International Association of Theoretical & Applied Limnology 29: 1178–1181.

Coyne, J. A., 2007. Sympatric speciation. Current Biology 17: R787–R788.

Coyne, J. A. & H. A. Orr, 2004. Speciation. Sinauer Associates, Sunderland, MA.

Crivelli, A. J., G. Catsadorakis, M. Malakou & E. Rosecchi, 1997. Fish and fisheries of the Prespa lakes. Hydrobiologia 351: 107–125.

Cvijic, J., 1906. Fundamentals of geography and geology of Macedonia and Old Serbia. Books I/II, Serbian Academy of Sciences, Special Edition, Beograd (in Serbian).

Cvijic, J., 1908. Grundlagen der Geographie und Geologie von Mazedonien und Altserbien. Petermans Mitteilungen Ergänzungsheft 162: VIII + 392.

Cvijic, J., 1911. Fundamentals of geography and geology of Macedonia and Old Serbia. Book III, Serbian Academy of Sciences, Special Edition, Beograd (in Serbian).

Decraemer, W. & A. Coomans, 1994. A compendium of our knowledge of the free-living nematofauna of ancient lakes. Advances in Limnology 44: 173–181.

Degens, E. T., R. P. von Herzen, H.-K. Wong, W. G. Deuser & H. W. Jannasch, 1973. Lake Kivu: Structure, chemistry and biology of an East African rift lake. International Journal of Earth Sciences 62(1): 245–277.

Denner, M., 2006. Bathymetrie, Sedimentverteilung und Tektonik im Ohridsee in Albanien und Mazedonien. Unpublished Diploma thesis, University of Leipzig.

Dobzhansky, T., F. J. Ayala, G. L. Stebbins & J. V. Valentine, 1977. Evolution. W. H. Freeman and Company, San Francisco.

Doflein, F., 1921. Mazedonien: Erlebnisse und Beobachtungen eines Naturforschers im Gefolge des deutschen Heeres. Fischer, Jena.

Dumurdzanov, N., T. Serofimovski & B. C. Burchfield, 2004. Evolution of the Neogene-Pleistocene Basins of Macedonia. Geological Society of America Digital Map and Chart Series 1. 1–20.

Dumurdzanov, N., T. Serafimovski & B. C. Burchfiel, 2005. Cenozoic tectonics of Macedonia and its relation to the South Balkan extensional regime. Geosphere 1: 1–22.

Eftimi, R., E. Micevski & A. Stamos, 2001. Geological and hydrogeological conditions of the Prespa Region. In Anovski, T. (ed.), Progress in the Study of Prespa Lake Using Nuclear and Related Techniques. Project Report, IAEA Regional Project RER/8/008, ISBN 9989-650-21-7. Skopje, Macedonia: 11–22.

Eftimi, R. & J. Zoto, 1997. Isotope study of the connection of Ohrid and Prespa lakes. Towards Integrated Conservation and Sustainable Development of Transboundary Macro and Micro Prespa Lakes. International Symposium. Korcha, Albania: 32–37.

Ernst Basler and Partners, 1995. Feasibility Study on the Lake Ohrid Conservation Project. The World Bank, Zollikon, Switzerland.

Fauna Europaea Web Service, 2004. Fauna Europaea version 1.1 [available online at http://www.faunaeur.org].

Field, M. S., 2002. A lexicon of cave and karst terminology with special reference to environmental karst hydrology. The National Center for Environmental Assessment—Washington Office, Washington, D.C.

Forrester, R. M., 1991. Pliocene-climate history of the western United States derived from lacustrine ostracods. Quaternary Science Reviews 10: 133–146.

Futyuma, D. J., 2005. Evolution. Sinauer, Sunderland, MA.

Gams, I., 2005. Tectonic impacts on poljes and minor basins (case studies of Dinaric Karst). Acta Carsologica 34: 25–41.

Gardner, J. V., L. A. Mayer & J. E. H. Clarke, 2000. Morphology and processes in Lake Tahoe (California-Nevada). GSA-Bulletin 5: 736–746.

Genner, M. J., P. Nichols, G. Carvalho, R. L. Robinson, P. W. Shaw, A. Smith & G. F. Turner, 2007. Evolution of a cichlid fish in a Lake Malawi satellite lake. Proceedings of the Royal Society B—Biological Sciences 274: 2249–2257.

Georgiev, M. L., 1957. Über die Hydrachniden des Ochridsee-Basins. Folia Balcanica 1: 39–49.

Gilbert, J. J. & S. Hadzisce, 1975. Sexual reproduction in the freshwater sponge, *Ochridospongia rotunda*. Verhandlungen der Internationalen Vereinigung Limnologie 19: 2785–2792.

Gilbert, J. J. & S. Hadzisce, 1977. Life cycle of the freshwater sponge *Ochridaspongia rotunda* Arndt. Archiv für Hydrobiologie 79: 285–318.

Gilbert, J. J. & S. Hadzisce, 1984. Taxonomic notes on the shallow-water endemic sponges of Lake Ohrid, Yugoslavia, with a description of two new species and a redescription of *Spongilla stankovici*. Archiv für Hydrobiologie 99: 331–339.

Gierlowski-Kordesch, E. H. & L. E. Park, 2004. Comparing lake species diversity in the modern and fossil record. Journal of Geology 112: 703–717.

Glaubrecht, M., 2008. Adaptive radiation of thalassoid Cerithioidean gastropods in Lake Tanganyika, East Africa: Morphology and systematization of a paludomid species flock in an ancient lake. Zoosystematics and Evolution 84(1): 71–122.

Glaubrecht, M. & T. von Rintelen, 2008. The species flocks of lacustrine gastropods: *Tylomelania* on Sulawesi as models in speciation and adaptive radiation. Hydrobiologia (this issue).

Glöer, P., C. Albrecht & T. Wilke, 2007. Enigmatic distribution pattern of the Bithyniidae in the Balkan Region (Gastropoda: Rissooidea). Mollusca 25: 13–22.

Goebel, F., 1919. Eine geologische Kartierung im mazedonisch-albanischen Grenzgebiet beiderseits des Ochrida-Sees. Sächsische Akademie der Wissenschaften 71: 257–276.

Golemansky, V. G., 1994. Thecamoebiens (Rhizopoda: Testacea) d'Hygropsammon littoral de trois lacs macedoniens d'origine tectonique: Ohrid, Prespa et Doiran. Archiv für Protistenkunde 144: 309–313.

Gorthner, A., 1992. Bau, Funktion und Evolution komplexer Gastropodenschalen in Langzeit-Seen. Mit einem Beitrag zur Paläobiologie von *Gyraulus „multiformes"* im Steinheimer Becken. Stuttgarter Beiträge zur Naturkunde Serie B 190: 1–173.

Gorthner, A., 1994. What is an ancient lake? In Martens K., B. Goddeeris & G. Coulter (eds), Speciation in Ancient Lakes. Archiv für Hydrobiologie 44: 97–100.

Gould, S. J., 1992. Punctuated equilibrium in fact and theory. In Somit, A. & S. Peterson (eds), The Dynamics of Evolution. Cornell University Press, New York: 54–84.

Greenwood, P. H., 1984. What is a species flock? In Echelle, A. A. & I. Kornfield (eds), Evolution of Fish Species Flocks. Orono Press, Maine: 13–19.

Griffiths, H. I. & M. R. Frogley, 2004. Fossil ostracods and faunistics: Implications for the evolution of regional biodiversity. In Griffiths, H. I. & B. Krystufek (eds), Balkan Biodiversity. Kluwer, Dordrecht: 261–272.

Guseska D., G. Kostoski & O. Tasevska, 2007. Seasonal successions of zooplankton in Lake Ohrid pelagic zone during 1996–2006. 1st Symposium for Protection of Natural Lakes in Republic of Macedonia, Ohrid: 163–170.

Haase, M. & P. Bouchet, 2006. The radiation of hydrobioid gastropods (Caenogastropoda, Rissooidea) in ancient Lake Poso, Sulawesi. Hydrobiologia 556: 17–46.

Hadzisce, S., 1953. Beitrag zur Kenntnis der Spongillidenfauna der großen mazedonischen Seen (Dojran, Prespa und Ohridsee). Recueil des Traveaux. Station Hydrobiologique Ohrid 1: 73–103.

Hadzisce, S., 1956. III. Beitrag zur Kenntnis der Gastropodenfauna des Ohridsees. Beschreibung der bis jetzt unbekannten Schnecken und Beispiele der Speciation bei den Gastropoden des Ohridsees. Recueil des Traveaux, Station Hydrobiologique Ohrid 4: 57–107.

Harzhauser, M. & O. Mandic, 2004. The muddy bottom of Lake Pannon—A challenge for dreissenid settlement (Late Miocene; Bivalvia). Palaeogeography, Palaeoclimatology, Palaeoecology 204: 331–352.

Harzhauser, M. & O. Mandic, 2008. Neogene lake systems of Central and South-Eastern Europe: Faunal diversity, gradients and interrelations. Palaeogeography, Palaeoclimatology, Palaeoecology 260: 417–434.

Haug, E., 1922. Traitéde géologie, I-II. Colin, Paris.

Hauswald, A.-K., C. Albrecht & T. Wilke, 2008. Testing two contrasting evolutionary patterns in ancient lakes: Species flock vs. species scatter in valvatid gastropods of Lake Ohrid. Hydrobiologia (this volume).

Herder, F., A. W. Nolte, J. Pfaender, J. Schwarzer, R. K. Hadiaty & U. K. Schliewen, 2006. Adaptive radiation and hybridization in Wallace's Dreamponds: Evidence from sailfin silversides in the Malili Lakes of Sulawesi. Proceedings of the Royal Society B 273: 2209–2217.

Hinderer, M. & G. Einsele, 2001. The world's large lake basins as denudation-accumulation systems and implications for their lifetimes. Journal of Paleolimnology 26: 355–372.

Hollis, G. E. & A. C. Stevenson, 1997. The physical basis of the Lake Mikri Prespa systems: Geology, climate, hydrology and water quality. Hydrobiologia 351: 1–19.

Hrabe, S., 1931. Die Oligochaeten aus den Seen Ochrida und Prespa (nach dem von Prof. S. Stankovic gesammelten Material bearbeitet). Zoologisches Jahrbuch, Abteilung für Systematik 61: 1–62.

Hubendick, B., 1960. The Ancylidae of Lake Ochrid and their bearing on intralacustrine speciation. Proceedings of the Zoological Society of London 133: 497–529.

Huxley, J., 1958. A logarithmic time scale for the evolution process. Nature 181: 1653–1654.

Itskovich, V. B., S. I. Belikov, S. M. Efremova, Y. Masuda, A. Krasko, H. C. Schroeder & W. E. G. Müller, 2006. Monophyletic origin of freshwater sponges in ancient lakes based on partial structures of COXI gene. Hydrobiologia 568(S): 155–159.

Kaiser, P., 1959. Über die Atmung von *Radix relicta* Pol. aus dem Ohrid-See. Recueil des travaux Ohrid 35: 1–5.

Karaman, S., 1953. Die Aselliden des Ohridsees. Periodicum Biologorum: Glasnik Bioloske Sekcije Ser II/B(4–6): 46–76.

Karaman, M., 1971. Zoogeographische Verhältnisse des Prespa- und Ohridsees. Izdanija 4: 1–21.

Karaman, G. S., 1977. Contribution to the knowledge of the Amphipoda. 77. *Gammarus ochridensis* Schaf. species complex of Ohrid Lake. Odjeljenja Prirodnikh Nauka Crnogorska Akademija Nauka i Umjetnosti Glasnik 2: 49.

Karaman, G. S., 1985. Contribution to the knowledge of the Amphipoda. 151. *Gammarus salemaai*, new species from Lake Ohrid (fam. Gammaridae). Fragmenta Balcanica Musei Macedonici Scientiarum Naturalium 12: 155–168.

Kenk, R., 1978. The Planarians (Turbellaria: Tricladida Paludicola) of Lake Ohrid in Macedonia. Smithsonian Contributions to Zoology 280: 1–56.

Kilikowska, A., A. Wysocka, J. Sell & M. Zietara, 2006. A note on genetic differentiation within the endemic *Proasellus* species group from Lake Ohrid. Biology Letters 43: 21–32.

Korniushin, A. V., 2004. The bivalve mollusc fauna of ancient lakes in the context of the historical biogeography of the Balkan region. In Griffith, H. I., et al. (eds), Balkan Biodiversity—Pattern and Process in the European Hotspot. Kluwer Academic Publishers, Dordrecht, Boston, London: 219–241.

Korniushin, A. V., Z. Krstanovski & G. Kostoski, 2000. Anatomical evidence of close affinity between endemic species of *Pisidium* (Bivalvia, Sphaeriidae) from some ancient lakes, and the widely distributed taxa. Journal of Zoological Systematics and Evolutionary Research 38: 81–86.

Kossmat, F., 1924. Geologie der zentralen Balkanhalbinsel. Borntraeger, Berlin.

Kostoski, G., D. Guseska, Z. Krstanovski, O. Tasevska, L. Velkova-Jordanoska, A. Kilikowska & T. Kretowics, 2007. The phylogenetic position of endemic *Cyclops ochridanus* from relict Lake Ohrid in relation to other non-endemic *Cyclops* species inferred from sequencing of the mtDNA cytochrome oxidase I gene. 1st Symposium for Protection of Natural Lakes in Republic of Macedonia, Ohrid: 72–73.

Kostoski, G., D. Guseska & O. Tasevska, 2004. I. Zooplankton Investigations in Lake Ohrid pelagic zone, II Investigations in Littoral Zone of Lake Ohrid. Lakes Ohrid and Prespa Monitoring Program 3rd Report, Ohrid: 45–56.

Kostoski, G., D. Guseska & O. Tasevska, 2005. Zooplankton research in the Lake Ohrid pelagic region. Limnological Investigations of Ohrid and Prespa Lakes 3/4: 79–87.

Krstanovski, Z., 1994. Taxonomy and ecology of Planaria from lakes Ohrid, Prespa, Dojran and Adjacent Water-bodies. PhD thesis, Sts. Cyril and Methodius University, Skopje (in Macedonian).

Laskarev, V., 1935. Aperçu de la paléogéographie néogène du lac d'Ohrid. Verhandlungen der Internationalen Vereinigung für Limnologie 7: 135–141.

Levkov, Z., S. Krstic, D. Metzeltin & T. Nakov, 2007. Diatoms from Lakes Prespa and Ohrid. Iconographia Diatomologica 16: 1–603.

Lomolino, M. V., B. R. Riddle & J. H. Brown, 2006. Biogeography, 3rd ed. Sinauer, Sunderland, MA.

Lukovic, M., 1935. A contribution to the geological history of the tertiary lakes in the Balkan Peninsula. Verhandlungen der Internationalen Vereinigung für Limnologic 7: 122–134.

Macdonald, K. S., L. Yampolsky & J. E. Duffy, 2005. Molecular and morphological evolution of the amphipod radiation of Lake Baikal. Molecular Phylogenetics and Evolution 35(2): 323–343.

Manconi, R. & R. Pronzato, 2002. Suborder Spongillina subord. nov: Freshwater sponges. In Hooper, J. N. A. & R. W. M. van Soest (eds), Systema Porifera: A Guide to the Classification of Sponges. Kluwer, New York: 921–1019.

Manconi, R. & R. Pronzato, 2008. Global diversity of sponges (Porifera: Spongillina) in freshwater. Hydrobiologia 595: 27–33.

Maric, D., B. Radujkovic & D. Sundic, 2007. Endemism of cyprinid fauna from Prespa-Ohrid-Skadar lakes system. 1st Symposium for Protection of Natural Lakes in Republic of Macedonia. Ohrid: 28.

Martens, K., 1997. Speciation in ancient lakes (review). Trends in Ecology and Evolution 12: 177–182.

Martens, K., G. Coulter & B. Goddeeris, 1994. Speciation in ancient lakes—40 years after Brooks. In Martens, K., B. Goddeeris & G. Coulter (eds), Speciation in Ancient Lakes. Archiv für Hydrobiologie 44: 75–96.

Martin, P., 1994. Lake Baikal. In Martens, K., B. Goddeeris & G. Coulter (eds), Speciation in Ancient Lakes. Archiv für Hydrobiologie Special Issues Advances in Limnology 44: 3–11.

Matzinger, A., M. Jordanoski, E. Veljanoska-Sarafiloska, M. Sturm, B. Müller & A. Wüest, 2006a. Is Lake Prespa jeopardizing the ecosystem of ancient Lake Ohrid? Hydrobiologia 553: 89–109.

Matzinger, A., M. Schmid, E. Veljanoska-Sarafiloska, S. Patceva, D. Guseska, B. Wagner, B. Müller, M. Sturm & A. Wüest, 2007. Eutrophication of ancient Lake Ohrid: Global warming amplifies detrimental effects of increased nutrient inputs. Limnology and Oceanography 52: 338–353.

Matzinger, A., Z. Spirkovski, S. Patceva & A. Wüest, 2006b. Sensitivity of ancient Lake Ohrid to local anthropogenic impacts and global warming. Journal of Great Lakes Research 32: 158–179.

Mayr, E., 1940. Speciation phenomena in birds. American Naturalist 74: 249–278.

Meixner, M. J., C. Lüter, C. Eckert, V. Itskovich, D. Janussen, T. von Rintelen, A. V. Bohne, J. M. Meixner & W. R. Hess, 2007. Phylogenetic analysis of freshwater sponges provide evidence for endemism and radiation in ancient lakes. Molecular Phylogenetics and Evolution 45: 875–886.

Michaloudi, E., M. Zarfdjian & P. S. Economidis, 1997. The zooplankton of Lake Mikri Prespa. Hydrobiologia 351: 77–94.

Michel, E., 1994. Why snails radiate: A review of gastropod evolution in long-lived lakes, both recent and fossil. In Martens, K., B. Goddeeris & G. Coulter (eds), Speciation in Ancient Lakes. Archiv für Hydrobiologie Special Issues Advances in Limnology 44: 285–317.

Miho, A. & H. Lange-Bertalot, 2003. Considerations on biodiversity and trophic state of Lake Ohrid (Albanian part) from a microscopic algae point of view. Journal of Environmental Protection and Ecology 4: 543–549.

Milevski, J., P. Ristevski & D. Tanusevska, 1997. Influence of rainfalls on the oscillations at Prespa Lake. Proceedings of International Symposium Towards Integrated Conservation and Sustainable Development of Transboundary Macro and Micro Prespa Lakes, PPNEA, ILAR Typography, Tirana, Albania: 52–57.

Ministry of Environment and Physical Planning, 2003. Country Study for Biodiversity of the Republic of Macedonia (First National Report), Skopje.

Molloy, D. P., L. Giamberini, L. E. Burlakova, A. Y. Karatayev, J. R. Cryan, S. L. Trajanovski & S. P. Trajanovska, 2008. Investigation of the endosymbionts of *Dreissena stankovici* with morphological and molecular confirmation of host species. In Van der Velde, G., S. Rajagopal & A. Bij de Vaate (eds), Zebra Mussels in Europe. Backhuys Publishers, Leiden (in press).

Müller, P., D. H. Geary & I. Magyar, 1999. The endemic molluscs of the Late Miocene Lake Pannon: Their origin, evolution, and family-level taxonomy. Lethaia 32: 47–60.

Naumoski, T. B., M. Jordanoski & E. Veljanoska Sarafiloska, 2007. Physical and chemical characteristics of Lake Ohrid. 1st International Symposium for Protection of the Natural Lakes in Republic of Macedonia: 103–112.

Noveska, V., T. Naumoski, D. Gjorgonoska & V. Mitic, 1985. Bibliography for Lake Ohrid. Special edition. Ohrid: 1–42.

Nowack, E., 1929. Geologische Übersicht von Albanien. Erläuterung zur geologischen Karte 1: 200000, Innsbruck.

O'Brien, S. J. & E. Mayr, 1991. Bureaucratic mischief: Recognizing endangered species and subspecies. Science 251: 1187–1188.

Papoutsi-Psychoudaki, S. & A. Psychoudakis, 2000. Agricultural externalities and policy for sustainable agriculture in the Greek part of Prespa. Proceedings of the International Symposium "Sustainable development of Prespa region" Otesehvo 2000: 174–185.

Patceva, S., 2005. Comparative analysis of the phytoplankton communities and the trophic state of Lake Ohrid and Prespa. PhD thesis, Faculty of Natural Sciences and Mathematics, Institute of Biology, University St. Cyril and Methodius, Skopje.

Pawlowski, L. K., 1936. Kurze Notiz über die Hirudineen des Ohrid-Sees. Annales Musei Zoologici Polonici 11: 371–378.

Pesic, V., 2003. Water mites (Acari: Hydrachnidia) from Macedonia. Part 2. Stagnant waters. Acta Zoologica Bulgarica 55(2): 29–42.

Petkovski, T. K., 1969. Einige neue und bemerkenswerte Candoninae aus dem Ohridsee und einigen anderen Fundorten in Europa (Crustacea-Ostracoda). Acta Musei Macedonici Scientiarum Naturalium 11: 81–111.

Petkovski, T. & D. Flössner, 1972. Eine neue *Mona*–Art (Crustacea: Cladocera) aus dem Ohridsee. Fragmenta Balcanica 9(10): 97–106.

Petkovski, T. K. & T. Karanovic, 1997. Two new copepod species (Crustacea: Copepoda) from the Ohrid Lake. Annales de Limnologie 33(4): 245–253.

Petkovski, T., B. Scharf & D. Keyser, 2002. New and little known species of the genus *Candona* (Crustacea, Ostracoda) from Macedonia and other Balkan areas. Limnologica 32: 114–130.

Phillips, R. B., M. P. Matsuoka, I. Konon & K. M. Reed, 2000. Phylogenetic analysis of mitochondrial and nuclear sequences supports inclusion of *Acantholingua ohridana* in the genus *Salmo*. Copeia 2: 546–550.

Polinski, V., 1929. Limnološka ispitivanja Balkanskog Poluostrva I. Reliktna fauna gasteropoda Ohridskog Jezera. Glas Srpska Kraljevske Akademije. Belgrade 137: 129–182.

Polinski, W., 1932. Die reliktäre Gastropodenfauna des Ohrida-Sees. Zoologische Jahrbücher Abteilung Systematik 62: 611–666.

Popov, S. V., F. Rögl, A. Y. Rozanov, F. F. Steininger, I. G. Shcherba & M. Kovac, 2004. Lithological-paleogeographic maps of the Paratethys. 10 maps Late Eocene to Pliocene. Courier Forschungsinstitut Senckenberg 250: 1–46.

Popovska, C. & O. Bonacci, 2007. Basic data on the hydrology of Lakes Ohrid and Prespa. Hydrological Processes 21: 658–664.

Por, F. D., 1964. The relict aquatic fauna of the Jordan Valley. Israel Journal of Zoology 12: 47–58.

Raabe, Z., 1965. The parasitic ciliates of gastropods in the Ohrid Lake. Acta Protozoologie 3: 311–320.

Raabe, Z., 1966. The parasitic ciliates of Dreissensia polymorpha and other Bivalvia in the Ohrid Lake. Acta Protozoologie 4: 1–14.

Radoman, P., 1973. New classification of fresh and brackish water Prosobranchia from the Balkans and Asia Minor. Posebna Izdanja, Prirodnjacki Musej u Beogradu 32: 1–30.

Radoman, P., 1983. Hydrobioidea a superfamily of Prosobranchia (Gastropoda), I. Systematics. Serbian Academy of Sciences and Arts, Belgrade: 1–256.

Radoman, P., 1985. Hydrobioidea a superfamily Prosobranchia (Gastropoda), II. Origin, Zoogeography, Evolution in the Balkans and Asia Minor. Monographs Institute of Zoology 1, Beograd.

Rakaj, M., F. Hindak & A. Hindakova, 2000. Phytoplankton diversity of the Albanian part of Lake Shkodra in 1998–1999. Biologia 55: 329–342.

Reed, J. M., B. Krystufek & W. J. Eastwood, 2004. The physical geography of the Balkans and nomenclature of place names. In Griffith, H. I. et al. (eds), Balkan Biodiversity. Kluwer Academic Publishers, Dordrecht, Boston, London: 9–22.

Remy, P., 1934. Sur quelques Hirudinees des Balkans. Annales de la Société Linnéenne de Lyon 77: 17–24.

Remy, P., 1937. Sangsues de Yougoslavie. Bulletin de la Societe Zoologique de France 62: 140–148.

Rensch, B., 1947. Neuere Probleme der Abstammungslehre. Enke, Stuttgart.

Rossiter, A. & H. Kawanabe (eds), 2000. Ancient Lakes: Biodiversity, Ecology and Evolution. Advances in Ecological Research, Vol. 31. Academic Press, London & New York.

Salemaa, H., 1994. Lake Ohrid. In Martens K., B. Goddeeris & G. Coulter (eds), Speciation in Ancient Lakes. Archiv für Hydrobiologie 44: 55–64.

Salemaa, H. & R. Kamaltynov, 1994. The chromosome numbers of endemic Amphipoda and Isopoda – an evolutionary paradox in the lakes Ohrid and Baikal. In, Martens, K., B. Goddeeris & G. Coulter (eds), Speciation in Ancient Lakes. Archiv für Hydrobiologie 44: 247–256.

Sanda, R., 2007. Endemism of cypriniform fishes of Lakes Ohrid and Prespa: A review. 1st Symposium for Protection of Natural Lakes in Republic of Macedonia, Ohrid: 40.

Sanda, R., V. Luskova & J. Vukic, 2005. Notes on the distribution and taxonomic status of *Gobio gobio* from the Moraca River basin (Montenegro). Folia Zoologica 54: 73–80.

Sapkarev, J., 1964. Fauna of Oligochaeta from Lake Ohrid. Godisen zbornik na Prirodno-matematickiot fakultet na Univerzitetot vo Skopje 15: 5–98 (in Serbocroatic).

Sapkarev, J., 1966. Fauna of Oligochaeta from Lake Ohrid. Godisen zbornik na Prirodno-matematickiot fakultet na Univerzitetot vo Skopje 16: 155–177 (in Serbocroatic).

Sapkarev, J. A., 1975. Sistematika i rasprostranjenje pijavica (Hirudinea) Makedonije (Taxonomy and distribution of leeches (Hirudinea) from Macedonia). Biosistematika 1: 87–99.

Sapkarev, J., 1990. New leeches (Hirudinea) of the ancient lake Ohrid. Fragmenta Balcanica 14: 155–162.

Sapkarev, J., Z. Krstanovski, N. Hristovski, G. Kostoski, Z. Spirkovski, T. Talevski, D. Gusevska & S. Stojanovski, 1998. Present knowledge of the living animal world diversity from the ancient Balkan Lake Ohrid. 2nd International Congress of the Biodiversity, Ecology and Conservation of the Balkan Fauna, Ohrid: 86–87.

Schliewen, U. K., D. Tautz & S. Pääbo, 1994. Sympatric speciation suggested by monophyly of crater lake cichlids. Nature 368: 629–632.

Schneider, W., 1943. Freilebende Nematoden aus dem Ochridsee. Bulletin de l'Academie Serbe des Sciences 136: 135–184.

Schön, I. & K. Martens, 2004. Adaptive, pre-adaptive and non-adaptive components of radiations in ancient lakes. A review. Organisms, Diversity and Evolution 4: 137–156.

Schultheiß, R., C. Albrecht, U. Bößneck & T. Wilke, 2008. The neglected side of speciation in ancient lakes: Phylogeography of an inconspicuous mollusk taxon in lakes Ohrid and Prespa. Hydrobiologia (this volume).

Seehausen, O., 2004. Hybridization and adaptive radiation. Trends in Ecology and Evolution 19: 198–207.

Seehausen, O., 2006. African cichlid fish: A model system in adaptive radiation research. Proceedings of the Royal Society of London B 273: 1978–1998.

Sell, J., A. Wysocka, G. Kostoski & S. Trajanovski, 2007. Genetic diversification of the endemic *Ochridagammarus* complex in Lake Ohrid explored with mtDNA sequencing. 1st Symposium for Protection of Natural Lakes in Republic of Macedonia, Ohrid: 72–73.

Sherbakov, D. Y., 1999. Molecular phylogenetic studies on the origin of biodiversity in Lake Baikal. Trends in Ecology and Evolution 14: 92–95.

Sherbakov, D. Y., R. M. Kamaltynov, O. B. Ogarkov & E. Verheyen, 1998. Patterns of evolutionary change in Baikalian gammarids inferred from DNA sequences (Crustacea, Amphipoda). Molecular Phylogenetics and Evolution 10: 160–167.

Shumka, S., 2002. Feeding relations of *Eudiaptomus gracilis* (SARS) and influence on some parameters of their life cycle in Lake Ohrid, Macedonia. Internationale Vereinigung für Theoretische und Angewandte Limnologie Verhandlungen 27(6): 3708–3711.

Siddall, M. E., 2002. Phylogeny of the leech family Erpobdellidae (Hirudinida: Oligochaeta). Invertebrate Systematics 16: 1–6.

Sitnikova, T. Y., 2006. Endemic gastropod distribution in Baikal. Hydrobiologia 568(S): 207–211.

Sket, B., 1968. Zur Kenntnis der Egelfauna (Hirudinea) Jugoslawiens. Academia Scientiarum et Artium Slovenica Cl. IV, 11(4): 127–178.

Sket, B., 1981. *Rhynchobdellid leeches* (Hirudinea, Rhynchobdellae) in the relic Ohrid Lake region. Biološki Vestnik 29: 67–90.

Sket, B., 1989. Intralacustrine speciation in the genus *Dina* (Hirudinea, Erpobdellidae) in Lake Ohrid (Yugoslavia). Hydrobiologia 182: 4–58.

Sket, B. & J. Sapkarev, 1992. Distribution of Hirudinea (Annelida) in the ancient Ohrid Lake region. Archiv für Hydrobiologie 124: 225–237.

Smiljkov, S., 2002. Ecology and dynamics of Chironomidae fauna larva (Diptera: Chironomidae) from Ohrid Lake. Makedonska Akademija na Naukite i Umetnostite Oddelenie za Bioloshki i Meditsinski Nauki Prilozi 22(1–2): 47–56.

Smiljkov, S. & J. Sapkarev, 1999. Taxonomical investigations of chironomid larvae (Diptera Chironomidae) from waters of the lowland region of Ohrid Valley. Godishen Zbornik Biologija Prirodno-Matematichki Fakultet na Univerzitetot "Sv. Kiril i Metodij", Skopje 51(52): 44–49.

Spirkovski, Z., O. Avramovski & A. Kodzoman, 2001. Watershed management in the Lake Ohrid region of Albania and Macedonia, lakes & reservoirs. Research and Management 6: 237–242.

Stankovic, S., 1932. Die Fauna des Ohridsees und ihre Herkunft. Archiv für Hydrobiologie 23: 557–617.

Stankovic, S., 1960. The Balkan Lake Ohrid and Is Living World. Monographiae Biologicae, Vol. IX. Uitgeverij Dr. W. Junk, Den Haag

Starobogatov, Y. I. (ed.), 1994. Freshwater Zebra Mussel, *Dreissena polymorpha* (Pall.) (Bivalvia: Dreissenidae): Systematics, Ecology, and Practical Meaning. Nauka, Moscow (in Russian).

Steindachner, F., 1892. Über einige neue und seltene Fischarten usw. Denkschriften der Akademie der Wissenschaften Wien, Mathematisch-naturwissenschaftliche Klasse 59: 357–384.

Stepien, C. A., C. D. Taylor, I. A. Grigorovich, S. V. Shirman, R. Wei, A. V. Korniushin & K. A. Dabrowska, 2003. DNA and systematic analysis of invasive and native dreissenid mussels: Is *Dreissena bugensis* really *D. rostriformis*? Aquatic Invaders: The Digest of the National Aquatic Nuisance Species Clearinghouse 14(1): 8–16.

Stojanovski, S., N. Hristovski, P. Cakic & M. Hristovski, 2004. Fauna of monogenean trematods—Parasites of some cyprinid fishes from the Lake Ohrid (Macedonia). 1st Symposium of Ecologists of the Republic of Montenegro. With International Participation. Abstract Book. Tivat, Serbia and Montenegro: 55.

Stojanovski, S., N. Hristovski, P. Cakic & M. Hristovski, 2005. Fauna of monogenean trematods—Parasites of some cyprinid fishes from the Lake Ohrid (Macedonia). Natura Montenegrina 4: 61–70.

Sturany, R., 1894. Zur Molluskenfauna der europäischen Türkei. Annalen des kaiserlich königlichen Naturhistorischen Hofmuseums 9: 369–394.

Susnik, S., I. Knizhin, A. Snoj & S. Weiss, 2006. Genetic and morphological characterization of a Lake Ohrid endemic, *Salmo (Acantholingua) ohridanus* with a comparison to sympatric *Salmo trutta*. Journal of Fish Biology 68(Suppl. A): 2–23.

Susnik, S., A. Snoj, I. Wilson, D. Mrdak & S. Weiss, 2007. Historical demography of brown trout (*Salmo trutta*) in the Adriatic drainage including the putative *S. letnica* endemic to lake Ohrid. Molecular Phylogenetics and Evolution 44: 63–76.

Swetlitz, M., 1995. Julian Huxley and the end of evolution. Journal of the History of Biology 28: 181–217.

Sywula, T., Z. Krstanovski, A. Biala, A. Wysocka, A. Kilikowska & J. Sell, 2006. Phylogenetic position of *Dendrocoelum lacteum* (Tricladida) from the Balcan Lake Ohrid evidenced by allozyme data. Biochemical Systematics and Ecology 34: 212–218.

Sywula, T., Z. Krstanovski, O. Tasevska, J. Sell & T. Kretowicz, 2003. Genetic differences among several species of Tricladida from the relict Lake Ohrid as revealed by enzyme electrophoresis. Folia Biologica 51: 105–109.

Talevska, M., 2007. Distribution of the genus *Potamogeton* in Lake Ohrid. 1st Symposium for Protection of Natural Lakes in Republic of Macedonia, Ohrid: 183–188.

Talevska, M. & S. Trajanovska, 2005. Lake Ohrid macrophyte vegetation. Limnological Investigations of Ohrid and Prespa Lakes 3/4: 91–102.

Trajanovska, S., 2005. Some basic investigations in the belt of stoneworths from Lake Ohrid. Chara belt. Limnological investigations of Ohrid and Prespa Lakes 3/4: 97–102.

Trajanovska, S. & M. Talevska, 2007. Depth distribution and colonization abilities of charophytes among the other submerged macrophytic vegetation in Lake Ohrid. 1st Symposium for Protection of Natural Lakes in Republic of Macedonia, Ohrid: 88–89.

Trajanovski, S., T. Wilke, B. Budzakoska-Djoreska & Z. Krstanovski, 2006. Evolution of the "*Dina*" ancient lake species flock (Hirudinae: Erpobdellidae) in lake Ohrid. Berliner Paläobiologische Abhandlungen 9: 63.

van Hinsbergen, D. J. J., C. G. Langereisa & J. E. Meulenkamp, 2005. Revision of the timing, magnitude and distribution of Neogene rotations in the western Aegean region. Tectonophysics 396: 1–34.

Viets, K., 1936. Hydracarinen aus Jugoslavien. (Systematische, ökologische, faunistische und tiergeographische Untersuchungen über die Hydrachnellae und Halacaridae des Süßwassers). Archiv für Hydrobiologie 29: 351–409.

von Rintelen, T., P. Bouchet & M. Glaubrecht, 2007. Ancient lakes as hotspots of diversity: A morphological review of an endemic species flock of *Tylomelania* (Gastropoda: Cerithioidea: Pachychilidae) in the Malili lake system on Sulawesi, Indonesia. Hydrobiologia 592(1): 11–94.

von Rintelen, T., A. B. Wilson, M. Meyer & M. Glaubrecht, 2004. Escalation and trophic specialization drive adaptive radiation of freshwater gastropods in ancient lakes on Sulawesi, Indonesia. Proceedings of the Royal Society of London B 271: 2842–2850.

Wagner, B., A. F. Lotter, N. Nowaczyk, J. M. Reed, A. Schwalb, R. Sulpizio, V. Valsecchi, M. Wessels & G. Zanchetta, 2008a. A 40,000-year record of environmental change from ancient Lake Ohrid (Albania and Macedonia). Journal of Paleolimnology. doi: 10.1007/s10933-008-9234-2.

Wagner, B., K. Reicherter, G. Daut, M. Wessels, A. Matzinger, A. Schwalb, Z. Spirkovski & M. Sanxhaku, 2008b. The potential of Lake Ohrid for long-term palaeoenvironmental reconstructions. Palaeogeogeography, Palaeoclimatology, Palaeoecology 259: 341–356.

Watzin, M. C., V. Puka & T. B. Naumoski (eds), 2002. Lake Ohrid and its watershed, state of the environment report. Lake Ohrid Conservation Project. Tirana, Albania and Ohrid, Macedonia.

West, K., E. Michel, J. Todd, D. Brown & J. Clabough, 2003. The Gastropods of Lake Tanganyika: Diagnostic key, classification and notes on the fauna. International Association of Theoretical and Applied Limnology, Occasional Publication 2: 1–132.

Wilke, T. & C. Albrecht, 2007. How to stop the creeping biodiversity crisis in Lake Ohrid? Suggestions for sustainable conservation strategies of biodiversity hotspots. Abstracts of the Symposium for protection of the natural lakes in Republic of Macedonia: 44–45.

Wilke, T., C. Albrecht, V. V. Anistratenko, S. K. Sahin & Z. Yildirim, 2007. Testing biogeographical hypotheses in space and time: Faunal relationships of the putative ancient lake Egirdir in Asia Minor. Journal of Biogeography 34: 1807–1821.

Wilke, T. & M. Pfenninger, 2002. Separating historic events from recurrent processes in cryptic species: Phylogeography of mud snails (*Hydrobia* spp.). Molecular Ecology 11: 1439–1451.

Williamson, P. G., 1981. Paleontological documentation of speciation of Cenozoic molluscs from the Turkana Basin. Nature 293: 437–443.

Wilson, A. B., M. Glaubrecht & A. Meyer, 2004. Ancient lakes as evolutionary reservoirs: Evidence from the thalassoid gastropods of Lake Tanganyika. Proceedings of the Royal Society B, Biological Sciences 271: 529–536.

Wysocka, A., G. Kostoski, A. Kilikowska, B. Wrobel & J. Sell, 2008. *Proasellus* (Crustacea, Isopoda) species group endemic to the Balkan Lake Ohrid: A case of ecological diversification? Fundamental and Applied Limnology (in press).

Zacharias, I., I. Bertachas, N. Skoulikidis & T. Koussouris, 2002. Greek lakes: Limnological overview. Lakes & Reservoirs: Research and Management 7: 55–62.

Hydrobiologia (2008) 615:141–156
DOI 10.1007/s10750-008-9553-3

SPECIATION IN ANCIENT LAKES

The neglected side of speciation in ancient lakes: phylogeography of an inconspicuous mollusc taxon in lakes Ohrid and Prespa

Roland Schultheiß · Christian Albrecht · Ulrich Bößneck · Thomas Wilke

Published online: 10 September 2008

Abstract The morphologically remarkable endemic fauna within ancient lakes has received much attention in the literature. More inconspicuous taxa, however, often lack detailed molecular and morphometrical examination, although their proportion of the endemic fauna of an ancient lake must not be underestimated. Consequently, a better understanding of evolutionary patterns and processes within these lakes requires more knowledge about the often-neglected inconspicuous taxa. In the present study, we focus on the notoriously cryptic pea clam genus *Pisidium* (Bivalvia: Sphaeriidae). Though the genus is widely distributed, most endemic species are reported only from ancient lakes, including the European ancient sister lake system of Ohrid and Prespa on the Balkan Peninsula. Here we test for the first time hypotheses on the evolution of the endemic pea clams in this European biodiversity hotspot by molecular means. Combining a broad 16S phylogeny (comprising most European pea clam species), network analyses and morphometrical analyses, we found interesting biogeographical patterns and provide evidence for cryptic species in both lakes. Furthermore, we confirmed the proposed sister-species relationship of the endemics *P. edlaueri* in Lake Ohrid and *P. maasseni* in Lake Prespa, and we suggest scenarios of the endemic pea clam evolution within both lakes. The patterns of speciation found in the genus *Pisidium* are compared to patterns in morphologically distinct molluscan groups in lakes Ohrid und Prespa.

Keywords Ancient lakes · *Pisidium* · *Pisidium edlaueri* · *Pisidium maasseni* · Phylogenetics · Morphometrics · Cryptic speciation

Guest editors: T. Wilke, R. Väinölä & F. Riedel
Patterns and Processes of Speciation in Ancient Lakes: Proceedings of the Fourth Symposium on Speciation in Ancient Lakes, Berlin, Germany, September 4–8, 2006

Electronic supplementary material The online version of this article (doi:10.1007/s10750-008-9553-3) contains supplementary material, which is available to authorized users.

R. Schultheiß (✉) · C. Albrecht · T. Wilke
Department of Animal Ecology & Systematics, Justus Liebig University Giessen, Heinrich-Buff-Ring 26-32 IFZ, 35392 Giessen, Germany
e-mail: Roland.Schultheiss@allzool.bio.uni-giessen.de

U. Bößneck
Schillerstr. 17, 99198 Erfurt-Vieselbach, Germany

Introduction

Ancient lakes as evolutionary hotspots are well known for their high degree of endemism (Brooks, 1950; Gorthner, 1994; Martens et al., 1994; Schön & Martens, 2004). Studying endemic species and species flocks within these old lakes have proven to be a powerful approach to reveal evolutionary pattern and processes in, for example, molluscs (Michel, 1994;

Rintelen & Glaubrecht, 2006). In particular, morphologically conspicuous taxa with a characteristic thalassoid shell shape have received much attention in the literature. Examples include the fascinating radiation of the genus *Tylomelania* Sarasin, 1897, in Sulawesi (Rintelen & Glaubrecht, 2005) and the evolution of the gastropod fauna of Lake Tanganyika, Africa (Michel, 2004; Wilson et al., 2004). In contrast to these examples of rapid morphological and ecological differentiation, many inconspicuous taxa still lack detailed molecular and morphological study within a biogeographical context (Korniushin, 2004). This is unfortunate because in many ancient lakes it is not the highly appraised morphologically diverse species flocks that primarily contribute to the often high degrees of biodiversity, but cryptic groups with little morphological differentiation (for example, the most species-rich family in Lake Ohrid—the Hydrobiidae; Radoman, 1983). As a consequence, we know relatively little about patterns and processes of speciation and radiation in cryptic ancient lake taxa.

A case in point is the notorious pea clam genus *Pisidium* C. Pfeiffer, 1821 (Bivalvia: Sphaeriidae). In contrast to many other molluscan taxa, particularly gastropod groups, pea clams are not known to form ancient lake species flocks (Korniushin, 2004; Rintelen & Glaubrecht, 2005). Worldwide ancient lakes appear to be one of the very few freshwater systems, where endemic *Pisidium* species occur (Kuiper, 1963). However, there is not a single lake worldwide with more than one endemic *Pisidium* species reported (Korniushin, 2004). A possible reason is the enormous passive dispersal capacity of pea clams. Hence, geographic isolation is thought to be of little meaning in speciation processes within this genus (Kuiper, 1963).

In this respect, the European ancient sister lake system of Ohrid (Macedonia and Albania) and Prespa (Macedonia, Albania, and Greece) (Fig. 2) harbors a remarkable degree of endemic pea clam taxa (for explanation of the term "sister lake system" see Albrecht et al., 2008; for details on physicogeographical settings of the two lakes see Albrecht & Wilke, 2008). One endemic *Pisidium* species is reported from each lake: *P. edlaueri* Kuiper, 1960, from Lake Ohrid and *P. maasseni* Kuiper, 1987, from Lake Prespa. Additionally, an endemic subspecies is known from Lake Ohrid, *P. subtruncatum recalvum* Kuiper, 1960 (Kuiper, 1960, 1987).

Previous studies established a biogeographical scenario for the origin of the endemic bivalve faunas of the ancient lakes Ohrid, Baikal (Russia) and Biwa (Japan). In particular, a common ancestor was suggested for all of them on the basis of shell characters (Stanković, 1960; Slugina & Starobogatov, 1999). Thus, the respective recent endemic species would be relicts of a widely distributed ancestor in the Tertiary. Korniushin et al. (2000), however, rejected this hypothesis on the basis of an anatomical analysis that indicates a convergent evolution of shell characters. Although he assumed similar processes for the two endemics of Lake Ohrid and Lake Prespa, *P. edlaueri* and *P. maasseni*, he also found anatomical evidence that supports the suggestion of Kuiper (1960, 1987), who regarded both taxa as vicariants of *P. nitidum* Jenyns, 1832, and, therefore, as sister species. Until now, these hypotheses have not been tested with molecular data.

It is not only this rare case of multiple endemism that calls for a detailed molecular-based examination of the pea clam fauna within these two lakes, but also a remarkable biogeographical feature: within an European context, the *Pisidium* species diversity of lakes Ohrid and Prespa is outstanding. While the temperate zone on the continent (i.e., Central Europe) has the highest species richness (17 species), the Mediterranean zone (where lakes Ohrid and Prespa are situated in) is generally poor in associated *Pisidium* species (Kuiper, 1960; Illies, 1978; Dyduch-Falniowska, 1989). Very often only *P. casertanum* (Poli, 1791) can be found there (Kuiper, 1963). This decline in species numbers from the temperate zone southwards is also reported by Angelov (1971). Although this might partly be due to fragmentary sampling, the observation is believed to have its basis in differences of climate and landscape (Angelov, 1971). In this context, lakes Ohrid and Prespa are water bodies with a remarkable *Pisidium* richness comprising altogether nine pea clam species (Kuiper, 1960 Angelov, 1971; Korniushin, 2004) (Table 1, Figs. 1 and 6).

In addition, a further biogeographical feature concerning the pea clam fauna of the two European ancient lakes is noteworthy: their partly disjunct distributions. Kuiper (1960) mentions two cases: first, *P. tenuilineatum* Stelfox, 1918, with the closest records to the population of Lake Ohrid being reported from Zeljesnica (Herzegovina) and Ribarić (Dalmatia)—both some 400 km north of Ohrid.

Table 1 Previous reports on *Pisidium* species composition within lakes Ohrid and Prespa (endemic taxa are marked in bold)

	P. amnicum	*P. casertanum* Type I	*P. casertanum* Type II	*P.*cf. *casertanum*	***P. edlaueri***	*P. henslowanum*	***P. maasseni***	*P. moitessierianum*	*P. personatum*	*P. subtruncatum*	***P. s. recalvum***	*P. tenuilineatum*	*P. parenzani*
Gambetta (1930)													O
Jaeckel et al. (1957)			O							O			O
Kuiper (1960)	O		O		O	O		O	O	O	O	O	X
Kuiper (1987)						P	P	P	P	P			X
Reischütz & Stummer (1989)			P				P			P			X
Present study	O	O	O P	O P	O	O P	P		O P	O P	O	O	X

According to Kuiper (1960), the report of *P. parenzani* was based on misidentification. Therefore, the column is marked with X. O indicates a record from Lake Ohrid (including watershed); P indicates a record from Lake Prespa. For explanation of the Types I and II of *P. casertanum* as well as *P.* cf. *casertanum* see text

A more recent work reports this species from Istria, Croatia (Bößneck, 2002). As a second species, Kuiper (1960) mentions *P. moitessierianum* Paladilhe, 1866, which closest record to Lake Ohrid is from Krain (Slovenia), about 700 km northwest of the lake. Kuiper (1960) furthermore states that *P. henslowanum* (Sheppard, 1823)—which is also reported from Lake Ohrid—is typically not found so far south in Europe.

Only a few attempts were made to analyze the remarkable pea clam fauna of these two European lakes: in addition to Kuiper's zoogeographical annotations in 1960, only Korniushin et al. (2000) and Korniushin (2004) worked in any detail on the phylogeography of *Pisidium* in Ohrid and Prespa. Their analyses, however, were based on morphological and anatomical data only. It should be noted that recent studies on North American sphaeriids revealed significant conflicts between morphological and molecular phylogenetic analyses (e.g., Lee, 2004). For instance, there is an ongoing debate on the monophyly of the genus. While morphological and anatomical phylogenies continue to support the traditional view of a monophyletic genus *Pisidium* with the genus *Sphaerium* Scopoli, 1777, as its sister taxon (Dreher-Mansur & Meier Brook, 1992; Korniushin & Hackenberg, 2000; Korniushin & Glaubrecht, 2002), molecular-based phylogenies suggest that *Pisidium* is a paraphyletic taxon (Cooley & Ó Foighil, 2000; Lee & Ó Foighil, 2002). Furthermore, the monophyly on the species level is doubted in at least one case: a paraphyletic pattern was discovered for the putative cosmopolitan species *P. casertanum* by molecular means. Hence, the suggestion was made that this taxon is not a single species, but represents a paraphyletic species complex (Lee & Ó Foighil, 2003). A conclusion of these findings is that the occurrence of enigmatic *Pisidium* species within lakes Ohrid and Prespa might be masked by phenotypic plasticity. Thus, molecular analyses are necessary to reveal such possible cases of cryptic speciation (Guralnick, 2005).

Considering the abovementioned conflicts and the reliance of previous analyses of the pea clam fauna in lakes Ohrid and Prespa on morphological and anatomical data, it appears to be essential to expand this work by analyses of DNA sequence data, a need already stressed by Korniushin (2004). The present study aims at closing this gap. For the first time, all pea clam species (except *P. moitessierianum* which

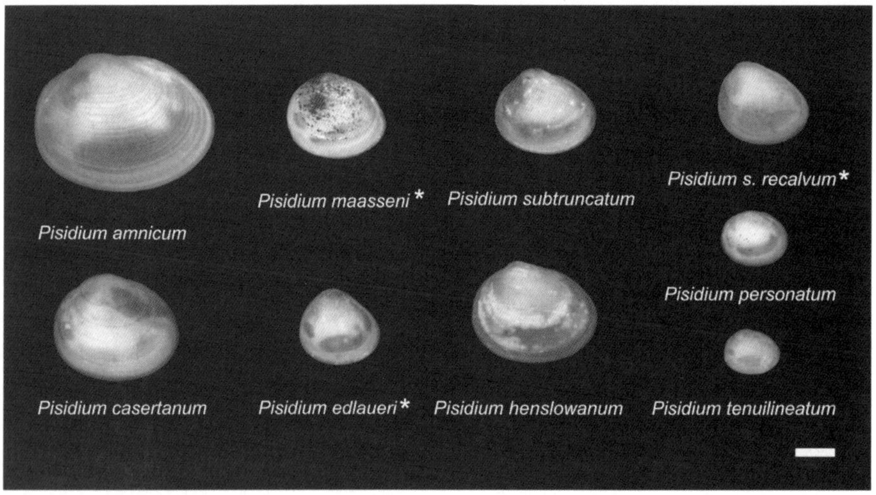

Fig. 1 *Pisidium* species of the Balkan lakes Ohrid and Prespa (with exception of *P. moitessierianum*; see discussion for details). Endemic taxa of the sister lake system are marked with *. Besides the here shown individual of *P. casertanum* Type II, the present study revealed within the study area a cryptic type that morphologically resembles *P. casertanum* but is genetically very distinct (see text for explanation). Scale bar is 1 mm

is, however, included as GenBank sequence in the present study) of lakes Ohrid and Prespa are analyzed genetically and embedded in a broad phylogeny comprising most European as well as several Asian and North American *Pisidium* species. Furthermore, we also examined these species using morphometrical methods and analyzed them in a combined approach. Therewith, we use a cryptic ancient lake group with little morphological differentiation to:

(1) test previous systematic and phylogenetic hypotheses for the pea clam fauna of lakes Ohrid and Prespa on the basis of molecular analyses with focus on (a) the proposed sister relationship of the two endemics *P. edlaueri* and *P. maasseni* and (b) the possible occurrence of cryptic species within the two ancient lakes;
(2) compare morphometrical and molecular analyses of selected *Pisidium* specimens and, therewith, test for significant phylogenetic signals in our morphometrical datasets;
(3) embed the new data into a phylogenetic context of the Pan-European pea clam fauna and discuss their biogeographical implications with respect to speciation patterns and processes in ancient lakes; and
(4) compare patterns of speciation found in the cryptic genus *Pisidium* with evolutionary patterns in morphologically distinct molluscan groups in lakes Ohrid und Prespa.

Materials and methods

Sampling design

The 78 specimens analyzed in this study were mainly collected in 2003, 2004, and 2005 within and in the surroundings of lakes Ohrid and Prespa (Fig. 2). Samples were collected from the littoral using sieves and forceps and in greater depths of up to 65 m using a dragnet. All specimens were immediately transferred into 80% ethanol. For comparison, additional specimens were collected in several European and Asian countries (see Supplementary material—Appendix I). In total, our sampling design comprises all species reported for Lake Ohrid and Lake Prespa (with exception of *P. moitessierianum*; see Discussion) as well as most European pea clam species (with exception of *P. conventus* Clessin, 1877; *P. dilatatum* Westerlund, 1897; *P. hinzi* Kuiper, 1975; *P. pulchellum* Jenyns, 1832; and *P. sogdianum* Izzatullaev & Starobogatov, 1986). DNA, shell and soft tissue

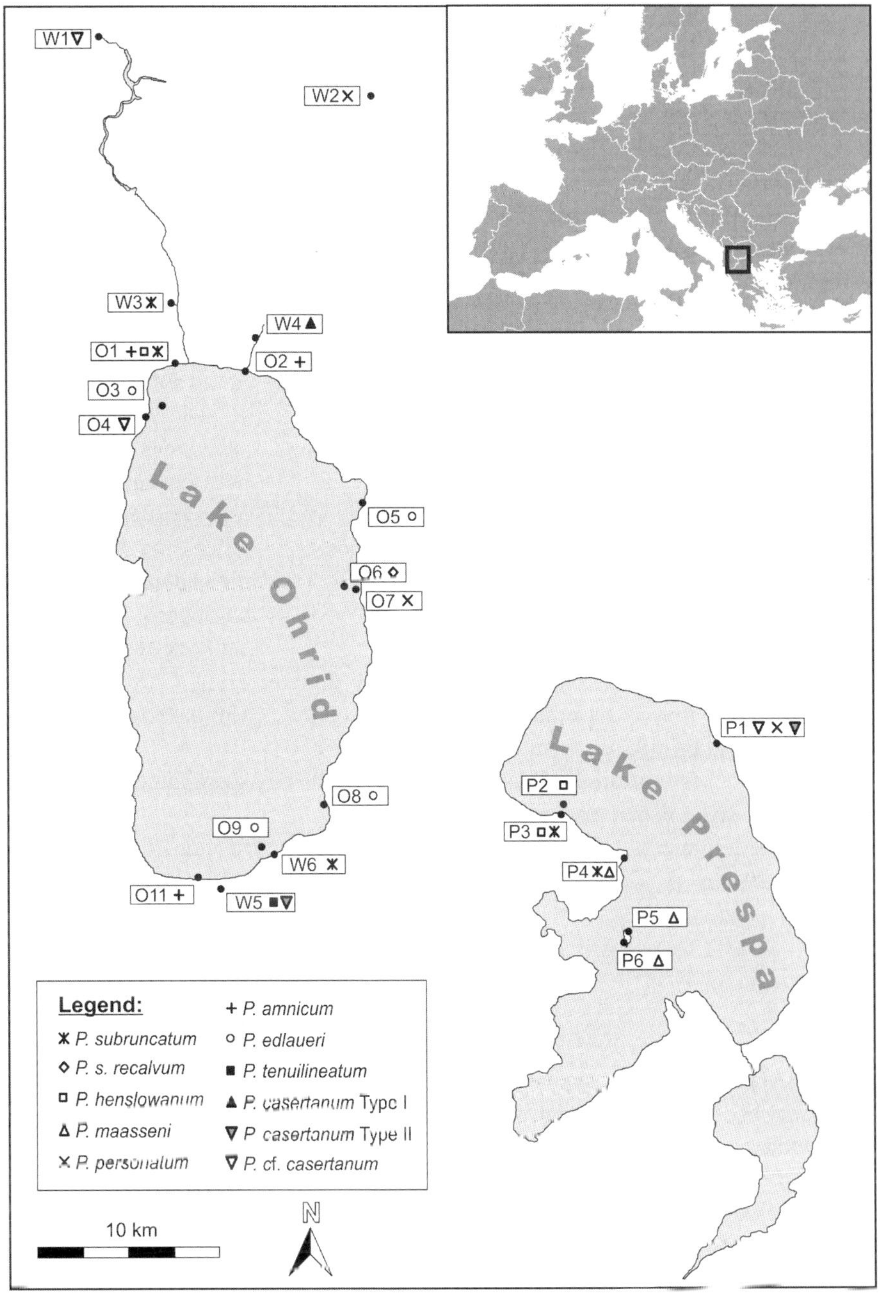

Fig. 2 Locality map showing ancient lakes Ohrid and Prespa on the Balkan Peninsula. Black dots mark 22 sampling points where specimens of *Pisidium* spp. were collected for the present study. Locality details are given in Supplementary material—Appendix I

vouchers are kept in the collection of the Department of Animal Ecology & Systematics at Justus Liebig University, Giessen, Germany.

DNA isolation and sequencing

DNA from individual pea clams was isolated using the CTAB protocol of Wilke et al. (2006). For amplification of the mitochondrial large ribosomal subunit (LSU rRNA or 16S), we used the primer-pair 16Sar-L and 16Sbr-H of Palumbi et al. (1991). A standard PCR mix included per sample a maximum amount of 10 ng genomic DNA (at a maximum volume of 1.5 μl), 1 μl reaction buffer, 0.7 μl dNTP, 0.2 μl of 100 mM $MgSO_4$, 0.7 μl forward primer, 0.7 μl reverse primer (both at a concentration of

10 μM), 6 μl high purity H_2O, 0.1 μl of 0.5 M TMAC, 0.6 μl BSA (10 mg/ml) and 0.2 μl *Taq* polymerase. The polymerase chain reaction conditions were as follows. An initial step of 95°C for 5 min was followed by 25 cycles of 30 s at 95°C, 30 s at 52°C and 30 s at 72°C. Reaction was terminated after a final step at 72°C for 4 min. PCR results were checked on a 1% agarose gel. Sequences (forward and reverse) were determined using the LI-COR DNA sequencer Long ReadIR 4200 (Lincoln, NE) and the Thermo Sequenase Fluorescent Labeled Primer Cycle Sequencing kit (Amersham Pharmacia Biotech, Piscataway, NJ). The alignment of the 16S fragment was done by using the default settings in ClustalW (version 1.4; Thompson et al., 1994) as implemented in BioEdit 7.0.5.3 (Hall, 1999). It showed three ambiguously aligned regions at positions 64 (seven base pairs), 275 (eight base pairs) and 333 (six base pairs). These regions were removed prior to subsequent analyses.

The first 35 base pairs behind the 3′ end of the forward and the first 31 base pairs behind the 3′ end of the reverse primer were difficult to read. We therefore cut off uniformly these bases of each sequence, leaving a 424 base pair long, completely overlapping fragment for the 16S gene. All sequences are available from GenBank (see Supplementary material—Appendix I).

Phylogenetic analyses

Only unique haplotypes were considered in subsequent analyses (except network analyses). Sequences from additional taxa were taken from GenBank (see Supplementary material—Appendix I). As outgroups served *Eupera cubensis* (Prime, 1865) and *Eupera platensis* Doello Juardo, 1921. In order to infer the 16S phylogeny, we used Bayesian inference as implemented in the software package MrBayes 3.0b3 (Huelsenbeck & Ronquist, 2001). Prior to the analysis, the computer program Modeltest 3.7 (Posada & Crandall, 1998) was used in order to find the optimal model of DNA substitution based on Akaike's information criterion as suggested by Posada & Buckley (2004). Hereby, the TVM+I+Γ model was selected with base frequencies of A = 0.3334, C = 0.0997, G = 0.2163, and T = 0.3506; substitution rates were from A to C 0.0000, A to G 5.1425, A to T 0.8553, C to G 0.3234, C to T 5.1425 and from G to T 1.0000. The proportion of invariable sites was 0.5226 and a gamma distribution shape parameter of $\alpha = 0.6438$ was chosen. Using these parameters in MrBayes, a 50% majority rule consensus tree was constructed from all 100,000 sampled trees (i.e., 1,000,000 generations) with the first 5,000 trees (50,000 generations) ignored as burn in.

In order to analyze the relationships between the two endemics *P. edlaueri* and *P. maasseni* and their proposed ancestor *P. nitidum*, we performed a median-joining network estimation as implemented in the software package NETWORK 4.2.0.1 (Bandelt et al., 1999). We included not only all sequenced specimens of the abovementioned three species in the network analysis, but also sequences of *P. pseudosphaerium* Favre, 1927, and *P. milium* Held, 1836. The latter ones were included since their relationship to the endemic species could not be resolved in the 16S phylogeny. All closely related sequences used for network estimation were unambiguously aligned using ClustalW resulting in a fragment length of 439 base pairs.

Morphometrical analyses

In total, 61 specimens were analyzed morphometrically. We acquired five shell parameters using the measurement environment as implemented in the software package analySIS FIVE (Soft Imaging Systems GmbH, Münster, Germany): shell length, width, height at the point of largest width, total height, and hinge height. In cases where the hinge was the highest point of the shell, the latter two values were equal (see Supplementary material—Appendix II).

Additionally to these parameters, the shape of the shell outline (rear and right side view) was examined using an Elliptic Fourier Analysis. Shell outlines of 100 equally distributed points were generated using the program tpsDig 1.40 (Rohlf, 2001). Elliptic Fourier Analysis was carried out using the software PAST (Hammer et al., 2001), resulting in a dataset of 40 variables per view (10 harmonics with sine x*n*, cosine x*n*, sine y*n* and cosine y*n*—whereby *n* is the number of the respective harmonic). This dataset was set to be invariant to starting point and size, leaving 39 variables per view. The combination of rear and side view datasets as well as the shell parameters yielded in two datasets with 78 and five variables, respectively. These datasets were processed twofold: first, we conducted Principle Component Analyses (PCA) for every dataset individually (shell parameters, rear and side view) and in combination using PAST. Second, we generated a distance matrix of the outline dataset based on Euclidean distances as implemented in PAST.

Correlation of genetic and morphometrical datasets

A pairwise distance matrix of the corresponding 16S dataset (see previous section; see Supplementary material—Appendix II) was created a priori with the program MEGA 3.1 (Kumar et al., 2004) based on the default Kimura-2-Parameter model. The setup of the 16S distance matrix dataset included all morphometrically examined specimens. This matrix was analyzed together with the morphometrical distance matrix (see previous section) using a Mantel test as implemented in the software package TFPGA 1.3 (Miller, 1997).

Results

Phylogenetic analyses

The Bayesian analysis revealed a tree with six, in part, well-supported major clades (Fig. 3), five of which correspond to subgeneric groups assigned by Lee & Ó Foighil (2003): *Afropisidium*, *Cyclocalyx*, *Odhneripisidium*, *Pisidium* s. str., *Sphaerium*, and *Musculium*.

The most basal cluster contains two species of the subgenus *Afropisidium* Kuiper, 1962: *P. clarkeanum* G. & H. Nevill, 1871, and *P. nevillianum* Theobald, 1876, with a Bayesian Posterior Probability (BPP) of 0.95. *Pisidium sterkianum* Pilsbry, 1897, supposedly belongs to this subgenus as well; however, it does not cluster within clade *Afropisidium* I, and the respective clade is therefore referred to as *Afropisidium* II. The subgenus *Pisidium* s. str. with the species *P. amnicum* (Müller, 1774) and *P. dubium* (Say, 1817) is supported by a BPP of 1.00 and forms the sister taxon to a clade that contains species of the nominal genera *Musculium* Link, 1807, and *Sphaerium* (BPP 0.91). Due to the position of this clade within the 16S phylogeny, the genus *Pisidium* appears to be paraphyletic. The largest clade within this analysis is formed by species of the subgenus *Cyclocalyx* Dall, 1903 (BPP 0.92). A fifth clade contains mainly members of the subgenus *Odhneripisidium* Kuiper, 1962 (with exception of *P. moitessierianum*).

All analyzed *Pisidium* species of lakes Ohrid and Prespa cluster within the clade *Cyclocalyx,* except for *P. amnicum* (*Pisidium* s. str.) and *P. tenuilineatum* (*Odhneripisidium*). Within this subgenus, the two endemic species *P. edlaueri* and *P. maasseni* are closely related. However, neither species is monophyletic in our analysis. The endemic subspecies *P. subtruncatum recalvum* forms a subclade within *P. subtruncatum* Malm, 1855 (BPP 1.00). Some specimens initially determined as *P. casertanum*, which have been collected from both lakes and their surroundings, are found in a sister relationship to *P. subtruncatum*. Since this group (henceforth referred to as *P.* cf. *casertanum*; see below for morphological description) is only distantly related to other members of *P. casertanum* and because it does not cluster with any other species described for both ancient lakes, its genetic determination suggests a new species for Lake Ohrid and Lake Prespa. Other specimens determined as *P. casertanum* from the two lakes form a paraphyletic group together with specimens of this taxon from Europe.

The median-joining network analysis shows two reticulations within the network (Fig. 4). Only *P. maasseni* is monophyletic, with its two haplotypes separated by two mutational steps. *P. edlaueri*, however, is genetically far more diverse, with up to 13 mutations between haplotypes; *P. nitidum* has over 20 mutational steps between haplotypes. Haplotypes of the latter two species do not form monophyletic clades but are distributed all over the network. In contrast, *P. milium* Held, 1836, and *P. pseudosphaerium* form more isolated cluster within the network.

Morphometrical analyses

The Principle Component Analysis (PCA) of the dataset containing the five shell parameters did not succeed in separating the analyzed specimens on a species level (Fig. 5A). About 97.64% of the variance within the dataset was expressed by the first Principle Component, 1.25% by the second. However, the Jolliffe cut-off value (2.10) as calculated in

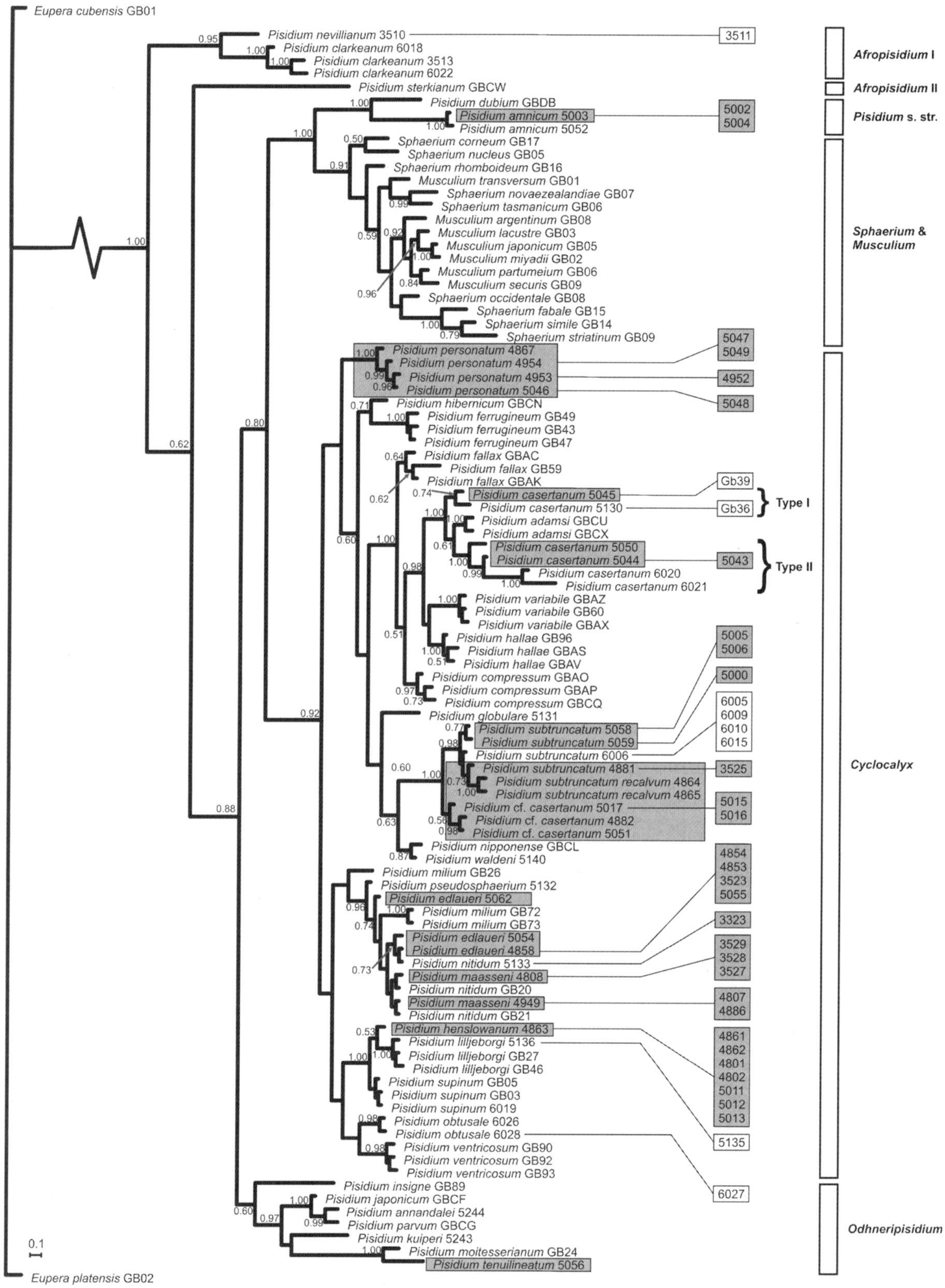
Eupera cubensis GB01
Pisidium nevillianum 3510
Pisidium clarkeanum 6018
Pisidium clarkeanum 3513
Pisidium clarkeanum 6022
Pisidium sterkianum GBCW
Pisidium dubium GBDB
Pisidium amnicum 5003
Pisidium amnicum 5052
Sphaerium corneum GB17
Sphaerium nucleus GB05
Sphaerium rhomboideum GB16
Musculium transversum GB01
Sphaerium novaezealandiae GB07
Sphaerium tasmanicum GB06
Musculium argentinum GB08
Musculium lacustre GB03
Musculium japonicum GB05
Musculium miyadii GB02
Musculium partumeium GB06
Musculium securis GB09
Sphaerium occidentale GB08
Sphaerium fabale GB15
Sphaerium simile GB14
Sphaerium striatinum GB09
Pisidium personatum 4867
Pisidium personatum 4954
Pisidium personatum 4953
Pisidium personatum 5046
Pisidium hibernicum GBCN
Pisidium ferrugineum GB49
Pisidium ferrugineum GB43
Pisidium ferrugineum GB47
Pisidium fallax GBAC
Pisidium fallax GB59
Pisidium fallax GBAK
Pisidium casertanum 5045
Pisidium casertanum 5130
Pisidium adamsi GBCU
Pisidium adamsi GBCX
Pisidium casertanum 5050
Pisidium casertanum 5044
Pisidium casertanum 6020
Pisidium casertanum 6021
Pisidium variabile GBAZ
Pisidium variabile GB60
Pisidium variabile GBAX
Pisidium hallae GB96
Pisidium hallae GBAS
Pisidium hallae GBAV
Pisidium compressum GBAO
Pisidium compressum GBAP
Pisidium compressum GBCQ
Pisidium globulare 5131
Pisidium subtruncatum 5058
Pisidium subtruncatum 5059
Pisidium subtruncatum 6006
Pisidium subtruncatum 4881
Pisidium subtruncatum recalvum 4864
Pisidium subtruncatum recalvum 4865
Pisidium cf. casertanum 5017
Pisidium cf. casertanum 4882
Pisidium cf. casertanum 5051
Pisidium nipponense GBCL
Pisidium waldeni 5140
Pisidium milium GB26
Pisidium pseudosphaerium 5132
Pisidium edlaueri 5062
Pisidium milium GB72
Pisidium milium GB73
Pisidium edlaueri 5054
Pisidium edlaueri 4858
Pisidium nitidum 5133
Pisidium maasseni 4808
Pisidium nitidum GB20
Pisidium maasseni 4949
Pisidium nitidum GB21
Pisidium henslowanum 4863
Pisidium lilljeborgi 5136
Pisidium lilljeborgi GB27
Pisidium lilljeborgi GB46
Pisidium supinum GB05
Pisidium supinum GB03
Pisidium supinum 6019
Pisidium obtusale 6026
Pisidium obtusale 6028
Pisidium ventricosum GB90
Pisidium ventricosum GB92
Pisidium ventricosum GB93
Pisidium insigne GB89
Pisidium japonicum GBCF
Pisidium annandalei 5244
Pisidium parvum GBCG
Pisidium kuiperi 5243
Pisidium moitesserianum GB24
Pisidium tenuilineatum 5056
Eupera platensis GB02
0.1
3511
5002
5004
5047
5049
4952
5048
Gb39
Gb36
Type I
5043
Type II
5005
5006
5000
6005
6009
6010
6015
3525
5015
5016
4854
4853
3523
5055
3323
3529
3528
3527
4807
4886
4861
4862
4801
4802
5011
5012
5013
5135
6027
Afropisidium I
Afropisidium II
Pisidium s. str.
Sphaerium & Musculium
Cyclocalyx
Odhneripisidium

◀ **Fig. 3** Bayesian phylogram (mitochondrial 16S gene) of Sphaeriidae (comprising the nominal genera *Pisidium*, *Sphaerium*, *Musculium*). The tree was rooted with *Eupera cubensis* and *Eupera platensis*. Bayesian posterior probabilities higher than 0.50 are given next to the respective node. The scale bar indicates the observed number of substitutions per site according to the applied model of sequence evolution. Highlighted clades in grey indicate specimens from lakes Ohrid and Prespa and their surrounding watershed. Major clades are marked with white bars on the right side; taxonomic classification follows Lee & Ó Foighil (2003). Braces indicate clades of the paraphyletic *P. casertanum* (see Discussion for details). Specimen details are given in Supplementary material—Appendix I

PAST was higher than the second Principle Component, indicating that the observed separation along the *y*-axis in Fig. 5A cannot be regarded as meaningful. All five parameters are strongly positively correlated with the first component (over 90%). Most of the specimens are found along the *x*-axis at a point below 6,000. Only members of *P. amnicum* are found completely above this value.

A separation on a species level using shell outline data did not succeed, either. Neither the single datasets of rear and right side view (data not shown), nor their combination (Fig. 5B) was sufficient for this task. The first Principle Component explains 76.58% of the variance within the dataset, the second 11.18%.

Mantel test results

The combination of genetic and morphometrical distances using a Mantel test revealed a weak positive correlation between both matrices. As original *Z* statistic, a value of 1.921 was computed. The average *Z* after 999 permutations was 1.762. The correlation was $r = 0.32$. Out of 999 permutations, no random dataset had a *Z*-value greater or equal to the original *Z* score. Therewith, the positive correlation is highly significant ($P = 0.001$).

Morphological description of *P.* cf. *casertanum*

The shell shape of *P.* cf. *casertanum* is trigonal-roundish to oval, the shell itself relatively thick, up to 5-mm long and resembles the external shell morphology of *P. casertanum*. The hinge structure, however, is more similar to *P. subtruncatum*, but in contrast to the latter, the cardinal teeth c2 and c4 are more curved and c4 is not overlapping. The hinge plate at the cardinal teeth is conspicuously thickened, which might be a critical morphological character for this taxon.

Discussion

Biodiversity and speciation of *Pisidium* in lakes Ohrid and Prespa

The present study provides the first molecular and morphometrical analyses of a cryptic molluscan group with little morphological differentiation in lakes Ohrid and Prespa using the pea clam fauna as a paradigm. In particular, the genetic analyses revealed novel insights into species composition and, therefore, shed new light on previous hypotheses on the phylogeography of *Pisidium* in these two European ancient lakes. Here we focus on three major topics concerning the biodiversity and speciation events of the pea clam fauna in lakes Ohrid and Prespa: (1) the phylogenetic relationship of the endemics *P. edlaueri* and *P. maasseni* in the light of the presented molecular data; (2) a newly discovered *Pisidium* species of Ohrid and Prespa, which we embed in a Pan-European context of the genus, and discuss not only the presence of a cryptic species, but also the potential loss of pea clam diversity; and (3) the phylogenetic signals within the morphometrical dataset and its correlation with genetic data.

(1) The close relationship for the endemics *P. edlaueri* (Lake Ohrid) and *P. maasseni* (Lake Prespa), as suggested by Kuiper (1960, 1987) and Korniushin et al. (2000), is now confirmed by molecular data. Furthermore, their proposed phylogenetic affinity to the widely distributed *P. nitidum* is also supported by the present study (Figs. 3, 4). A clear differentiation on species level, however—as it can be found in other clades of the 16S phylogeny—could neither be achieved for the two endemic species nor for their proposed sister taxon, *P. nitidum*, or for the likewise closely related *P. milium*. This might be due to incomplete lineage sorting for the analyzed 16S fragment, indicating a comparatively young evolutionary age of the endemic branches, comparatively low substitution rates in the 16S gene, or a combination of both. Therewith, it could be argued, that none of the two endemic species is actually a "good species" because they form no

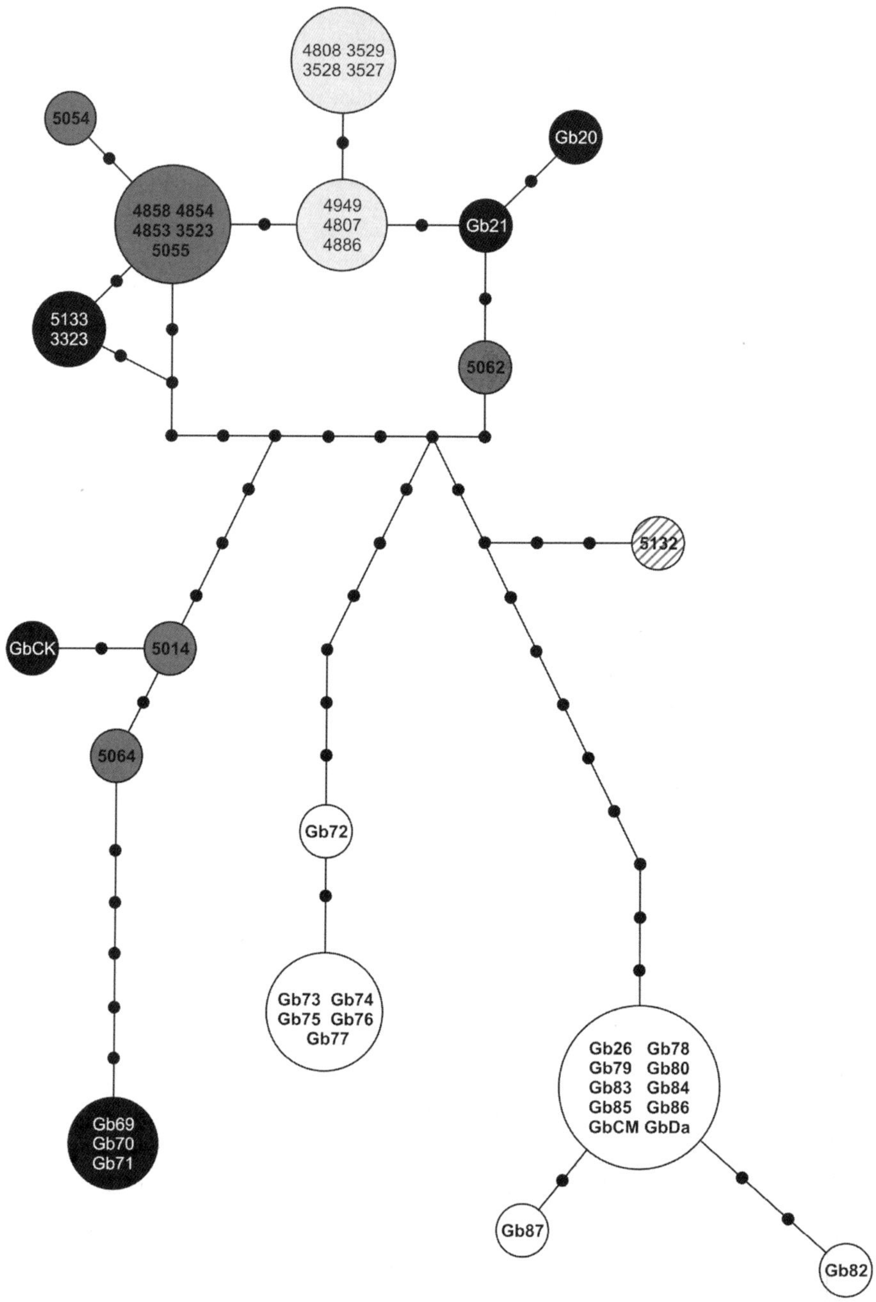

Fig. 4 Median-joining network comprising all analyzed specimens of the endemics *P. edlaueri* (marked in dark grey) and *P. maasseni* (marked in light grey). In addition, specimens of the closely related *P. nitidum* (black), *P. milium* (white) and *P. pseudosphaerium* (striped) are included. Each dot indicates a mutational step. Specimen details are given in Supplementary material—Appendix I

monophyletic clades. The line of argument would be similar to the suggestions of Lee & Ó Foighil (2003), who assumed cryptic species within a paraphyletic clade of *P. casertanum*. By accepting the paraphyly of *P. edlaueri*, *P. maasseni* and *P. nitidum* as a reflection of the "true" phylogenetic relationship (like the previously mentioned authors did for *P. casertanum* spp.), the species status of both Balkan endemics must be doubted. However, the analyses of more rapidly evolving markers, such as mitochondrial cytochrome *c* oxidase I or AFLP loci, are necessary to answer this question.

(2) The present study also revealed new insights into the species composition of lakes Ohrid and Prespa and their surroundings. Using molecular methods, the occurrence of a cryptic clade was

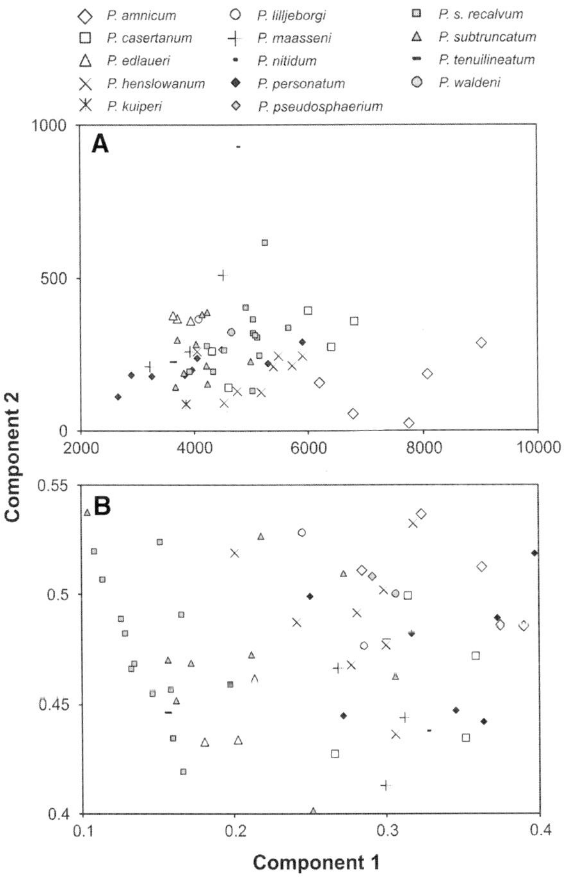

Fig. 5 Results of PCA of five shell parameters (**A**; for details see text, measurements are given in Supplementary material—Appendix II) and of the combined dataset resulting from the Elliptic Fourier Analyses of the rear- and right side view, respectively (**B**)

discovered—a clade which members were previously determined as *P. casertanum* (in Figs. 2 and 3 labeled as *P.* cf. *casertanum*). In this context, it is noteworthy that determination on species level is difficult in the genus *Pisidium* (Ellis, 1978). Due to the scarcity of sufficient morphological characters, "classical" determination requires considerable experience (Kuiper, 1983). However, our species determination was approved by molecular means in all other cases. A mistake in determination of the individuals of *P.* cf. *casertanum* can be excluded since all reported species from Lake Ohrid and Lake Prespa are analyzed in our study. The only exception is *P. moitessierianum*, which is, however, also included in the 16S phylogeny as a sequence from GenBank. Therefore, *P.* cf. *casertanum* likely represents a new species for Lake Prespa (Fig. 2, Locality P1) and for Lake Ohrid (Fig. 2, O4). We have also found three specimens of this clade within the River Drim, north of Lake Ohrid (W1). Moreover, it should be noted that the present study comprises nearly all known European pea clam species with exception of the boreal *P. hinzi*, and of *P. conventus*, *P. dilatatum*, *P. pulchellum* and *P. sogdianum*. The morphological difference between *P.* cf. *casertanum* and these five species is, however, considerable. Therefore, *P.* cf. *casertanum* is also likely to represent a new pea clam species for the European continent (Fig. 6), although genetic analyses of the five missing European species are necessary to confirm this.

A paraphyletic species complex of *P. casertanum* was already suggested by Lee & Ó Foighil (2003). We have found a similar pattern within the European clade of members of *P. casertanum* (braces Type I and Type II in Fig. 3). It can be shown that specimens of *P. casertanum* from the sister lake system Ohrid and Prespa (including the surrounding watershed) belong to both clades: one specimen (5045; W4 in Fig. 2) clusters with a *P. casertanum* from Germany (5130), while other specimens from the feeding springs of Lake Ohrid (5043 & 5045; W5 in Fig. 2) and from Lake Prespa (5050; P1 in Fig. 2) cluster with specimens from Picos de Europa, Spain (6020 & 6021). We want to stress, however, that the observed paraphyletic pattern within *P. casertanum* as seen in Types I and II might also be due to incomplete lineage sorting. We intended to include *P. casertanum* from the type locality in Caserta, Italy, in our study. However, none of the several dozen specimens we determined from two ponds within Caserta obtained in 2006 was *P. casertanum* (see specimen 6005 in Supplementary material—Appendix I for locality information). Instead we found *P. subtruncatum* only.

Considering the faunal composition of lakes Ohrid and Prespa, a further annotation is noteworthy: we did not find live specimens of *P. moitessierianum*. Despite the extensive fieldwork of three years, only a single valve was found (Lake Prespa, 2003). In contrast, Kuiper (1960) determined over 400 specimens out of a total of 1,100 individuals of pea clams from lakes Ohrid and Prespa as *P. moitessierianum*. Kuiper (1987) reported again the occurrence of *P. moitessierianum* in Lake Prespa. He refers, however, to samples collected by Maassen in 1975 and 1976 (Maassen, 1980). Since this date, we know

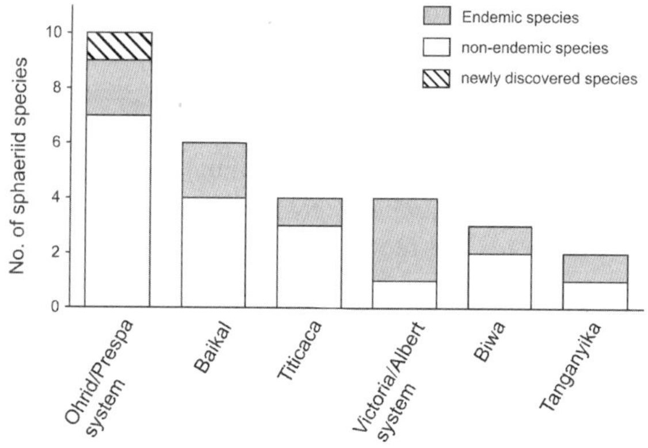

Fig. 6 Comparison of endemic sphaeriid species from selected ancient lakes of the world: The sister lake system Ohrid and Prespa harbors two endemic species *P. edlaueri* and *P. maasseni* and one endemic subspecies, *P. subtruncatum recalvum*. Lake Baikal harbors the endemic *P. raddei* Dybowski, 1902 and *Sphaerium baicalense* Dybowski, 1902. Lake Titicaca harbors the endemic *Musculium titicacense* (Pilsbry, 1924) and the Victoria-Albert Lake System the endemic *P. fistulosum* Mandahl-Barth, 1954, *M. stuhlmanni* (Martens, 1897), and *M. nyanzae* (Smith, 1892). Lake Biwa harbors the endemic species *P. kawamurai* Mori, 1938 and the endemic subspecies *P. casertanum lacustre* Woodward in Mori, 1938. From Lake Tanganyika the endemic species *P. giraudi* Bourguignat, 1885, is known (data from Korniushin, 2004). The present study revealed a so far unknown species within the sister lake system Ohrid and Prespa (striped column)

of no published report of this species in neither lake (Reischütz & Stummer, 1989; Korniushin, 2004; Slugina, 2006). Although the absence of this species within our collection is not a conclusive evidence for its local extinction, the obvious change in overall abundance is noteworthy.

(3) The analysis of the five shell parameters failed to provide sufficient data for species separation (Fig. 5A). This confirms the view of Holopainen & Kuiper (1982), who found their parameter set overlapping and too simple to describe the shape of *Pisidium* well enough. Despite the fact that the Elliptic Fourier Analyses of shell outline data provided better separation on the species level (Fig. 5B), this approach was also not sufficient in order to identify phylogenetic informative morphometrical characters. Although the sample size of individuals per species used for Principle Component Analyses was generally low, the overlap as seen in Fig. 5B is striking and will unlikely be diminished by including more specimens into the analysis.

The result of the Mantel test, however, shows a positive and strongly significant correlation between morphometrical and molecular distances. Therewith, the present study confirms the findings of Guralnick (2005), who suggested that lineage specific shell patterns are not completely masked by eco-phenotypic variance. Consequently, it can be assumed that the inclusion of more morphometrical parameters, for example, hinge characteristics, into the dataset could increase the resolution of this approach. However, *Pisidium* remains a notoriously difficult genus of freshwater molluscs in respect to morphological based species determination (Dance & Kuiper, 2002). Similar difficulties can be found in gastropods as well (Wilke & Falniowski, 2001; Wilke et al., 2002). In this context, the present study stresses the need and the potential of molecular analyses for identifying cryptic speciation not only in ancient lakes.

Biogeographical implications

The occurrence of two endemic sister-species of *Pisidium* within the neighboring lakes Ohrid and Prespa is a unique biogeographical phenomenon. As the present study shows, both taxa are more closely related in respect to each other than most other species within the broad pea clam phylogeny (Fig. 3). If the species status of both endemics and their proposed sister species relationship will be confirmed by population genetics, three possible biogeographical hypotheses can be assumed (Fig. 7):

(a) A parallel dispersal scenario: a common ancestor of *P. edlaueri* and *P. maasseni* colonized simultaneously lakes Ohrid and Prespa, and evolved independently in geographical isolation.
(b) A vicariance scenario: it has been suggested, that both lakes once formed a single watershed (Karaman, 1971; but see Albrecht & Wilke, 2008). According to this scenario, a common ancestor colonized the united watershed of the later lakes Ohrid and Prespa. When the lakes became separated, the equally separated populations of the common ancestor started to evolve independently into the two endemic species.
(c) An "Ohrid-stepping stone scenario": a common ancestor colonized Lake Ohrid and, thence, Lake Prespa. This pattern might have happened several times. Likewise, possibly due to a desiccation of Lake Prespa, the endemic pea clam population therein might have gone extinct

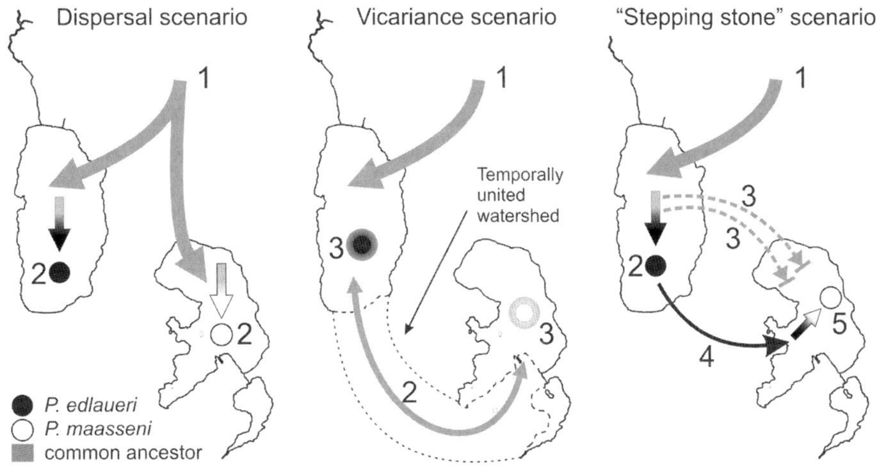

Fig. 7 Three possible scenarios of the evolution of the endemic pea clam species *P. edlaueri* (black circles) and *P. maasseni* (white circles) from a common ancestor (grey arrows) within the lakes Ohrid and Prespa (see text for explanations)

several times (indicated by the dotted arrows in the right image in Fig. 7). According to this scenario, the endemic pea fauna of Lake Prespa has its origin in Lake Ohrid and is, therefore, younger.

Examination of the median-joining network (Fig. 4) shows that *P. edlaueri* is far more diverse than *P. maasseni*, most likely indicating a greater phylogenetic age of the former. This observation contradicts the idea of a parallel colonization of both water bodies (dispersal scenario, Fig. 7, left image) as well as the hypothesis of a single invasion with a subsequent separation of a formerly united watershed into lakes Ohrid and Prespa (vicariance scenario, same figure, center image). In fact, the present data suggests a primary colonization of Lake Ohrid and a subsequent colonization of Lake Prespa by the endemic *P. edlaueri* (Fig. 7, right image). This assumption would explain the differences in genetic diversity as revealed by the present study. Such an "Ohrid-stepping stone" pattern—indicated by a relatively high genetic diversity of a taxon in Lake Ohrid and a much lower diversity of this taxon in adjacent water bodies—is also known from other molluscan taxa, for example, within the genus *Dreissena* on the Balkan Peninsula (Albrecht et al., 2007).

This scenario, however, raises the question of timing and mode of isolation between the putative source population within Lake Ohrid and the population in Lake Prespa. If, for example, bird- or insect-mediated colonization of Prespa happened (Mackie, 1978; Jensen et al., 2002), did gene flow ever stop? The observed paraphyly shown in Fig. 4 may indicate—as discussed in the previous sections—the presence of a single species with its geographical variants (*edlaueri* and *maasseni*). Thus, both taxa would still exchange genes and the observed phenotypic divergence would be either environmentally induced or genetically based. A genetically based phenotypic divergence, however, could only be maintained by strong selection in the presence of gene flow (Funk & Omland, 2003).

If, on the other hand, gene flow has stopped, it seems for two reasons questionable that the resulting reproductive isolation was due to geographical barriers. First, specimens of *P. henslowanum* from both lakes share the same 16S haplotype (Figs. 2, 3), indicating appropriate opportunities for some pea clams to overcome the geographic barriers between these water bodies. Second, according to the "Ohrid-stepping stone" scenario, specimens of *P. edlaueri* might have been able to overcome the geographic barrier between the sister lakes at least once (and, therewith, perhaps the Galičica mountains). Therefore, the subsequent colonization of Lake Prespa was probably followed by the occupation of an ecological niche, driven by environmental settings that constituted a strong diverging selection between *P. edlaueri* and *P. maasseni*.

General patterns of speciation in lakes Ohrid and Prespa

Despite of the partly unresolved phylogeographical relationships of the abovementioned taxa, which could be, in part, due to the performance of the gene fragment used, the data presented here are interesting in comparison to the better studied morphologically distinct molluscan groups in lakes Ohrid und Prespa, i.e., the genus *Ancylus* (Albrecht et al., 2006), the genus *Valvata* (Hauswald et al., 2008), the genus *Radix* (Albrecht et al., 2008), the subfamily Pyrgulinae (Wilke et al., 2007), and the genus *Dreissena* (Albrecht et al., 2007).

General patterns include:

(A) the confirmed presence of endemic species;
(B) the presence of putatively undescribed species;
(C) a generally high biodiversity of the respective taxa; and
(D) a close phylogeographical relationship of Ohrid and Prespa taxa, where applicable.

However, there are also striking differences in speciation patterns of pea clams in lakes Ohrid and Prespa compared to other taxa. These differences comprise:

(A) the absence of ancient lake species flock on the subgenus or higher taxonomic level (but see *Valvata*; Hauswald et al., 2008);
(B) the presence of widespread European representatives in lakes Ohrid and Prespa; and
(C) a close phylogenetic relationships of some endemic taxa to taxa outside the Balkan area (but see *Radix*, Albrecht et al., 2008).

Future studies involving more cryptic groups in lakes Ohrid and Prespa may show whether the interesting evolutionary patterns presented here are specific for *Pisidium* or whether morphologically distinct groups generally show differences compared to the largely neglected cryptic groups in lakes Ohrid and Prespa.

Conclusion

The present study stresses the need of shifting the focus in studying speciation in ancient lakes. Besides dealing with eye-catching, "thalassoid" gastropod species, an understanding of the evolutionary history within these remarkable biological hotspots also requires a closer look at the so-far often-neglected biodiversity of inconspicuous taxa. Within these potentially cryptic speciations, which possibly contribute significantly to ancient lake endemic faunas (Radoman, 1983), lies an important key for tracing unbiased the origin and evolution of biodiversity within ancient lakes. Moreover, the presence of cryptic species within these lakes—and, therefore, their evolutionary history therein—must be considered as normal case, while the conspicuous taxa might constitute the morphological and evolutionary exception to this rule. Focusing on merely one of both groups bears not only the risk of biased research, but also of an incomplete and, therefore, potentially erroneous understanding of speciation patterns and processes within ancient lakes.

Our study also stresses the importance of taxonomic expertise, particular in such notorious taxa as *Pisidium*. Since modern morphometrical analyses fail to provide relatively simple characters for species determination, a long-term taxonomic experience is required to work with such difficult taxa. This expertise—combined with ecological familiarity with the taxon in question—forms the framework for further research on the subject. Only on that basis the application of modern approaches like DNA sequencing is reasonable and efficient.

Acknowledgments We are very grateful to the colleagues of the Hydrobiological Institute Ohrid (HBI) for their hospitality, immense support, and for issuing the permits to collect in Lake Ohrid. G. Kostoski, B. Budzakoska, S. Trajanovski and S. Trajanovska provided valuable information and took part in field trips. D. Georgiev is gratefully acknowledged for supporting us with his local expertise and equipment. P. Glöer and V. Pešić as well as V. Fiorentino kindly provided material from Lake Skutari and from Caserta, Italy. S. Nachtigall is gratefully acknowledged for her help in the lab. We are thankful to two anonymous reviewers for their valuable comments that significantly enhanced an earlier version of the manuscript. This work was supported in part by a research grant of The Malacological Society of London to RS. Last not least, the authors would like to thank J. G. J. Kuiper for supporting us with literature and for encouraging our work.

References

Albrecht, C., R. Schultheiß, T. Kevrekidis, B. Streit & T. Wilke, 2007. Invaders or endemics? Molecular phylogenetics, biogeography and systematics of *Dreissena* in the Balkans. Freshwater Biology 52: 1525–1536.

Albrecht, C., S. Trajanovski, K. Kuhn, B. Streit & T. Wilke, 2006. Rapid evolution of an ancient lake species flock: freshwater limpets (Gastropoda: Ancylidae) in the Balkan Lake Ohrid. Organisms. Diversity & Evolution 6: 294–307.

Albrecht, C. & T. Wilke, 2008. Ancient Lake Ohrid: biodiversity and evolution. Hydrobiologia. doi:10.1007/s10750-008-9558-y.

Albrecht, C., C. Wolf, P. Glöer & T. Wilke, 2008. Concurrent evolution of ancient sister lakes and sister species: The freshwater gastropod genus *Radix* in lakes Ohrid and Prespa. Hydrobiologia. doi:10.1007/s10750-008-9555-1.

Angelov, A., 1971. Beitrag zur Kenntnis der Pisidienfauna (Molluska-Sphaeriidae) Mazedoniens. Fragmenta Balcanica 7: 53–55.

Bandelt, H. J., P. Forster & A. Rohl, 1999. Median-joining networks for inferring intraspecific phylogenies. Molecular Biology and Evolution 16: 37–48.

Bößneck, U., 2002. Lebendnachweise von *Microcondylaea bonellii* (A. Ferussac 1827) und *Pisidium tenuilineatum* Stelfox 1918 aus Istrien (Kroatien) (Bivalvia: Unionidae et Sphaeriidae). Malakologische Abhandlungen 20: 313–317.

Brooks, J. L., 1950. Speciation in ancient lakes. The Quaterly Review of Biology 25: 131–176.

Cooley, L. R. & D. Ó Foighil, 2000. Phylogenetic analysis of the Sphaeriidae (Mollusca: Bivalvia) based on partial mitochondrial 16S rDNA gene sequences. Invertebrate Biology 119: 299–308.

Dance, S. P. & J. G. J. Kuiper, 2002. B.B. Woodward and the '*Pisidium* affair'. Journal of Conchology 37: 635–650.

Dreher-Mansur, M. C. & C. Meier Brook, 1992. Morphology of *Eupera* Bourguignat, 1854 and *Byssanodonta* Orbigny, 1846 with contributions to the phylogenetic systematics of Sphaeriidae and Corbiculidae (Bivalvia, Veneroida). Archiv für Molluskenkunde 128: 1–59.

Dyduch-Falniowska, A., 1989. Remarks on the Sphaeriidae of Greece (Mollusca, Bivalvia). Basteria 53: 15–18.

Ellis, A. E., 1978. British freshwater bivalve Mollusca. The Linnean Society of London, London.

Funk, D. J. & K. E. Omland, 2003. Species-level paraphyly and polyphyly: frequency, causes, and consequences, with insights from animal mitochondrial DNA. Annual Review of Ecology Evolution and Systematics 34: 397–423.

Gambetta, L., 1930. Molluschi terrestri e lacustri d'Albania. Atti della Accademia Scientifica Veneto – Trentino – Istriana. Padova 21: 51–74.

Gorthner, A., 1994. What is an ancient lake? In Martens, K., B. Goddeeris & G. Coulter (eds), Speciation in Ancient Lakes. Archiv für Hydrobiologie, Stuttgart: 97–100

Guralnick, R. P., 2005. Combined molecular and morphological approaches to documenting regional biodiversity and ecological patterns in problematic taxa: a case study in the bivalve group *Cyclocalyx* (Sphaeriidae, Bivalvia) from western North America. Zoologica Scripta 34: 469–482.

Hall, T. A., 1999. BioEdit: a user-friendly biological sequence alignment editor and analysis program for Windows 95/98/NT. Nucleic Acids Symposium Series 41: 95–98.

Hammer, Ø., D. A. T. Harper & P. D. Ryan, 2001. PAST: Palentological statistics software package for education and data analysis. Palaeontologia Electronica 4: 9.

Hauswald, A.-K., C. Albrecht & T. Wilke, 2008. Testing two contrasting evolutionary patterns in ancient lakes: species flock vs. species scatter in valvatid gastropods of Lake Ohrid Hydrobiologia. doi:10.1007/s10750-008-9556-0.

Holopainen, I. J. & J. G. J. Kuiper, 1982. Notes on the morphometry and anatomy of some *Pisidium* and *Sphaerium* species (Bivalvia, Sphaeriidae). Annales Zoologici Fennici 19: 93–107.

Huelsenbeck, J. P. & F. Ronquist, 2001. MrBayes: Bayesian inference of phylogeny. Bioinformatics 17: 754–755.

Illies, J., 1978. Limnofauna Europaea. Gustav Fischer Verlag, Stuttgart, New York.

Jaeckel, S. G., W. Klemm & W. Meise, 1957. Die Land- und Süsswasser-Mollusken der nördlichen Balkanhalbinsel. Abhandlungen und Berichte des Staatliche Museums für Tierkunde. Dresden 23: 141–205.

Jensen, C., J. G. J. Kuiper & K. D. Vorren, 2002. First postglacial establishment of forest trees: early Holocene vegetation, mollusc settlement and climate dynamics in central Troms, North Norway. Boreas 31: 285–301.

Karaman, M., 1971. Zoogeographische Verhältnisse des Prespa- und Ochridasees. Izdanija 4: 1–21.

Korniushin, A. V., 2004. The bivalve mollusca fauna of ancient lakes in the context of the historical biogeography of the Balkan region. In Griffiths, H. I., B. Krystufek & J. M. Reed (eds), Balkan Biodiversity – Pattern and Process in the European Hotspot. Kluwer Academic Publishers, Dordrecht/Bosten/London: 219–241.

Korniushin, A. V. & M. Glaubrecht, 2002. Phylogenetic analysis based on the morphology of viviparous freshwater clams of the family Sphaeriidae (Mollusca, Bivalvia, Veneroida). Zoologica Scripta 31. 415–459.

Korniushin, A. V. & E. Hackenberg, 2000. Verwendung konchiologischer und anatomischer Merkmale für die Bestimmung mitteleuropäischer Arten der Familie Shaeriidae (Bivalvia), mit neuem Bestimmungsschlüssel und Diagnosen. Malakologische Abhandlungen – Staatliches Museum für Tierkunde Dresden 20: 45–72.

Korniushin, A. V., Z. Krstanovski & G. Kostoski, 2000. Anatomical evidence of close affinity between endemic species of *Pisidium* (Bivalvia, Sphaeriidae) from some ancient lakes, and the widely distributed taxa. Journal of Zoological Systematics and Evolutionary Research 38: 81–86.

Kuiper, J. G. J., 1960. Die Pisidien des Ohridsees, Mazedonien, nebst Bemerkungen über die Verbreitung der Pisidien in der Balkanhalbinsel und den Donauländern. Beaufortia 7: 219–231.

Kuiper, J. G. J., 1963. Hauptzüge der Verbreitung des Genus *Pisidium* in Europa. Archiv für Molluskenkunde 92: 247–252.

Kuiper, J. G. J., 1983. The Sphaeriidae of Australia. Basteria 47: 3–52.

Kuiper, J. G. J., 1987. *Pisidium maasseni* n. sp., a new species from Lake Prespa, Jugoslavia (Bivalvia, Sphaeriidae). Basteria 51: 163–165.

Kumar, S., K. Tamura & M. Nei, 2004. MEGA3: integrated software for molecular evolutionary genetics analysis and sequence alignment. Briefings in Bioinformatics 5: 150–163.

Lee, T. W., 2004. Morphology and phylogenetic relationships of genera of North American Sphaeriidae (Bivalvia, veneroida). American Malacological Bulletin 19: 1–13.

Lee, T. & D. Ó Foighil, 2002. 6-phosphogluconate dehydrogenase (PGD) allele phylogeny is incongruent with a recent origin of polyploidization in some North American Sphaeriidae (Mollusca, Bivalvia). Molecular Phylogenetics and Evolution 25: 112–124.

Lee, T. & D. Ó Foighil, 2003. Phylogenetic structure of the Sphaeriinae, a global clade of freshwater bivalve molluscs, inferred from nuclear (ITS-1) and mitochondrial

(16S) ribosomal gene sequences. Zoological Journal of the Linnean Society 137: 245–260.
Maassen, W. J. M., 1980. De zoetwatermollusken van het Ohrid/en Prespameer en hun omgeving. De Kreukel 16: 81–98.
Mackie, G. L., 1978. Larval growth in fingernail and pill clams (Bivalvia: Sphaeriidae). Bulletin of the American Malacological Union: 6–13.
Martens, K., G. Coulter & B. Goddeeris, 1994. Speciation in Ancient Lakes – 40 years after Brooks. Archiv für Hydrobiologie – Advances in Limnology 44: 75–96.
Michel, E., 1994. Why snails radiate – A review of gastropod evolution in long lived lakes, both recent and fossil. In Martens, K., B. Goddeeris & G. Coulter (eds), Speciation in Ancient Lakes. Archiv für Hydrobiologie, Stuttgart: 285–317.
Michel, E., 2004. *Vinundu*, a new genus of gastropod (Cerithioidea: 'Thiaridae') with two species from Lake Tanganyika, East Africa, and its molecular phylogenetic relationships. Journal of Molluscan Studies 70: 1–19.
Miller, M. P., 1997. Tools for population genetic analyses (TFPGA) 1.3: a Windows program for the analysis of allozyme and molecular population genetic data. Northern Arizona University, Flagstaff.
Palumbi, S., A. Martin, S. Romano, W. O. Mcmillian, L. Stice & G. Grabowski, 1991. The simple fool's guide to PCR. University of Hawaii, Honolulu.
Posada, D. & T. R. Buckley, 2004. Model selection and model averaging in phylogenetics: advantages of Akaike information criterion and Bayesian approaches over likelihood ratio tests. Systematic Biology 53: 793–808.
Posada, D. & K. A. Crandall, 1998. MODELTEST: testing the model of DNA substitution. Bioinformatics 14: 817–818.
Radoman, P., 1983. Hydrobiodea, a Superfamily of Prosobranchia (Gastropoda). Faculty of Science, Department of Biology, Beograd.
Reischütz, P. L. & B. Stummer, 1989. Ein Beitrag zur Molluskenfauna der dessaretischen Seen. Malakologische Abhandlungen 14: 105–109.
Rintelen, T. v. & M. Glaubrecht, 2005. Anatomy of an adaptive radiation: a unique reproductive strategy in the endemic freshwater gastropod Tylomelania (Cerithioidea : Pachychilidae) on Sulawesi, Indonesia and its biogeographical implications. Biological Journal of the Linnean Society 85: 513–542.
Rintelen, T. v. & M. Glaubrecht, 2006. Rapid evolution of sessility in an endemic species flock of the freshwater bivalve Corbicula from ancient lakes on Sulawesi, Indonesia. Biology Letters 2: 73–77.
Rohlf, F. G., 2001. Comparative methods for the analysis of continuous variables: geometric interpretation. Evolution 55: 2143–2160.
Schön, I. & K. Martens, 2004. Adaptive, pre-adaptive and non-adaptive components of radiations in ancient lakes: a review. Organisms, Diversity & Evolution 4: 137–156.
Slugina, Z. V., 2006. Endemic Bivalvia in ancient lakes. Hydrobiologia 568: 213–217.
Slugina, Z. V. & Y. I. Starobogatov, 1999. Guides and keys to identification of fauna and flora of Lake Baikal. Izdalelstvo SO RAN, Novosibirsk.
Stanković, S., 1960. The Balkan Lake Ohrid and Its Living World. Junk, Den Haag.
Thompson, J. D., D. G. Higgins & T. J. Gibson, 1994. CLUSTAL W: improving the sensitivity of progressive multiple sequence alignment through sequence weighting, positions-specific gap penalties and weight matrix choice. Nucleic Acids Research 22: 4673–4680.
Wilke, T., C. Albrecht, V. V. Anistratenko, S. Kosal & M. Z. Yildirim, 2007. Testing biogeographical hypotheses in space and time: faunal relationships of the putative ancient Lake Egirdir in Asia Minor. Journal of Biogeography Journal of Biogeography 34: 1807–1821.
Wilke, T., G. M. Davis, D. Qiu & R. C. Spear, 2006. Extreme mitochondrial sequence diversity in the intermediate schistosomiasis host *Oncomelania hupensis robertsoni*: another case of ancestral polymorphism? Malacologia 48: 143–157.
Wilke, T. & A. Falniowski, 2001. The genus *Adriohydrobia* (Hydrobiidae: Gastropoda): polytypic species or polymorphic populations? Journal of Zoological Systematics and Evolutionary Research 39: 227–234.
Wilke, T., M. Pfenninger & G. M. Davis, 2002. Anatomical variation in cryptic mudsnail species: statistical discrimination and evolutionary significance. Proceedings of the Academy of Natural Sciences of Philadelphia 152: 45–66.
Wilson, A. B., M. Glaubrecht & A. Meyer, 2004. Ancient lakes as evolutionary reservoirs: evidence from the thalassoid gastropods of Lake Tanganyika. Proceedings of the Royal Society of London Series B-Biological Sciences 271: 529–536.

Hydrobiologia (2008) 615:157–167
DOI 10.1007/s10750-008-9555-1

Concurrent evolution of ancient sister lakes and sister species: the freshwater gastropod genus *Radix* in lakes Ohrid and Prespa

Christian Albrecht · Christian Wolff · Peter Glöer · Thomas Wilke

Published online: 11 September 2008

Abstract Ancient sister lakes are considered to be ancient lakes lying in close geographic proximity, sharing a related origin and significant time of co-existence, usually having hydrological connection as well as a balanced degree of faunal overlap and distinctness. A paradigm for studying sister lake relationships are the ancient lakes Ohrid and Prespa in the Balkans, which are characterized by high degrees of endemicity. Three general patterns of endemic species can be distinguished for these lakes: (1) taxa that are endemic to either lake, with no close relatives in the respective sister lake, (2) closely related but distinct endemic taxa in both lakes (sister species) and (3) shared endemic taxa occurring in both lakes. In the present paper, two endemic freshwater pulmonate gastropod species, *Radix relicta* (Lake Ohrid) and *R. pinteri* (Lake Prespa), are used to study the evolution of presumed sister species based on biogeographical and comparative DNA data from world-wide *Radix* taxa. Phylogenetic, phylogeographical and parametric bootstrap analyses all suggest a sister group relationship of *R. relicta* and *R. pinteri* (pattern 2 of endemic diversity). Sister to these two taxa is the widespread *R. ampla*, which does not occur in the vicinity of lakes Ohrid and Prespa. The southern feeder spring complexes of Lake Ohrid are inhabited by another lineage (*Radix* sp. 1), which resembles *Radix relicta* in morphology/anatomy. For Lake Prespa, the widespread *R. auricularia* was reported in addition to the endemic *R. pinteri*. Comparative phylogenetic data favour a western Adriatic zoogeographical affinity of lakes Ohrid and Prespa over an Aegean-Anatolian faunal connection. The status of lakes Ohrid and Prespa as sister lakes is evaluated in the light of current knowledge on gastropod speciation and endemism in these hotspots of biodiversity.

Keywords Ancient lake · Lake Ohrid · Lake Prespa · *Radix* · Balkan · Sister lakes

Guest editors: T. Wilke, R. Väinölä & F. Riedel
Patterns and Processes of Speciation in Ancient Lakes: Proceedings of the Fourth Symposium on Speciation in Ancient Lakes, Berlin, Germany, September 4–8, 2006

Electronic supplementary material The online version of this article (doi:10.1007/s10750-008-9555-1) contains supplementary material, which is available to authorized users.

C. Albrecht (✉) · C. Wolff · T. Wilke
Department of Animal Ecology and Systematics, Justus Liebig University Giessen, Heinrich-Buff-Ring 26-32 (IFZ), 35392 Giessen, Germany
e-mail: Christian.Albrecht@allzool.bio.uni-giessen.de

P. Glöer
Schulstraße 3, 25491 Hetlingen, Germany

Introduction

Ancient lakes are usually seen as compact and isolated geographic entities that have undergone unique histories in a certain hydrological setting (e.g. Gorthner, 1994). It is clear that over geological time scales, these lakes do not remain constant; rather, some of these lakes became repeatedly subdivided and reconnected in the past (e.g. Lake Tanganyika, Coulter, 1991). Even some extant lakes show distinct subdivision like the two parts of Lake Titicaca (Lago Grande and Lago Pequeno). Such conditions have been suggested to trigger allopatric speciation in ancient lakes (e.g. Martens, 1997). Ancient sister lakes, such as the Malili Lake system on the Indonesian island Sulawesi (e.g. Rintelen & Glaubrecht, 2005, 2006), are here considered to be ancient lakes lying in close geographic proximity, sharing a related origin and significant time of co-existence, usually having hydrological connection as well as a balanced degree of faunal overlap and distinctness. A paradigm for ancient sister lakes is the European lake group called Dessaretes located on the Balkan Peninsula. Originally, the Dessaretes consisted of Lake Ohrid (Macedonia, Albania), Lake Prespa (Macedonia, Greece, Albania), Lake Mikri Prespa (Greece, Albania), and Lake Maliq (Albania). The latter was drained in the middle of the last century (Denèfle et al., 2000). Some workers even argued that the Dessarete basin was filled in with water and all of these lakes were interconnected during evolutionary time (reviewed in Albrecht & Wilke, 2008).

Today, Lake Ohrid (693 m a.s.l.) constitutes the oldest lake in Europe, with an estimated age ranging from 2 or 3 my (Stanković, 1960) to 10 my (Spirkovski et al., 2001), and a high degree of endemism (Albrecht & Wilke, 2008). It has a surface area of 358 km^2 and a maximum depth of 289 m (Matzinger et al., 2007). Lake Prespa, located at an altitude of 849 m a.s.l., is the highest large lake in the Balkans. It covers 254 km^2 (Matzinger et al., 2006) and has a maximum depth of 58 m (Albrecht & Wilke, 2008). Lake Mikri Prespa (approx. 853 m a.s.l.) is 53 km^2 in size and 8.4 m deep (Zacharias et al., 2002). The latter lake is directly connected to Lake Prespa and its water level is regulated through the outflow to Lake Prespa. No reliable data is available for age and origin of the two Prespa lakes, though many workers consider Lake Prespa to have a similar age as Lake Ohrid.

While the sister lake status of Lake Prespa and Lake Mikri Prespa is obvious, it is, on first sight, less clear for Lake Ohrid and Lake Prespa. Though they

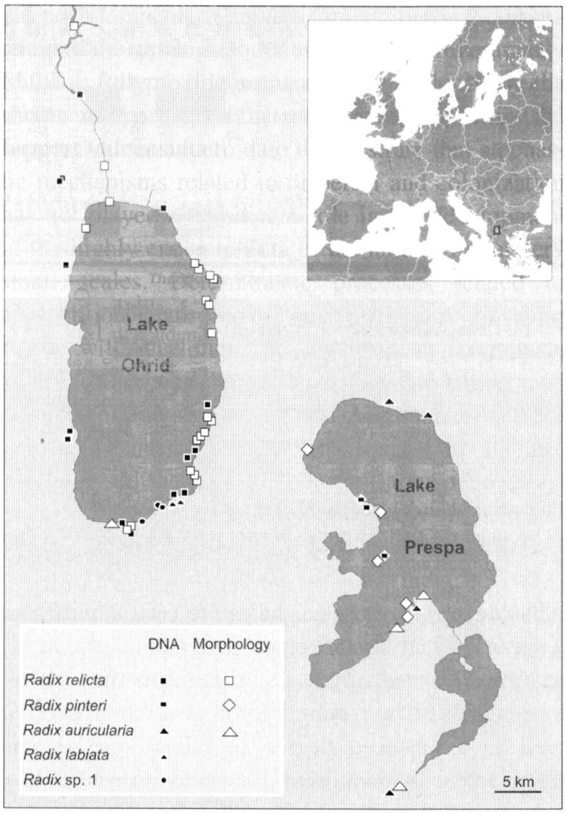

Fig. 1 Distribution pattern of *Radix* spp. in the Ohrid/Prespa region. Populations studied by morphology/anatomy (*white symbols*) or (additionally) DNA sequences (*black symbols*) are shown; for details of sampling localities, see Appendix 1

are less than 10 km apart, a high mountain chain (Galičica Mountains) separates the two basins (Fig. 1).

Subterranean karstic channels, however, are known to connect lakes Ohrid and Prespa (Amataj et al., 2007), and according to Matzinger et al. (2006), Lake Prespa contributes approx. 20% to the water balance of Lake Ohrid. Although the age of Lake Prespa remains unknown, they presumably co-existed for several hundreds of thousand years (Radoman, 1985). In fact, some workers even suggest that the fauna of Lake Prespa is older than that of Lake Ohrid, at least according to fish endemism data (e.g. Karaman, 1971). While fish endemism in both lakes has attracted many researchers, the lakes are arguably most famous for their extraordinary degree of mollusc endemism, which has been studied for more than 100 years (e.g. Sturany, 1894; Polinski, 1932; Hadžišče, 1956; Radoman, 1985; Burch & Hadžišče,

1974). There are three general patterns of endemic species that can be distinguished for lakes Ohrid and Prespa:

1) taxa that are endemic to either lake with no close relatives in the respective sister lake,
2) closely related but distinct endemic taxa in both lakes (sister species) and
3) shared endemic taxa occurring in both lakes.

Of particular interest for evolutionary biologists is the presence of sister species in ancient lakes, as these taxa cannot only help unravelling processes of speciation in space in time; they can also help to establish hypotheses about the hydrological or even geological histories of the respective lakes (e.g. Wilke et al., 2007). Although sister species are often postulated for related water bodies, they are sometimes difficult to identify, particularly when dealing with inconspicuous taxa. In fact, robust data showing sister species relationships in molluscs of lakes Ohrid and Prespa are still sparse, for example, some species of the genus *Pisidium* (Schultheiß et al., 2008).

A candidate group for studying sister species in these lakes is the pulmonate family Lymnaeidae. The group is best known for some of its members acting as an intermediate host for trematodiases (Bargues et al., 2001). Many representatives are widespread (Remigio, 2002) or even invasive (Cowie, 2001), and narrow-range endemism is unusual in these gastropods. However, for lakes Ohrid and Prespa, one endemic lymnaeid species each has been described. In Lake Ohrid, *Radix relicta* Polinski, 1929, occurs in depths up to 30 m. This relatively large species (up to 3 cm in length) has a special physiological adaptation to life in greater depths (Kaiser, 1959). Its pneumostome is completely closed, tissues are less vascularised, and respiration takes places exclusively through the body surface. From Lake Prespa, *Radix pinteri* Schütt, 1974, is believed to be a 'relictary' Pontic species (Schütt, 2006). It lives on hard substrata and is characterized by its expanded last whorl and extraordinarily large aperture. While the two species can clearly be distinguished by their shell morphology, there are only relatively few diagnostic anatomical characters separating the two taxa. They include, among others, a cylindrical versus sacciform bursa copulatrix in *R. relicta* and *R. pinteri*, respectively (Fig. 2).

Several family-level molecular phylogenies for the Lymnaeidae have been published. These studies include nuclear 18S rDNA (Bargues & Mas-Coma, 1997), ITS2 (Bargues et al., 2001) and mitochondrial 16S rDNA data (Remigio & Blair, 1997; Remigio, 2002). The specific composition and relationships within the genus (or subgenus) *Radix*, however, remain unknown or confused. Even the monophyly of *Radix* was recently doubted (Remigio, 2002). Some attempts have been made to clarify the phylogenetics of western and northwestern European *Radix* populations (Bargues et al., 2001; Pfenninger et al., 2006). Molecular data, however, are almost entirely absent for Balkan taxa in general and for the ancient lake taxa in particular.

Given that only two endemic *Radix* taxa are known for lakes Ohrid and Prespa, with one taxon occurring in each lake, we here use biogeographical and ecological data, as well as DNA sequences from the mitochondrial cytochrome oxidase *c* subunit I (COI) gene to:

a) develop a phylogenetic and biogeographical framework for Balkan *Radix* spp. and discuss their mode of diversification in European ancient lakes,
b) test the sister species relationship of Lake Ohrid *Radix relicta* and Lake Prespa *R. pinteri* and study their degree of differentiation as well as their putative relictary status and
c) discuss the status of lakes Ohrid and Prespa as sister lakes.

Materials and methods

Materials

Materials from the Dessarete Lakes and other water bodies of the Balkan Peninsula, including ponds, springs, artificial lakes, streams and rivers, were mainly collected during several field trips carried out between May 2003 and September 2005 (Fig. 2). In addition to the Balkan samples, *Radix* populations from Germany, Austria, Corsica and Asia Minor were included in the analyses. Individuals were obtained by hand collecting from hard substrata in shallow waters or from stones and rocks lifted from depths to 5 m by snorkelling. Deeper parts of the littoral and sublittoral to 60 m were sampled using a dredge from small boats or from the research vessel of the

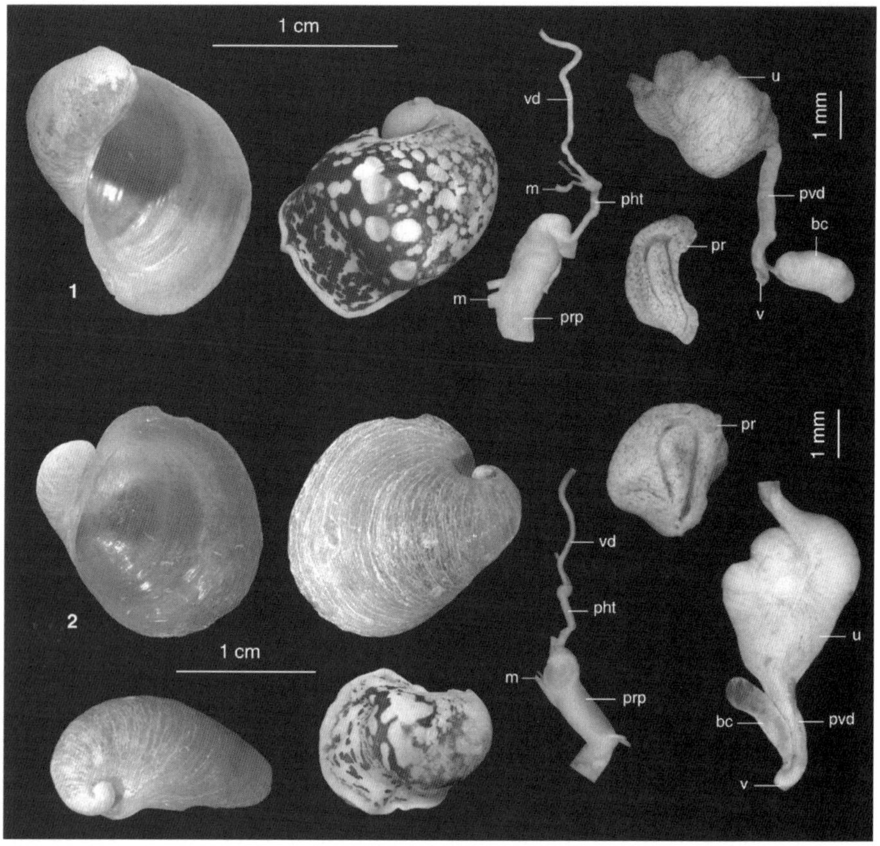

Fig. 2 Shell morphology and anatomy of *Radix relicta* (1) and *Radix pinteri* (2) with shell, mantle pigmentation, male copulatory organ and female genital organs. *Abbreviations*: bc = bursa copulatrix, bd = bursa duct, cp = corpus pyriforme, m = protractor and retractor muscle, pht = phallotheca, pr = prostate, prp = praeputium, pvd = provaginal duct, v = vagina, vd = vas deferens

Hydrobiological Institute in Ohrid (HBI). Locality information, collecting details and GenBank accession numbers are provided in Appendix 1. All materials are deposited at the permanent DNA and tissue collection of the University of Giessen, Systematics and Biodiversity Group (UGSB).

Several outgroup taxa of the family Lymnaeidae representing all European genera with their respective type species were included in the phylogenetic analyses to account for the potential paraphyly of *Radix*. Representatives of the family Physidae, *Haitia acuta* and *Physa fontinalis*, were used as additional outgroups. As primary outgroup for rooting the trees, *Planorbarius corneus* (family Planorbidae) was used.

DNA isolation and sequencing

The method described in Wilke et al. (2006) was used for isolating DNA from individual snails. The primers for amplifying a fragment of the COI gene with a target length of 655 base pairs (excluding 51 bp primer sequence) were LCO1490 and HCO2198, as described by Folmer et al. (1994). Sequences (forward and reverse) were determined using the LI-COR (Lincoln, NE) DNA sequencer Long ReadIR 4200 and the Thermo Sequenase Fluorescent Labeled Primer Cycle Sequencing kit (Amersham Pharmacia Biotech, Piscataway, NJ). The protein-coding COI sequences were aligned unambiguously by eye using BioEdit 7.0.4.1 (Hall, 1999).

The first base pairs (bp) behind the 3′ end of each primer were difficult to read. We therefore uniformly cut these parts, leaving an up to 600 bp-long overlapping fragment for the COI gene. Sequences of the non-*Radix* taxa were derived using the methodology described in Albrecht et al. (2004, 2007). All sequences are available at GenBank (see Appendix 1).

Phylogenetic analyses

Two commonly employed approaches were chosen for phylogenetic reconstructions, Bayesian inference (BI) and maximum parsimony (MP). Prior to the phylogenetic analyses, the computer program

Modeltest 3.6 (Posada & Crandall, 1998) was used to find the optimal model of DNA substitution based on the Akaike information criterion. For the COI dataset, the K81uf + I + G model was selected with base frequencies of A = 0.3023, C = 0.1204, G = 0.1328 and T = 0.4445; a rate matrix of [A – C] = 1.0000, [A – G] = 9.6093, [A – T] = 2.7121, [C – G] = 2.7121, [C – T] = 9.6093, [G – T] = 1.000; a proportion of invariable sites of 0.3225; and a gamma distribution shape parameter of 0.7060.

BI was performed utilizing MrBayes 3.0b4 (Huelsenbeck & Ronquist, 2001). For a general understanding of phylogenetic relationships within the genus *Radix*, BI analysis of 40 ingroup specimens and four lymnaeid as well as three non-lymnaeid outgroup species was performed based on the best-fit model. During a preliminary run, the log-likelihoods started at around −8,600 and quickly converged on a stable value of about −4,900 after approximately 6,500 generations. The final run was then carried out with four chains (one cold, three heated) and 1,000,000 sampled generations with the current tree saved at intervals of ten generations. A 50% majority rule tree was constructed from all sampled trees with the first 2,000 trees (= 20,000 generations) ignored as burn in.

The MP analysis was conducted using PAUP* v. 4.0b10 (Swofford, 2002) with branch-and-bound search, 500 random-addition-sequence replications and TBR branch swapping. Statistical robustness was assessed by non-parametric bootstrapping (Felsenstein, 1985) with 500 replications.

Parametric bootstrapping

For testing the sister species relationship of Lake Ohrid *Radix relicta* and Lake Prespa *R. pinteri* (i.e. the monophyly of the two taxa), we used the parametric bootstrapping approach (Hillis et al., 1996; Huelsenbeck & Crandall, 1997). First, we conducted maximum likelihood (ML) searches in PAUP under the topological constraint of non-monophyly (null hypothesis) and the best-fit model of sequence evolution. The resulting tree was, together with the aligned sequences, imported into Seq-Gen 1.3.1. (Rambaut & Grassly, 1997) to generate 100 random datasets based on the model suggested by Modeltest. We then analyzed in PAUP the ratios of the likelihood differences in tree lengths between the constrained and unconstrained trees for each of the 100 replicates. The frequency of likelihood ratios of the simulated datasets then was plotted and compared to the likelihood-ratio (constrained vs. unconstrained topologies) of the original dataset. Finally, we estimated how likely it is that this difference could have been observed randomly, that is, whether the null hypothesis can be rejected.

Phylogeographical analysis

We utilized a statistical parsimony (SP) network approach to demonstrate phylogeographical relationships among the putative *Radix* sister species from lakes Ohrid and Prespa. The SP network was generated in TCS 1.18 (Clement et al., 2000) and a reduced dataset (14 sequences, 365 bp long).

Genetic distances within and between *R. relicta and R. pinteri* were calculated using MEGA 3.1 (Kumar et al., 2004) by employing the Kimura-2-Parameter model. Standard errors were calculated with 500 bootstrap replicates and a random seed.

Results

Phylogenetic and phylogeographical relationships

The phylogenetic analyses of the mitochondrial COI dataset resulted in a mostly well-supported phylogeny comprising ten major *Radix* clades (Fig. 3; MP consensus tree with CI of 0.866). The topologies derived from both analyses were compatible. The major clades correspond to six recognized *Radix* species and four lineages for which most likely no names are available and that are here numbered consecutively for convenience. The endemic *Radix relicta* (Lake Ohrid) and *R. pinteri* (Lake Prespa) together form a well-supported monophylum (100% bootstrap support [BS], 0.99 Bayesian posterior probabilities [BPP]) with *Radix ampla* being its sister species. However, their respective relationships could not be resolved. An unnamed lineage (*Radix* sp. 1) is endemic to the feeder springs south of Lake Ohrid and clustered as sister taxon to the clade *R. relicta* + *R. pinteri* + *R. ampla*. These four species form a well-supported monophylum (100% BS, 0.96 BPP). A species from the Turkish Burdur Province (*Radix* sp. 3) was the sister species to the latter clade. The remaining species in the dataset

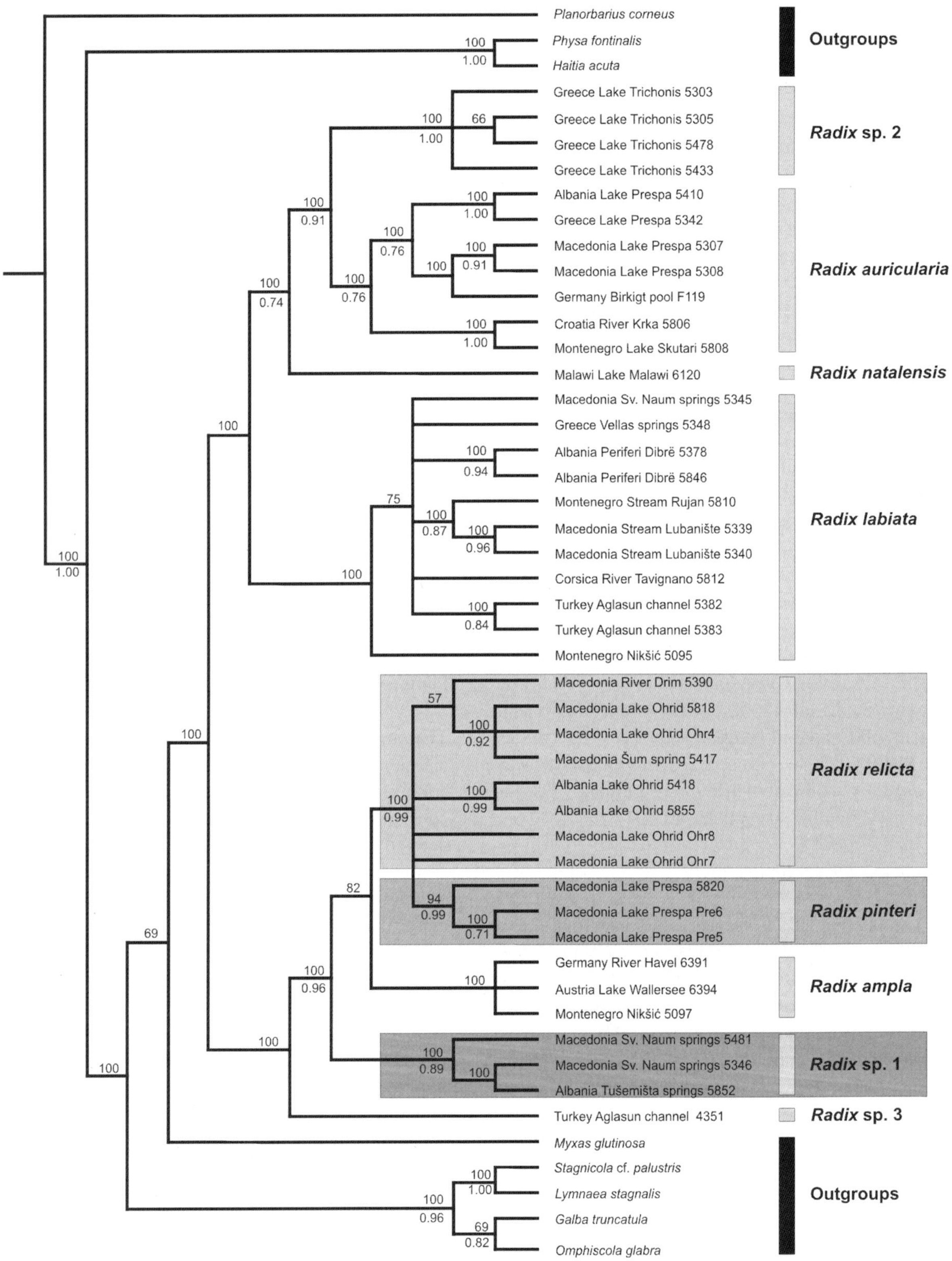
Planorbarius corneus
Physa fontinalis
Haitia acuta
Outgroups
Greece Lake Trichonis 5303
Greece Lake Trichonis 5305
Greece Lake Trichonis 5478
Greece Lake Trichonis 5433
Radix sp. 2
Albania Lake Prespa 5410
Greece Lake Prespa 5342
Macedonia Lake Prespa 5307
Macedonia Lake Prespa 5308
Germany Birkigt pool F119
Croatia River Krka 5806
Montenegro Lake Skutari 5808
Radix auricularia
Malawi Lake Malawi 6120
Radix natalensis
Macedonia Sv. Naum springs 5345
Greece Vellas springs 5348
Albania Periferi Dibrë 5378
Albania Periferi Dibrë 5846
Montenegro Stream Rujan 5810
Macedonia Stream Lubanište 5339
Macedonia Stream Lubanište 5340
Corsica River Tavignano 5812
Turkey Aglasun channel 5382
Turkey Aglasun channel 5383
Montenegro Nikšić 5095
Radix labiata
Macedonia River Drim 5390
Macedonia Lake Ohrid 5818
Macedonia Lake Ohrid Ohr4
Macedonia Šum spring 5417
Albania Lake Ohrid 5418
Albania Lake Ohrid 5855
Macedonia Lake Ohrid Ohr8
Macedonia Lake Ohrid Ohr7
Radix relicta
Macedonia Lake Prespa 5820
Macedonia Lake Prespa Pre6
Macedonia Lake Prespa Pre5
Radix pinteri
Germany River Havel 6391
Austria Lake Wallersee 6394
Montenegro Nikšić 5097
Radix ampla
Macedonia Sv. Naum springs 5481
Macedonia Sv. Naum springs 5346
Albania Tušemišta springs 5852
Radix sp. 1
Turkey Aglasun channel 4351
Radix sp. 3
Myxas glutinosa
Stagnicola cf. palustris
Lymnaea stagnalis
Galba truncatula
Omphiscola glabra
Outgroups

◀ **Fig. 3** Maximum parsimony majority-rule consensus tree of *Radix* spp. based on 600 nucleotide positions of the COI gene. The tree was rooted with planorbid and physid outgroups. Additionally, all European lymnaeid genera are represented by their respective type species in the dataset. Maximum parsimony bootstrap values (>70%) are provided above and Bayesian posterior probabilities (>0.80) below the branches. For explanations of species assignments, see text. New species-level clades are consecutively numbered for convenience. Endemics of Lake Ohrid (including its southern feeder springs) and Lake Prespa are highlighted in grey boxes

form a second major clade, containing, among others, the widespread *Radix labiata*. This species was represented by populations from Corsica, Greece, Turkey, Montenegro and Albania, and the Lake Ohrid inflow stream Lubanište and the feeder spring complex Sv. Naum. The sister clade to *R. labiata* comprises a possibly undescribed species from Lake Trichonis (Greece), *Radix natalensis* from Lake Malawi (Malawi) and *Radix auricularia* (Germany, Croatia, Montenegro [Lake Skutari] and many parts of Lake Prespa).

Given the inability of phylogenetic methods used here to fully resolve the closely related sister species *R. relicta* and *R. pinteri*, a SP network of 14 specimens was generated (Fig. 4). The network analysis resulted in a total of 11 haplotypes and a clear separation of the two species. Acknowledging that only three specimens of *R. pinteri* could be included in the analysis, it is obvious that genetic diversity is more pronounced in *R. relicta*. The SP network suggested one *R. relicta* haplotype occurring in the spring complex Sv. Naum and the eastern shore of the lake as having the highest probability of being the ancestral haplotype of the analysed specimens. The average K2P diversity within *R. relicta* was 0.009 (±0.004) and within *R. pinteri* 0.005 (±0.003), whereas the divergence between *R. relicta* and *R. pinteri* was calculated to be 0.024 (±0.008).

Testing the sister species null hypothesis

As the phylogenetic analyses suggest *R. relicta* and *R. pinteri* to form a monophyletic group, but could not resolve the relationship between the two species, we tested the null hypothesis that these taxa are not monophyletic using parametric bootstrapping.

From the 100 sampled trees from the simulated datasets, all had differences in log likelihood units of ≤5.1. As the observed difference in the original dataset was 34.0 log likelihood units, the null hypothesis is rejected at $P < 0.01$, and the alternative hypothesis—monophyly of *R. relicta* and *R. pinteri*—can be accepted.

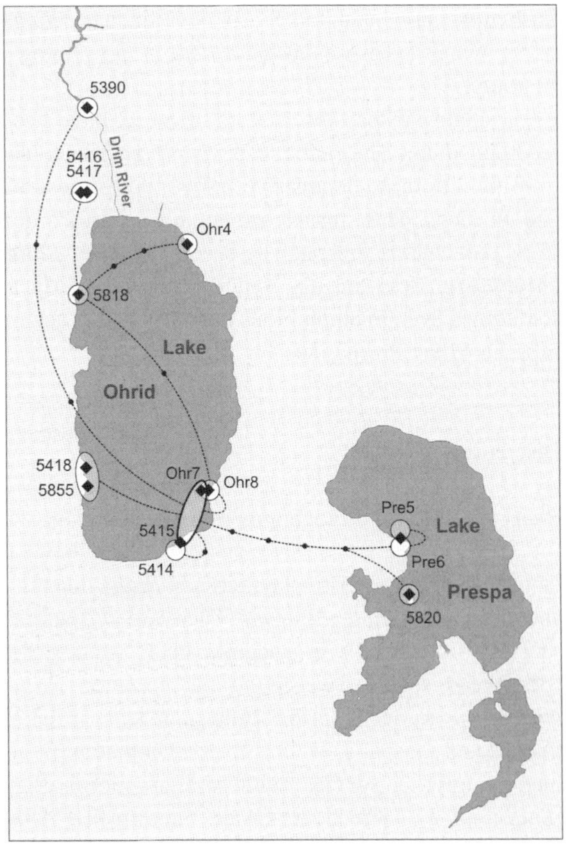

Fig. 4 COI area cladogram of *Radix pinteri* and *R. relicta* based on a statistical parsimony network. Missing haplotypes are indicated by small black dots. The haplotype with the highest outgroup probability is indicated by a bold outline

Distribution and ecology

Radix relicta is not restricted to Lake Ohrid but occurs in the Drim System at least upstream to the dam Lake Globočičko (approximately 15 km N of Lake Ohrid). In the lake proper of Lake Ohrid, no other *Radix* species than *R. relicta* was found in depths from 0 m to 35 m. It lives mainly on hard substrata (e.g. rocks, *Dreissena* bundles) or on algae of the *Chara* belt.

Radix labiata was recorded in one stream approximately 0.5 km SE of Lake Ohrid, as well as in the feeder spring complex Sv. Naum. These springs and the sister complexes in Albania (Zagorican and

Tushemist) are inhabited by another lineage (*Radix* sp. 1), which resembles *Radix relicta* in morphology/anatomy.

For Lake Prespa, two species of *Radix* have been recorded so far. The endemic *Radix pinteri* was found from the littoral to depths of 6 m on muddy substrata and on rocks. This species has not yet been recorded from Lake Mikri Prespa. The second species, *Radix auricularia*, was mainly found near the surface, sometimes even in amphibious habitats. It also occurs in Lake Mikri Prespa (see Fig. 2).

Discussion

Radix diversity and biogeography in the Balkans

Our study revealed distinct patterns of *Radix* diversity and distribution on the Balkans. There are widespread species contrasted with endemic taxa. Among the widespread species, we found the common *Radix auricularia* to occur in the Balkans at least in lakes Skutari and Prespa, as well as in one Dalmatian river. Interestingly, a species with similar morphology (*Radix* sp. 2; Albrecht & Glöer, unpublished data) so far occurs exclusively in Lake Trichonis (Greece). Hubendick's (1951) superspecies concept of '*Radix auricularia*' becomes less convincing given the new data indicate the existence of distinct phylogenetic units within '*R. auricularia*'. Another widespread species is *Radix labiata* with occurrences throughout the Balkans and southwest into Turkey (Aglasun). *Radix labiata* is a characteristic species of springs and small rivers. Bargues et al. (2001) also mentioned this taxon from Turkey as well as an undetermined *Radix* species, indicating the need for more comprehensive sampling in Asia Minor.

Concerning the more restricted *Radix* species on the Balkan Peninsula, it is an unexpected outcome of this study to find two geographically localized clades (*Radix* sp. 1 from the Lake Ohrid feeder springs and *Radix* sp. 2 from Lake Trichonis) besides the well-known Ohrid and Prespa endemic species. The latter, *R. relicta* and *R. pinteri*, are closely related to another widespread species with comparably low genetic diversity, *R. ampla*. Such a relationship was already suggested by Polinski (1932). *Radix ampla* itself was neither found in lakes Ohrid and Prespa nor in their surroundings. The nearest confirmed records are from Nikšić (Montenegro) as well as from Romania (Glöer & Sîrbu, 2006). This pattern of occurrence of endemics in lakes Ohrid and Prespa in combination with rather long geographical distances to the next known locality of the respective widespread sister species was also found for some other taxa. A good example is the endemic pea clam species studied by Schultheiß et al. (2008). The absence of *R. balthica* in the investigated waters of the Balkans is remarkable, despite a long list of published records for the area. This can probably be attributed to a persistent taxonomic confusion of European *Radix* spp. in general (discussed by Bargues et al., 2001) and a mere adoption of a central European taxonomic framework for the Balkan species in particular. Such a complex situation was already demonstrated for other gastropod groups on the Balkans such as the Bithyniidae (Glöer et al., 2007). Given our results, we would expect to find more distinct lineages on the Balkans, a tendency already shown for northwestern Europe (Pfenninger et al., 2006).

Sister species *Radix relicta* and *R. pinteri*

The phylogenetic relationship between these two endemic species from lakes Ohrid and Prespa is close as shown in the phylogenetic analysis (Fig. 3). The fact that these sister taxa are not fully resolved is, however, not surprising. Part of the problem is the high diversity within these presumably old populations, which may obscure species-level divergences (Albrecht et al., 2006, discuss a similar example in Lake Ohrid *Ancylus*), as well as the fact that most phylogenetic packages using traditional tree-building approaches may perform poorly at and below the species level (Posada & Crandall, 2001). Thus, the dedicated phylogeographical network analysis performed here (Fig. 4) can clearly separate the two sister species. Moreover, parametric bootstrapping rejected the null hypothesis of non-monophyly of *R. relicta* and *R. pinteri* and the two species can be assigned to pattern 2 of endemicity discussed in the Introduction.

As mentioned before, Lake Ohrid *Radix relicta* is adapted to live in greater depths. Although adaptation to greater depth and bathymetric segregation is a common feature of the Ohrid gastropod fauna (Radoman, 1985), it is not at all common for lymnaeids. Interestingly, Lake Prespa *R. pinteri* was also found down to 6 m. It has been suggested that *R. pinteri* occurs on soft substrata (Schütt, 1974) and that therefore ecological constraints would favour its extraordinary shell shape. We could, however, also regularly find it on stones and rocks. If, indeed, ecophenotypical modulation is responsible for the shell shape pattern seen in *R. pinteri*, other factors should be considered as well, such as wave action (at least during certain times of the year). The existence of fossil lymnaeids with shell forms having enlarged apertures and expanded last whorls led Schütt (1974, 2006) to conclude that both species represent 'relictary' species. The phylogenetic results of our study and the close relationship to *Radix ampla* raise doubts as to their ancientness. It has to be noted that *R. ampla* has not been considered a distinct species for a long time (e.g. Jackiewicz, 1998). Our mitochondrial data supports findings based on nuclear sequences, suggesting specific distinctness of *R. ampla* as previously proposed (Falkner, 1990). Schütt (1974) doubted a close relationship of the endemic *Radix* with amploid forms and could not address the question of the relationship to *R. auricularia*. We here showed that *Radix auricularia* is a distant lineage.

Among the most obvious adaptive zones in the Ohrid basin are spring habitats both within the actual lake or adjacent to it. The most important spring complexes are the southern feeder springs at Sv. Naum (Macedonia) and the less well-studied neighbouring spring complex Zagorican/Tushemist (Albania). These springs have long been recognized as a habitat for endemic species that have close relatives in the actual lake (e.g. Radoman, 1985). We found a distinct *Radix* lineage (*Radix* sp. 1) in both springs systems that morphologically resembles *Radix relicta*. Its specific character had already been suspected based on an immunotaxonomic study (Burch & Hadžišče, 1974). Other examples of closely related species in the feeder spring systems and the actual lake proper include *Carinogyraulus* spp. (Hubendick & Radoman, 1959; Gorthner & Meier-Brook, 1985) and *Ancylus* spp. (Albrecht et al., 2006). However, in contrast to *Carinogyraulus* and *Ancylus*, *Radix* did not radiate in Lake Ohrid (a pattern also found in *Valvata* spp.; Hauswald et al., 2008). Nonetheless, the statistical parsimony network (Fig. 4) shows substructuring of *R. relicta* throughout the lake as well as in surrounding springs and rivers that could lead to diversification. Although, it is not clear from the available data whether populations of *R. relicta* in the Drim River (i.e. the outflow of Lake Ohrid) and adjacent waters in the northern Ohrid Basin recently dispersed there or whether they constitute remnant populations from times when all these waters belonged to the former, larger lake and thus were sublacustrine. The pond at the Šum Spring, however, is artificial, and *R. relicta* might therefore have been introduced there recently (possibly with eggs for the salmon hatchery that uses the pond).

Considering the diversity within *R. relicta* and the divergence between Lake Ohrid *R. relicta* and Lake Prespa *R. pinteri*, the genetic differentiation in the COI gene between the two taxa is with ≤ 5 mutational steps (Fig. 4) or $\leq 1.7\%$ K2P distance relatively low, yet distinct. Acknowledging that detailed molecular clock analyses are beyond the scope of the paper and that the limited data set presented here (i.e. missing data) and the likely presence of ancestral polymorphism might result in an overestimation of divergence times, the most recent common ancestor of the two species would be ≤ 0.75 My old (Albrecht et al., unpublished data).

Sister lakes Ohrid and Prespa

Given the close phylogenetic relationship of endemic taxa from Lake Ohrid and Lake Prespa found in the present study as well as by other workers, it is worth discussing the putative sister lake status of these lakes. According to our understanding of ancient sister lakes, they have to lie in close geographic proximity. This is undoubtedly the case with extant lakes Prespa and Ohrid, although the altitudinal difference is considerable, approximately 156 m. As members of the Dessarete Lake group, both lakes likely share a related origin as graben systems that formed during the Uppermost Meotioan-Pontian (Dumurdzanov et al., 2004), and therefore, share a significant time of co-existence. The hydrological connection criterion is easily applicable for lakes Prespa and Ohrid due to their karstic underground connection (Cvijic, 1911), as demonstrated by modern trace methods (Amataj et al., 2007). As far as a balanced degree of faunal overlap and distinctness is concerned, there are only few examples of species that occur both in lakes Prespa and Ohrid (Radoman, 1985). One case would be *Dreissena presbensis* (Albrecht et al., 2007). However, a pattern similar to that in *Radix* was found in the genus

Pisidium, of which the respective sister species live in lakes Prespa and Ohrid (Schultheiß et al., 2008). Future phylogenetic analyses of other zoobenthic taxa will unravel the degree of recent faunal overlap between lakes Ohrid and Prespa and give a more complete picture of the biogeographic history of these lakes. Preliminary data for several hydrobiid groups (e.g. Wilke et al., 2007) already challenge Radoman's (1985) hypothesis of a partial Aegean–Anatolian faunal connection of Lake Prespa. Rather, a western Adriatic zoogeographical affinity (also assumed by Radoman, 1985) appears more likely and supports our sister lake hypothesis for lakes Ohrid and Prespa. As demonstrated here for *Radix*, these sister lakes provide prime models for studying in situ evolutionary processes.

Acknowledgements We are very grateful to the colleagues of the Hydrobiological Institute Ohrid for their immense support. S. Trajanovski, B. Budzakoska and S. Trajanovska provided valuable information and took part in some field trips. D. Georgiev and B. Streit generously supported our field work. T. Kevrekidis and A. Mogias perfectly organized field work in Greece. Z. Yildirim and S. Koşal Şahin sent material from Turkey, whereas Z. Fehér provided Albanian material, M. Zettler material from Corsica, R. Patzner from Wallersee, and V. Pešić from Montenegro and Serbia. The comments of two anonymous reviewers enhanced an earlier version of the manuscript. We wish to express our deepest gratitude to all of them.

References

Albrecht, C., R. Schultheiß, T. Kevrekidis, B. Streit & T. Wilke, 2007. Invaders or endemics? Molecular phylogenetics, biogeography and systematics of *Dreissena* in the Balkans. Freshwater Biology 52: 1525–1536.

Albrecht, C., S. Trajanovski, K. Kuhn, B. Streit & T. Wilke, 2006. Rapid evolution of an ancient lake species flock: freshwater limpets (Gastropoda: Ancylidae) in the Balkan lake Ohrid. Organisms Diversity & Evolution 6: 294–307.

Albrecht, C. & T. Wilke, 2008. Lake Ohrid: biodiversity and evolution. Hydrobiologia. doi:10.1007/s10750-008-9558-y.

Albrecht, C., T. Wilke, K. Kuhn & B. Streit, 2004. Convergent evolution of shell shape in freshwater limpets. The African genus Burnupia. Zoological Journal of the Linnean Society 140: 577–588.

Amataj, S., T. Anovski, R. Benischke, R. Eftimi, L. L. Gourcy, L. Kola, I. Leontiadis, E. Micevski, A. Stamos & J. Zoto, 2007. Tracer methods used to verify the hypothesis of Cvijic about the underground connection between Prespa and Ohrid Lake. Environmental Geology 51: 749–753.

Bargues, M. D. & S. Mas-Coma, 1997. Phylogenetic analysis of lymnaeid snails based on 18S rDNA sequences. Molecular Biology and Evolution 14: 569–577.

Bargues, M. D., M. Vigo, P. Horak, J. Dvorak, R. A. Patzner, J. P. Pointier, M. Jackiewicz, C. Meier-Brook & S. Mas-Coma, 2001. European Lymnaeidae (Mollusca: Gastropoda), intermediate hosts of trematodiases, based on nuclear ribosomal DNA ITS-2 sequences. Infection. Genetics and Evolution 1: 85–107.

Burch, J. B. & S. Hadžišče, 1974. Immunological relationships of two species of Lymnaeidae (Mollusca: Basommatophora) from Macedonia, Yugoslavia. Malacological Review 7: 25–32.

Clement, M., D. Posada & K. Crandall, 2000. TCS. A computer program to estimate gene genealogies. Molecular Ecology 9: 1657–1660.

Coulter, G. W. (ed.), 1991. Lake Tanganyika and Its Life. Oxford University Press, New York.

Cowie, R. H., 2001. Invertebrate invasions on Pacific Islands and the replacement of unique native faunas: a synthesis of the land and freshwater snails. Biological Invasions 3: 119–136.

Cvijic, J., 1911. Fundamentals of geography and geology of Macedonia and Old Serbia. Book III, Serbian Academy of Sciences, Special Edition, Beograd (in Serbian).

Denèfle, M., A.-M. Lézine, E. Fouache & J.-J. Dufaure, 2000. A 12, 000-year pollen record from Lake Maliq, Albania. Quaternary Research 54: 423–432.

Dumurdzanov, N., T. Serofimovski & B. C. Burchfield, 2004. Evolution of the Neogene-Pleistocene Basins of Macedonia. Geological Society of America Digital Map and Chart Series 1: 1–20.

Falkner, G., 1990. Binnenmollusken. In Fechter, R. & G. Falkner (eds), Weichtiere. Europäische Meeres- und Binnenmollusken. Mosaik Verlag, München.

Felsenstein, J., 1985. Confidence limits on phylogenies: an approach using the bootstrap. Evolution 39: 783–791.

Folmer, O., M. Black, W. Hoeh, R. Lutz & R. Vrijenhoek, 1994. DNA primers for amplification of mitochondrial cytochrome c oxidase subunit I from diverse metazoan invertebrates. Molecular Marine Biology & Biotechnology 3: 294–299.

Glöer, P., C. Albrecht & T. Wilke, 2007. Enigmatic distribution pattern of the Bithyniidae in the Balkan Region (Gastropoda: Rissooidea). Mollusca 25: 13–22.

Glöer, P. & I. Sîrbu, 2006. Freshwater molluscs species, new fort he Romanian fauna. Heldia 6: 207–216.

Gorthner, A., 1994. What is an ancient lake? Archiv für Hydrobiologie 44: 97–100.

Gorthner, A. & C. Meier-Brook, 1985. The Steinheim Basin as a paleo-ancient lake. In Bayer, U. & A. Seilacher (eds), Sedimentary and Evolutionary Cycles. Springer, Berlin, Heidelberg, New York, Tokyo: 322–334.

Hadžišče, S., 1956. III. Beitrag zur Kenntnis der Gastropodenfauna des Ohridsees. Beschreibung der bis jetzt unbekannten Schnecken und Beispiele der Speciation bei den Gastropoden des Ohridsees. Recueil des Traveaux, Station Hydrobiologique. Ohrid 4: 57–107.

Hall, T. A., 1999. BioEdit. A user-friendly biological sequence alignment editor and analysis program for Windows 95/98/NT. Nucleic Acids Symposium Series 41: 95–98.

Hauswald, A.-K., C. Albrecht & T. Wilke, 2008. Testing two contrasting evolutionary patterns in ancient lakes: species flock vs. species scatter in valvatid gastropods of Lake Ohrid. Hydrobiologia. doi:10.1007/s10750-008-9556-0.

Hillis, D. M., B. K. Mable & C. Moritz, 1996. Applications of molecular systematics. In Hillis, D. M. C. Moritz & B. K. Mable (eds), Molecular Systematics, 2 edn. Sinauer Associates Inc., Sunderland, MA: 515–543.

Hubendick, B., 1951. Recent Lymnaeidae. Their variation, morphology, taxonomy, nomenclature, and distribution. Kunglinska Svenska Vetenskapsacedemiens Handlingar, Ser. 4 3: 1–223.

Hubendick, B. & P. Radoman, 1959. Studies on the Gyraulus species of Lake Ochrid. Morphology. Archiv för Zoologi 12: 223–243.

Huelsenbeck, J. P. & K. A. Crandall, 1997. Phylogeny estimation and hypothesis testing using maximum likelihood. Annual Review of Ecology and Systematics 28: 437–466.

Huelsenbeck, J. P. & F. Ronquist, 2001. MRBAYES. Bayesian inference of phylogeny. Bioinformatics 17: 754–755.

Jackiewicz, M., 1998. European species of the family Lymnaeidae (Gastropoda: Pulmonata: Basommatophora). Genus 9: 1–93.

Karaman, M., 1971. Zoogeographische Verhältnisse des Prespa- und Ohridsees. Izdanija 4: 1–21.

Kaiser, P., 1959. Über die Atmung von *Radix relicta* Pol. aus dem Ohrid-See. Recueil des travaux Ohrid 35: 1–5.

Kumar, S., K. Tamura & M. Nei, 2004. MEGA3: integrated software for molecular evolutionary genetics analysis and sequence. Alignment Briefings in Bioinformatics 5: 150–163.

Martens, K., 1997. Speciation in ancient lakes (review). Trends in Ecology and Evolution 12: 177–182.

Matzinger, A., M. Jordanoski, E. Veljanoska-Sarafiloska, M. Sturm, B. Müller & A. Wüest, 2006. Is Lake Prespa jeopardizing the ecosystem of ancient Lake Ohrid? Hydrobiologia 553: 89–109.

Matzinger, A., M. Schmid, E. Veljanoska-Sarafiloska, S. Patceva, D. Guseska, B. Wagner, B. Müller, M. Sturm & A. Wüest, 2007. Eutrophication of ancient Lake Ohrid: global warming amplifies detrimental effects of increased nutrient inputs. Limnology and Oceanography 52: 338–353.

Pfenninger, M., M. Cordellier & B. Streit, 2006. Comparing the efficacy of morphologic and DNA-based taxonomy in the freshwater gastropod genus *Radix* (Basommatophora, Pulmonata). BMC Evolutionary Biology 6: 100.

Polinski, W., 1932. Die reliktäre Gastropodenfauna des Ohrida-Sees. Zoologische Jahrbücher Abteilung Systematik 62: 611–666.

Posada, D. & K. A. Crandall, 1998. Modeltest: testing the model of DNA substitution. Bioinformatics 14: 817–818.

Posada, D. & K. A. Crandall, 2001. Selecting the best-fit model of nucleotide substitution. Systematic Biology 50: 580–601.

Radoman, P., 1985. Hydrobioidea, a superfamily Prosobranchia (Gastropoda), II. Origin, zoogeography, evolution in the Balkans and Asia Minor. Monographs Institute of Zoology 1, Beograd.

Rambaut, A. & N. C. Grassly, 1997. Seq-Gen: an application for the Monte Carlo simulation of DNA sequence evolution along phylogenetic trees, Version 1.1. Computer Applications in the Biosciences 13: 235–238.

Remigio, E. A., 2002. Molecular phylogenetic relationships in the aquatic snail genus Lymnaea, the intermediate host of the causative agent of fascioliasis: insights from broader taxon sampling. Parasitology Research 88: 687–696.

Remigio, E. A. & D. Blair, 1997. Molecular systematics of the freshwater snail family Lymnaeidae (Pulmonata: Basommatophora) utilising mitochondrial ribosomal DNA sequences. Journal of Molluscan Studies 63: 173–185.

von Rintelen, T. & M. Glaubrecht, 2005. Anatomy of an adaptive radiation: a unique reproductive strategy in the endemic freshwater gastropod *Tylomelania* (Cerithioidea: Pachychilidae) on Sulawesi, Indonesia, and its biogeographic implications. Biological Journal of the Linnean Society 85: 513–542.

von Rintelen, T. & M. Glaubrecht, 2006. Rapid evolution of sessility in an endemic species flock of the freshwater bivalve *Corbicula* from ancient lakes on Sulawesi, Indonesia. Biology Letters 2: 73–77.

Schultheiß, R., C. Albrecht, U. Bößneck & T. Wilke, 2008. The neglected side of speciation in ancient lakes: phylogeography of an inconspicuous mollusk taxon in lakes Ohrid and Prespa. Hydrobiologia. doi:10.1007/s10750-008-9553-3.

Schütt, H., 1974. Zwei neue reliktäre Süßwassermollusken der Dinariden. Annalen des Naturhistorischen Museums in Wien 78: 473–480.

Schütt, H., 2006. Die Entrollung des Gewindes bei Lymnaeiden (Eine Literaturübersicht). Mitteilungen der deutschen malakozoologischen Gesellschaft 75: 5–7.

Spirkovski, Z., O. Avramovski & A. Kodzoman, 2001. Watershed management in the Lake Ohrid region of Albania and Macedonia. Lakes & Reservoirs: Research and Management 6: 237–242.

Stanković, S., 1960. The Balkan Lake Ohrid and Its Living World. Monographiae Biologicae Vol. IX, Uitgeverij Dr. W. Junk, Den Haag.

Sturany, R., 1894. Zur Molluskenfauna der europäischen Türkei. Annalen des kaiserlich königlichen Naturhistorischen Hofmuseums 9: 369–394.

Swofford, D. L., 2002. PAUP. Phylogenetic analysis using parsimony (and other methods). Version 410b. Sinauer Associates, Sunderland, Mass.

Wilke, T., C. Albrecht, V. V. Anistratenko, S. K. Sahin & Z. Yildirim, 2007. Testing biogeographical hypotheses in space and time: faunal relationships of the putative ancient lake Egirdir in Asia Minor. Journal of Biogeography 34: 1807–1821.

Wilke, T., G. M. Davis, Q. Dongchuan & R. C. Spear, 2006. Extreme mitochondrial sequence diversity in the intermediate schistosomiasis host *Oncomelania hupensis robertsoni*: another case of ancestral polymorphism? Malacologia 48: 143–157.

Zacharias, I., I. Bertachas, N. Skoulikidis & T. Koussouris, 2002. Greek lakes: limnological overview. Lakes & Reservoirs: Research and Management 7: 55–62.

Hydrobiologia (2008) 615:169–179
DOI 10.1007/s10750-008-9556-0

SPECIATION IN ANCIENT LAKES

Testing two contrasting evolutionary patterns in ancient lakes: species flock versus species scatter in valvatid gastropods of Lake Ohrid

Anne-Kathrin Hauswald · Christian Albrecht · Thomas Wilke

Published online: 13 September 2008

Abstract Ancient lakes have long been recognized as "hot spots of evolution" and "evolutionary theatres" and they have significantly contributed to a better understanding of speciation and radiation processes in space and time. Yet, phylogenetic relationships of many ancient lake taxa, particularly invertebrate groups, are still unresolved. Also, the lack of robust morphological, anatomical, and phylogeographical data has largely prevented a rigorous testing of evolutionary hypotheses. For the freshwater gastropod genus *Valvata*—a group with a high degree of endemism in several ancient lakes—different evolutionary scenarios are suggested for different ancient lakes. Lake Baikal, for example, is inhabited by several endemic *Valvata* taxa that presumably do not form a monophyletic group. For such an evolutionary pattern, the term *ancient lake species scatter* is introduced here. In contrast, for the Balkan Lake Ohrid, workers previously suggested the presence of a monophyletic group of endemic *Valvata* species, that is, an *ancient lake species flock*. Sequence data of the mitochondrial cytochrome oxidase *c* subunit I gene (COI) from worldwide taxa, with a strong emphasis on Balkan species, are here used to test whether the putative Ohrid *Valvata* endemics represent an ancient lake species flock and to study patterns of speciation both on the Ohrid and the Balkan scale. The study reveals three distinct clades of endemic *Valvata* in Lake Ohrid. Monophyly of these taxa, however, is rejected, and they therefore do not represent an ancient lake species flock, but rather an ancient lake species scatter. Also, in contrast to many other gastropod groups in Lake Ohrid, the valvatids apparently did not radiate. Many *Valvata* taxa in ancient lakes are characterized by enhanced levels of shell complexity. However, it remains unclear whether these patterns are associated with ancient lake environments per se. It is here suggested that similarities in shell structure between North American and Balkan taxa might simply be due to convergent evolution.

Keywords Ancient lakes · Lake Ohrid · Lake Prespa · Ancient lake species flock · Ancient lake species scatter · *Valvata*

Guest editors: T. Wilke, R. Väinölä & F. Riedel
Patterns and Processes of Speciation in Ancient Lakes: Proceedings of the Fourth Symposium on Speciation in Ancient Lakes, Berlin, Germany, September 4–8, 2006

Electronic supplementary material The online version of this article (doi:10.1007/s10750-008-9556-0) contains supplementary material, which is available to authorized users.

A.-K. Hauswald · C. Albrecht (✉) · T. Wilke
Department of Animal Ecology and Systematics, Justus Liebig University Giessen, Heinrich-Buff-Ring 26-32 (IFZ), 35392 Giessen, Germany
e-mail: christian.albrecht@allzool.bio.uni-giessen.de

Introduction

Phylogenetic relationships of many major endemic freshwater taxa living in ancient lakes are still unresolved. These lakes are worldwide represented only by a small number of water bodies, characterized by several features that distinguish them from widespread post-glacial lakes (Martens, 1997). During their histories, an astonishing fauna has often formed within these lakes, typically reflected in high species diversities and outstanding degrees of endemism. This earned them attributes like "hot spots of evolution" (Martens et al., 1994) or "evolutionary theatres" (Wilson et al., 2004).

For Europe, Lake Ohrid is considered to be the most outstanding extant ancient lake (Salemaa, 1994; Frogley & Preece, 2004). This trans-boundary lake belongs to approximately two-thirds to Macedonia and one-third to Albania. For Lake Ohrid, often an age of 2–3 million years (my) is suggested (reviewed in Albrecht & Wilke, 2008). Lake Ohrid is known for its high degree of endemism, with one of the highest percentages being found in gastropods (Stanković, 1932, 1960; Radoman, 1985).

A key gastropod group previously recognized for its speciation patterns in ancient lakes is the family Valvatidae Gray, 1840 (Michel, 1994). This family belongs to the heterobranch molluscs clade (Bouchet & Rocroi, 2005) and comprises small freshwater snails common throughout eastern and central Europe, eastern Asia, and North America. Their general anatomy has been well studied (Rath, 1988). Prominent characters are a secondary gill and an unique osphradial morphology (Haszprunar, 1985). The family is comparably species poor, for example, with a total of only 11 species in North America (Burch, 1989). However, in some ancient lakes (i.e., Baikal, Ohrid and Prespa), the Valvatidae are comparably species rich, with an average of 76% of its taxa being endemic (calculated from Michel, 1994). Though valvatid species diversity in Lake Baikal is outstanding, the highest degree of endemicity as calculated in endemic species per area is found in Lake Ohrid (Table 1).

In Lake Ohrid the family Valvatidae is represented by the single genus *Valvata* and the three subgenera *Cincinna* Hübner, 1810, *Costovalvata* (Polinski, 1929), and *Tropidina* H. Adams & A. Adams, 1854 (Fig. 1). Four endemic species of the genus *Valvata* were described from Lake Ohrid—*V. stenotrema* Polinski, 1929; *V. rhabdota* Sturany, 1894; *V. hirsutecostata* Polinski, 1929; and *V. relicta* (Polinski, 1929), the latter with the nominal subspecies *V. relicta relicta* (Polinski, 1929) and *V. relicta interlithonis* Hadžišče, 1956. *Valvata relicta* occurs from the littoral to 100 m depth (Hadžišče, 1956). *Valvata stenotrema* can be found in the littoral and sublittoral, whereas *V. rhabdota* (littoral) and *V. hirsutecostata* (sublittoral) are thought to be vertically separated (Radoman, 1985).

In contrast, only one species, *Valvata "piscinalis,"* is mentioned for Lake Prespa, and it is also assumed to be common throughout Europe, where it is represented by several nominal subspecies (Falkner et al., 2001; Glöer, 2002). A *Valvata* sp., formerly attributed to *V. piscinalis* is known from Lake Pamvotis, together with *Valvata cristata* (Frogley & Preece, 2004, 2007). The Skutari Basin is inhabited by *V. piscinalis* and *V. cristata* as well as by an additional species, *Valvata montenegrina* Glöer & Pešić, 2008. For Lake Trichonis, Schütt (1962) described the occurrences of *V. piscinalis* and of the sculptured *Valvata klemmi* Schütt, 1962. Unfortunately, no information on phylogenetic relationships are available from any of the ancient lakes, and molecular work on the genus *Valvata* in general is apparently restricted to a few studies on two rare North American species (Miller et al., 2006).

Nonetheless, two contrasting scenarios are suggested for the evolution of ancient lake species:

(A) the presence of an *ancient lake species flock*, that is, a group of species that is characterized by speciosity, endemicity, and monophyly (Greenwood, 1984); and

Table 1 Present knowledge on endemism in Valvatidae of worldwide ancient lakes

Lake	Total number of species	Number of endemics	Endemism [%]	Index [endemic species/log km^2]	References
Baikal	16	11	69	2.45	Sitnikova et al. (2004)
Biwa	1	1	100	0.35	Nishino & Watanabe (2000)
Ohrid	4	4	100	1.57	Radoman (1985)

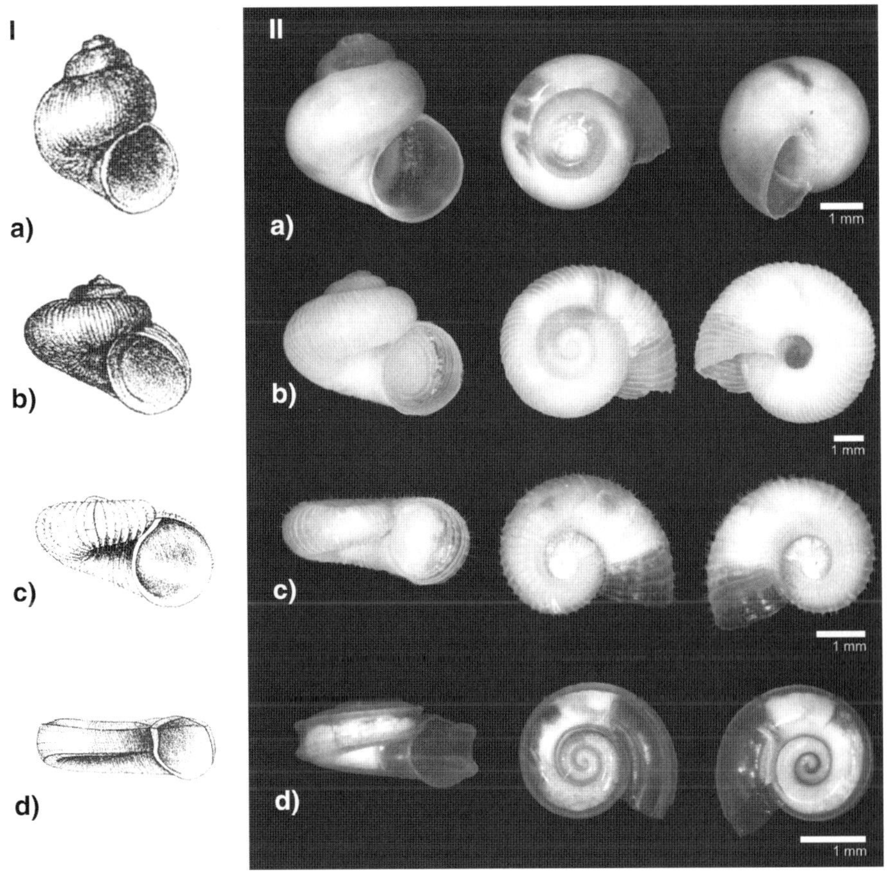

Fig. 1 Endemic *Valvata* species of Lake Ohrid (I: figures adopted from original descriptions, II: apertural, spire and umbilicus views of studied specimens). (**a**) *V. stenotrema* Polinski, 1929; (**b**) *V. rhabdota* Sturany, 1894; (**c**) *V. hirsutecostata* Polinski, 1929 and (**d**) *V. relicta* (Polinski, 1929). For the assignment of studied specimens of *V. hirsutecostata* see discussion

(B) a group of closely related taxa that are characterized by the criteria speciosity and endemicity, but which does not represent a monophyletic entity. Since no term is apparently available for the latter group, we here call it *ancient lake species scatter*. Processes leading to such a species scatter pattern regularly include multiple colonizations.

An example for an ancient lake species scatter would be the presumably diverse relic *Valvata* lineages known for Lake Baikal (e.g., Sitnikova, 1994), whereas Radoman (1985) suggested for Lake Ohrid *Valvata* to represent an ancient lake species flock. However, as two of the criteria for species flocks—endemicity and monophyly—are difficult to test in *Valvata* due to high degrees of homoplasy and considerable inter- and intraspecific variability in shell characters, we here use DNA sequence data of the mitochondrial cytochrome *c* oxidase subunit I (COI) gene to:

(1) determine the phylogenetic status of the known Ohrid and Prespa species of the genus *Valvata* within the framework of other Balkan as well as worldwide taxa;
(2) investigate whether the putative Ohrid *Valvata* endemics represent an ancient lake species flock;
(3) discuss the mode of diversification of the genus *Valvata* in Lake Ohrid; and
(4) assess the possible relationship between enhanced levels of shell complexity in *Valvata* and the occurrence in ancient lakes.

Materials and methods

Materials

The sampling is based on field trips to lakes Ohrid and Prespa conducted in September 2004 and 2005

Fig. 2 Distribution of sequenced populations of *Valvata* spp. in Lake Ohrid (OH), Lake Prespa (PR) and adjacent waters. Black squares indicate *V. rhabdota*, blank square *V.* cf. *hirsutecostata*, circles *V. stenotrema*, triangles *V. relicta*, black asterisks *Valvata* sp. 1, blank asterisk *Valvata* sp. 2

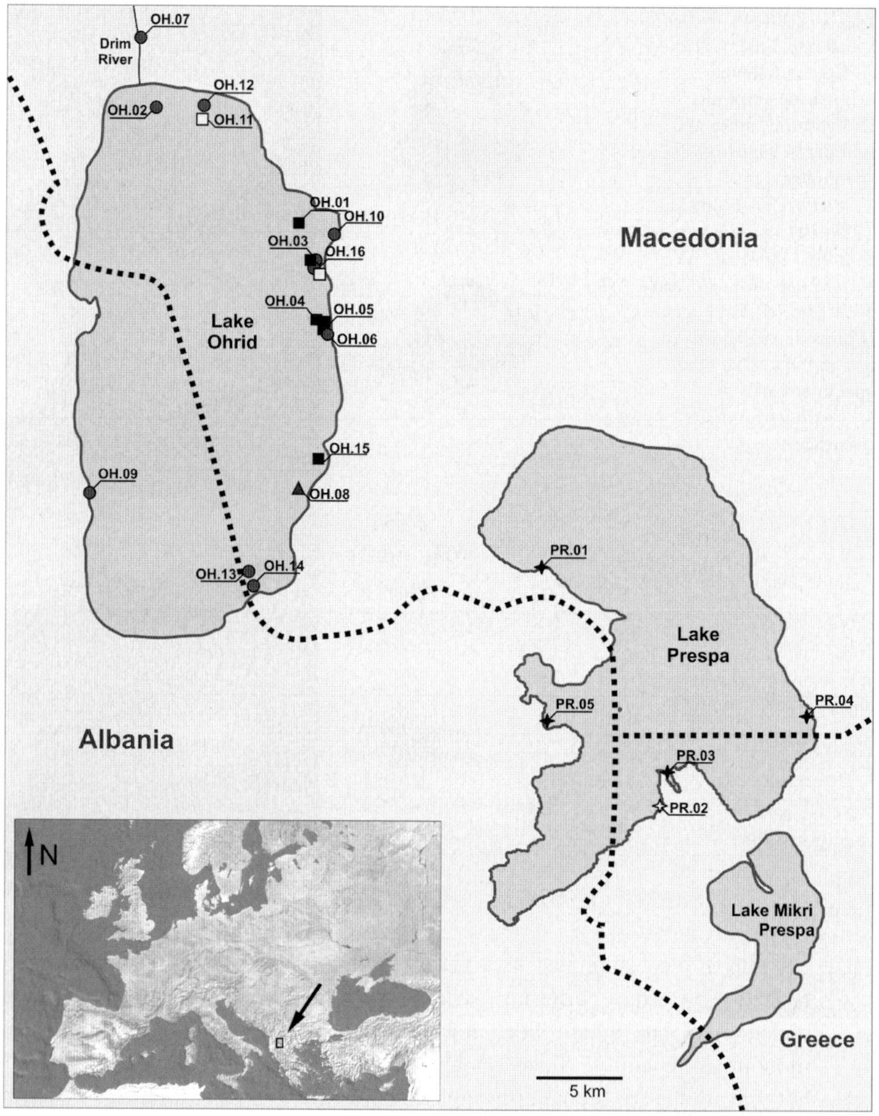

(Fig. 2). Individuals were obtained by hand collecting from hard substrata in shallow waters or from stones and rocks lifted from depths down to 5 m by snorkelling. Soft substrata were sampled with the help of metal sieves. Deeper parts of the littoral and sublittoral down to 60 m were sampled using a dredge from either a small boat or from the research vessel of the Hydrobiological Institute in Ohrid (HBI).

The study comprised specimens of Lake Ohrid *V. stenotrema*, *V. rhabdota*, *V.* cf. *hirsutecostata*, *V. relicta* and Lake Prespa *Valvata* spp. For phylogenetic analyses, additional materials from other European lakes and rivers were included: *V. montenegrina* from the Skutari Basin, *Valvata klemmi* from the Greek Lake Trichonis, *Valvata* sp. from the Greek lakes Vegoritis and Kastorias, *V. cristata* O. F. Müller, 1773, from the Greek Lake Pamvotis, *Valvata* cf. *piscinalis* from the German Lake Hohennauen and *Valvata* sp. from the German River Gülper Havel.

DNA extraction, amplification, sequencing, and alignment

Genomic DNA was extracted either from parts of the foot or from entire individuals depending on specimen size. For DNA isolation, the proteinase K protocol following Wilke et al. (2006) was used.

The final amount of DNA was dissolved in 35–40 μl of ddH_2O. DNA quantity and quality were assessed by electrophoresis using 1% agarose gel in 0.5× TBE buffer.

For amplification of the COI gene, for each sample a polymerase chain reaction mix was prepared containing: 2 μl each of 10 μM universal primers LCO1495 and HCO2198 (Folmer et al., 1994); 10–20 ng template; 3 μl of 10× ThermoPol buffer; 0.3 μl of 0.5 M TMAC; 2 μl of 10 mg/ml BSA; 2 μl of each 2.5 mM dNTPs; 0.6 μl of 5 U/μl Taq polymerase (New England Biolabs, Inc, MA, USA), and ddH_2O to a final volume of 30 μl. PCR amplifications were performed with an initial denaturation step at 95°C for 1 min, followed by 29 cycles of denaturing at 95°C for 30 s, annealing at 52°C for 30 s, and elongation at 72°C for 30 s. The reaction was terminated after a final elongation step of 4 min at 72°C. DNA fragments were purified in a 1% low melting point agarose gel in TAE buffer followed by a Wizard Prep treatment (Promega GmbH, Germany). Sequences (forward and reverse) were determined using the LI-COR (Lincoln, NE) DNA sequencer Long ReadIR 4200 and the Thermo Sequenase Fluorescent Labeled Primer Cycle Sequencing kit (Amersham Pharmacia Biotech, Piscataway, New Jersey).

The protein-coding COI sequences were aligned unambiguously by eye using BioEdit 7.0.4.1 (Hall, 1999), resulting in a 638-bp-long overlapping fragment. All sequences are available from Genbank (See Electronic supplementary material).

Phylogenetic analyses

For phylogenetic reconstruction, neighbor joining (NJ), maximum parsimony (MP) and Bayesian inference (BI) based methods were used. A NJ analysis with 1,000 bootstrap replicates under the K2P model was conducted in MEGA 3.1 (Kumar et al., 2004).

The MP analysis was performed utilizing PAUP* v4.0b10 (Swofford, 1998). Bootstrapping with 1,000 replicates and 10,000 sampled trees was performed and branch swapping was implemented using the tree bisection–reconnection (TBR) option.

The BI analysis was conducted using the phylogenetic software MRBAYES v3.0b4 (Huelsenbeck & Ronquist, 2001). Prior to the BI run, the optimal model of sequence evolution was determined using log-likelihood ratio tests implemented in the program MODELTEST 3.7 (Posada & Crandall, 1998). The best-fit model selected based on the Akaike Information Criterion (Akaike, 1974) was GTR + I + G.

As outgroups, sequences of three different species from GenBank were used: *Pyramidella dolabrata* (AY345054), *Littorina littorea* (DQ093525), *Theodoxus fluviatilis* (AY765317).

Molecular clock analyses

Molecular clock analyses were performed utilizing the average local Kimura-2-parameter (K2P) COI clock for non-saturated Protostomia taxa suggested by Wilke (2003) and Albrecht et al. (2006). Prior to the analyses, the COI dataset was tested for significant levels of saturation using the test of Xia et al. (2003), implemented in the software package DAMBE 5.0.7. (Xia & Xie, 2001). The test did not reveal a significant degree of saturation ($I_{SS} = 0.106$, $I_{SS.C} = 0.715$, $P = 0.000$) and the dataset was therefore considered to be suitable for subsequent analyses. Then, a log likelihood ratio test (LRT, Huelsenbeck & Rannala, 1997) was performed in the R environment (R Development Core Team, 2008) for statistical computing utilizing the log likelihoods from the best BI trees generated in MRBAYES v3.0b4 (1,000,000 generation with the current tree saved at intervals of 100 generations) under the K2P distance model both with and without the clock assumption. The test rejected the molecular clock null hypothesis of unequal substitution rates for the COI dataset as $-2 \log \Lambda$ (19.37) did not exceed the critical value of 59.3 (df = 43, $P \leq 0.05$).

In order to estimate divergence times for two major phylogenetic events for the endemic *Valvata* species of Lake Ohrid, the 1,000 best clock-enforced K2P trees of the BI analyses were used to determine mean node depth and respective standard deviations (details: Wilke et al., 2007). For estimating selected node ages, the average COI clock rate with an K2P distance of 2.23 ± 0.22% my^{-1} was applied due to the absence of an adequate calibration point for *Valvata* spp. Total error estimations (error of node depths + error of calibration point) were obtained by additive variance analyses. The clock estimates were not corrected for ancestral polymorphism (Edwards & Beerli, 2000), because the applied average Protostomia clock is not corrected, either.

Results

The COI dataset included a total of 42 ingroup specimens of *Valvata* spp. and comprised 196 variable sites, of which 175 were parsimony-informative. The NJ, MP and BI trees exhibited compatible topologies. All methods, however, partly failed to fully resolve some tip clades. The NJ tree is shown in Fig. 3 with the respective NJ, MP and BI support values added.

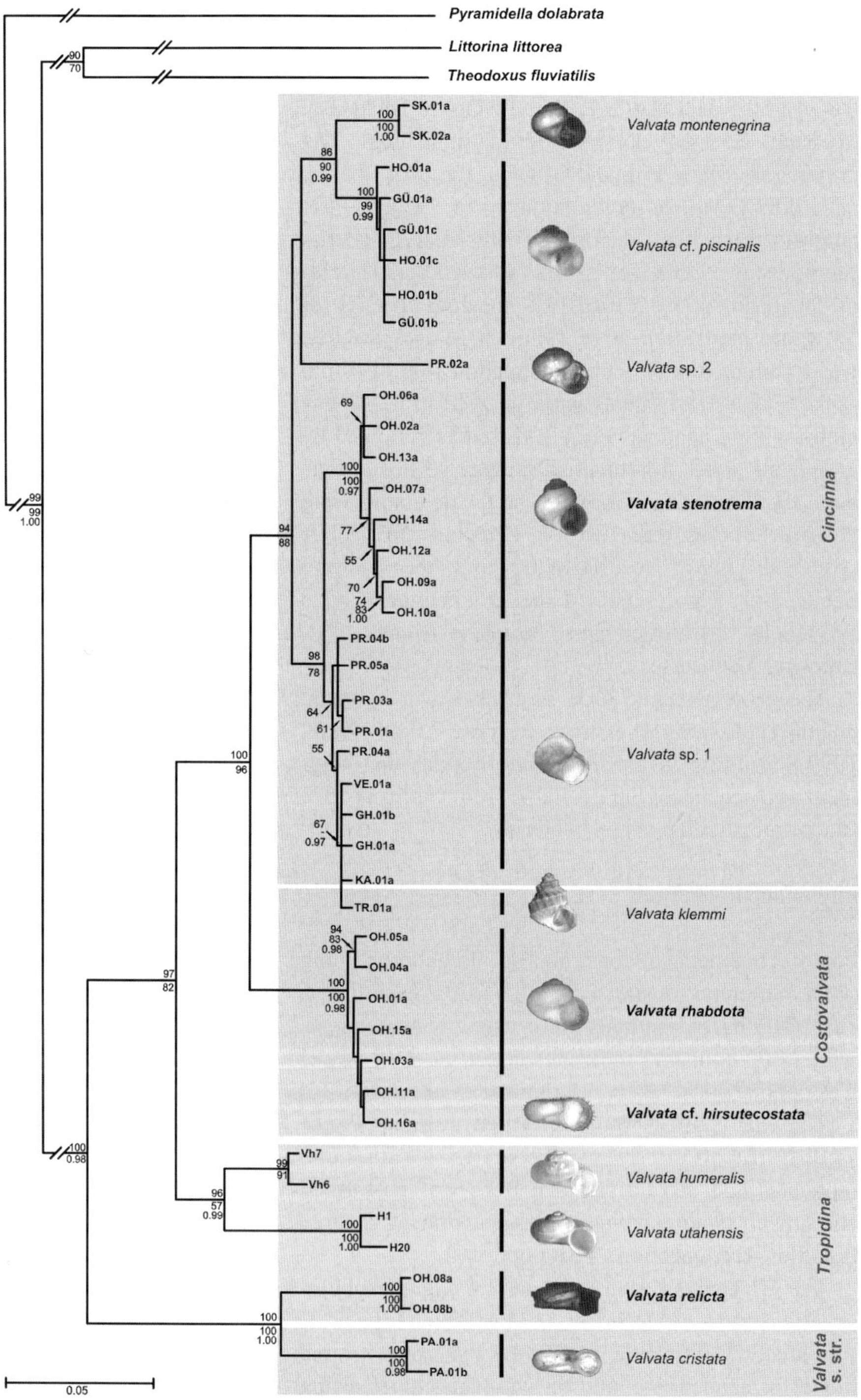

Fig. 3 Neighbor joining tree of *Valvata* spp. based on COI mtDNA sequencing data. Bootstrap values with more than 50% support for neighbor joining and maximum parsimony are presented above and directly below nodes. Bayesian posterior probabilities (>0.95) of branches are given second below nodes. The scale bar indicates the expected number of substitutions per site according to the model of sequence evolution applied. Representative shells of sequenced specimens are displayed. Subgeneric assignment follows Falkner et al. (2001) and Hadžišče (1956). Endemic taxa for lakes Ohrid and Prespa are shown in bold

All analyses indicated the presence of four major clades of *Valvata*, largely corresponding to the four subgenera previously described: clade 1 (=subgenus *Cincinna*) comprises the endemic Lake Ohrid *V. stenotrema*, *Valvata* sp. 1 from Germany and Balkan Lakes (including Lake Prespa but not Lake Ohrid), *Valvata* sp. 2 from Lake Prespa, *Valvata* cf. *piscinalis* from Germany, and the endemic *Valvata montenegrina* from the Skutari Basin. Interestingly, one Lake Trichonis *V. klemmi* specimen adheres to this clade. Note that this species has a costate shell.

Clade 2 (=subgenus *Costovalvata*) includes the Lake Ohrid endemic *V. rhabdota* and one putative specimen of *V.* cf. *hirsutecostata*. Clade 3 comprises the North American *V. humeralis* and *V. utahensis*, and clade 4 the Ohrid endemic *V. relicta*, as well as the Balkan *V. cristata*. The latter forms the most distinct *Valvata* clade in our analysis. Also, clades 3 and 4 do not fully correspond to the nominal subgenera *Tropidina* and *Valvata* s. str., respectively, as the Ohrid endemic *V. relicta* does not cluster with the American *Valvata* species of the subgenus *Tropidina* but with the Balkan *V. cristata* (subgenus *Valvata* s. str.).

In Lake Ohrid, three monophyletic *Valvata* subclades exist, which largely correspond to the four described endemic *Valvata* species: the subclade *V. stenotrema*, the subclade *V. relicta*, and the subclade *V. rhabdota*, including the putative specimen of *V.* cf. *hirsutecostata*. However, these three subclades do not form a monophyletic group.

Interestingly, whereas shells and live animals of *V. relicta* could only be found at Lake Ohrid's southeastern shore "biodiversity hot spots" near Veli Dab and north of Sv. Naum, specimens of *V. rhabdota* and *V. stenotrema* were found at all sampling points along the eastern shore and many parts of the lakes northwestern shore. *Valvata stenotrema* was found in the River Drim as well. Populations of Lake Prespa *Valvata* sp. 1 occurred in all major parts of the lake (Fig. 2).

Molecular clock analyses estimated the basal split of the endemic *V. relicta* from all other *Valvata* taxa to be 10.5 ± 1.2 my old (Fig. 4, split A). The second split was calculated for the separation of the most recent common ancestor of Lake Ohrid *V. rhabdota* from *V. stenotrema*, in particular, and resulted in a divergence time of 4.6 ± 0.6 my (Fig. 4, split B).

Discussion

Phylogenetic status of the known Ohrid and Prespa species of the genus *Valvata*

Our study represents the first molecular phylogeny of European valvatids in general and of ancient lake taxa in particular. It revealed distinct clades that

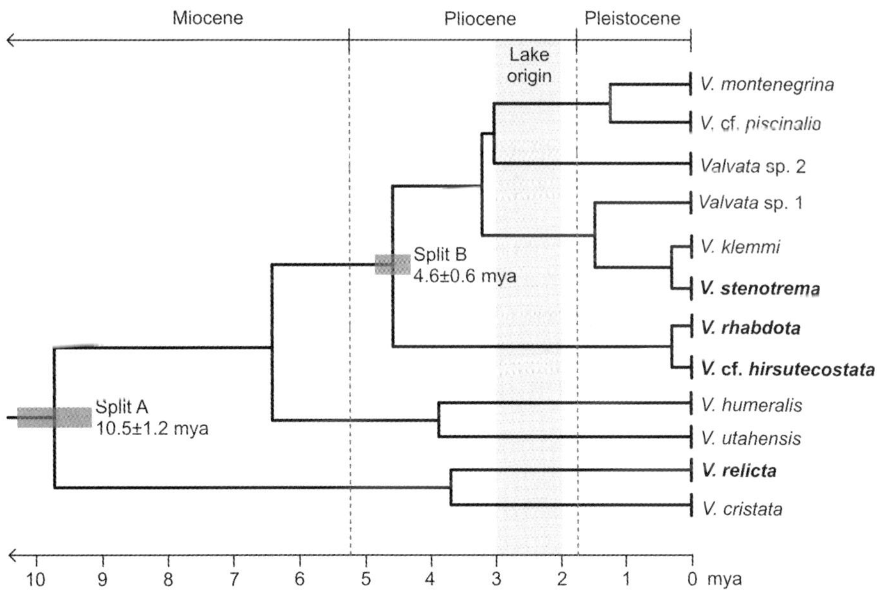

Fig. 4 Schematic clock-enforced Bayesian tree based on COI mtDNA sequencing data of *Valvata* spp. under the K2P model. Two evolutionary events of interest are marked as splits A and B. Standard deviations are plotted onto the respective nodes. Endemic taxa for lakes Ohrid and Prespa are shown in bold

largely correspond to previous taxonomic assignments both at the subgenus and species level. The comparative character of the study also allows for a discussion of biogeographical and evolutionary patterns among Balkan valvatids.

Three distinct clades of endemic *Valvata* were found within Lake Ohrid, corresponding to the species *V. stenotrema*, *V. relicta*, and *V. rhabdota*/*V.* cf. *hirsutecostata*. Note that our preliminary data suggests possible conspecificity of *V.* cf. *hirsutecostata* and *V. rhabdota*. This is also supported by a continuous shell morphospace in our collection throughout Lake Ohrid. Interestingly, Rath (1988) found strong differences in the gill anatomy between *V. hirsutecostata* and *V. rhabdota*. *Valvata rhabdota* is also described as being significantly larger than *V. hirsutecostata* (Polinski, 1929; Sturany, 1894), whereas gross anatomy and shell features suggest close relationship. Final conclusions must, however, be postponed until more data becomes available.

Our study also shows that the Ohrid endemic *V. relicta* is not a member of the American subgenus *Tropidina*. Instead, it is genetically distinct with *V. cristata* (subgenus *Valvata* s. str.) being its sister taxon. Hadžišče (1953) established for *V. relicta* the taxon *Ohridotropidina* as a section within the subgenus *Tropidina*. Based on our genetic data, it might be justifiable to raise this taxon to subgenus level.

Ancient lake species flock versus ancient lake species scatter

The endemic taxa of Lake Ohrid meet the criteria I (speciosity) and II (endemicity) of the definition of an ancient lake species flock. However, the third criterion, monophyly, is rejected for the *Valvata* endemics in Lake Ohrid. Therefore, they do not represent an ancient lake species flock but rather an ancient lake species scatter. This is interesting because monophyly has been recognized before for several groups of endemic species in Lake Ohrid (Albrecht & Wilke, 2008), and the current study represents the first evidence for an ancient lake species scatter in molluscs of this lake.

Radoman (1985) suggested the endemic Lake Ohrid valvatids to be the result of intralacustrine speciation. Our findings, however, indicate that no radiation occurred among those taxa.

Mode of diversification of the genus *Valvata* in Lake Ohrid

Acknowledging possible missing taxa in our phylogeny, molecular clock estimates for the split of Lake Ohrid endemics from their respective sister taxa outside the lake indicate largely different divergence times. Whereas the split of a *V. relicta* ancestor from all other *Valvata* taxa is less than 10 my, the split leading to the evolutionary lineages of *V. rhabdota* and *V. stenotrema* is less than 4 my old.

Assuming a time frame of 2–3 my for the origin of Lake Ohrid, the most recent common ancestor of *V. relicta* likely predates the origin of Lake Ohrid. This is also true for the lineages leading to *V. rhabdota* and *V. stenotrema*. The evolution of *V. stenotrema* may either be associated with the origin of Lake Ohrid or it may even be younger than the lake. In any event, the patterns found suggest at least three independent immigration events of Lake Ohrid *Valvata* ancestors.

If one considers the subgeneric level, it would mean that in each of the three subgenera, one ancestor invaded Lake Ohrid independently. This pattern resembles the assumed immigration of Lake Baikal by ancestors of the endemic valvatids (Sitnikova, 1994).

Furthermore, the Lake Prespa and endemic Lake Ohrid valvatids appear to be closely related to species from surrounding Balkan water bodies. *Valvata relicta*, for example, turned out to be the sister to Lake Pamvotis *V. cristata* and not to the other Lake Ohrid endemics. Interesting is also the status of Lake Prespa *Valvata*. In the literature (Maassen, 1980), it is generally listed as *V. piscinalis*. However, our phylogenetic studies indicate that it belongs to two different clades; here referred to as *Valvata* sp. 1 and *Valvata* sp. 2. The former also occurs in the Greek lakes Vegoritis and Kastorias. Our preliminary data suggests that *V. piscinalis* most likely represents a cryptic species complex with possibly more species to be discovered. It is also interesting that the genetic diversity within Lake Ohrid *V. stenotrema* and *V. rhabdota* appears to be more pronounced than in the more widespread clades *V.* cf. *piscinalis* and *Valvata* sp. 1, suggesting old populations to occur in Lake Ohrid. Such a pattern was also found in *Radix relicta* (Albrecht et al., 2008) and *Dreissena presbensis* (Albrecht et al., 2007).

Table 2 Summarized diversity and distribution of *Valvata* in Balkan lakes

	Dojran	Kastorias	Mikri Prepsa	Ohrid	Pamvotis	Prespa	Skutari	Trichonis	Vegoritis
V. piscinalis	x		x		x[a]	x[b]	x	x	x[b]
V. montenegrina							**x**		
Valvata sp. 1		x				x			x
Valvata sp. 2						x			
V. stenotrema				x					
V. rhabdota				**x**					
V. hirsutecostata				**x**					
V. klemmi								**x**	
V. cristata					x		x	x	x
V. relicta				**x**					
Sum	1	1	1	4	2	2[c]	3	3	2[c]

Information are based on own data and literature survey (Falniowski et al., 1988; Frogley & Preece, 2004, 2007; Glöer & Pešić, 2008; Hadžišče, 1956; Koussouris & Pugh-Thomas, 1982; Maassen, 1980; Petridis & Sinis, 1995; Radoman, 1983, 1985; Reischütz & Reischütz, 2003; Schütt, 1962; Stankovic, 1985). Species with complex shell structures are in bold

[a] Frogley & Preece (2004) listed the species as *V. piscinalis* while new data (2007) raises doubts as to the status of this taxon

[b] It is most likely that *Valvata* sp. 1 and the published record of *V. piscinalis* refer to the same species

[c] The sum number does not include the published records of *Valvata piscinalis* (see footnote b)

Enhanced levels of shell complexity in *Valvata* and occurrence in ancient lakes

Another pattern found in many of the Lake Ohrid endemic gastropods is enhanced levels of shell complexity (Gorthner, 1992). Two of three *Valvata* lineages show complex shell structures (costation, keels). Considering *V. hirsutecostata* as a valid species, the share of complex shelled valvatids would be 75%. On a European scale, costate forms are also found in the Balkan lakes Trichonis and Skutari (Table 2). On a worldwide scale, *Valvata biwaensis* Preston, 1916, is a keeled and costate species from Lake Biwa, presumably belonging to the subgenus *Cincinna*. In Lake Baikal, the share of complex forms among the endemics is 18% (compiled from Starobogatov & Sitnikova, 1998). If one considers Nearctic species (not living in ancient lakes), complex shells are very common and can even differ strikingly in single populations of, for example, *V. tricarinata* (Burch, 1982). From the survey given in Table 2, it becomes clear that lakes Ohrid, Prespa and Skutari have the highest valvatid species diversity in the Balkans. They also harbor a high percentage of *Valvata* species with complex shell forms.

Although a tendency toward enhanced levels of shell complexity is obvious, it remains unclear whether complex shell structures are a typical pattern for ancient lake valvatids. In previous studies, some authors mentioned conchological similarities between Nearctic and Balkan valvatids. Hadžišče (1956) noted a strong similarity of *V. relicta* and the North American *V. (Tropidina) tricarinata*, and Schütt (1962) a relation of the Greek *V. klemmi* and *V. tricarinata*. *Valvata klemmi*, a species treated as *V. piscinalis* by some authors (Falniowski et al., 1988), clusters in the present study within the unsculptured *Valvata* sp. 1 complex. Therefore, conchological similarities between North American and Balkan taxa might be due to convergent evolution. Such processes have often obscured phylogenetic relationships in freshwater gastropods in general (Albrecht et al., 2004) and in Lake Ohrid in particular (Albrecht et al., 2006).

Acknowledgments We are very grateful to the colleagues of the HBI for their hospitality, immense support and interest in our joint project. B. Budzakoska, S. Trajanovski and S. Trajanovska provided valuable information and took part in field trips. D. Georgiev is gratefully acknowledged for providing us with vast local expertise, support, and equipment. M. Colling, C. Renker, A. Falniowski, C. Meier-Brook, S. Koşal Şahin, V. Pešić, and Z. Feher kindly provided important material for this study. D. Lohfink, R. Schultheiß, and T. Geertz were excellent field fellows. We thank S. Nachtigall for technical assistance in the lab and R. Schultheiß for help with the R-analysis. The comments of P. Glöer, T. Sitnikova, and an anonymous reviewer on an earlier version are also acknowledged.

References

Akaike, H., 1974. A new look at the statistical model identification. IEEE Transaction on Automatic Control 19: 716–723.

Albrecht, C., R. Schultheiß, T. Kevrekidis & T. Wilke, 2007. Invaders or endemics? Molecular phylogenetics, biogeography and systematics of *Dreissena* in the Balkans. Freshwater Biology 52: 1525–1536.

Albrecht, C., S. Trajanovski, K. Kuhn, B. Streit & T. Wilke, 2006. Rapid evolution of an ancient lake species flock: freshwater limpets (Gastropoda: Ancylidae) in the Balkan lake Ohrid. Organisms, Diversity & Evolution 6: 294–307.

Albrecht, C. & T. Wilke, 2008. Lake Ohrid: biodiversity and evolution. Hydrobiologia (this volume).

Albrecht, C., T. Wilke, K. Kuhn & B. Streit, 2004. Convergent evolution of shell shape in freshwater limpets: the African genus *Burnupia*. Zoological Journal of the Linnean Society 140: 577–586.

Albrecht, C., C. Wolff, P. Glöer & T. Wilke, 2008. Concurrent evolution of ancient sister lakes and sister species: the freshwater gastropod genus *Radix* in lakes Ohrid and Prespa. Hydrobiologia (this volume).

Bouchet, P. & J.-P. Rocroi, 2005. Classification and nomenclator of gastropod families. Malacologia 47: 1–397.

Burch, J. B., 1982. North American freshwater snails: identification keys, generic synonymy, supplemental notes, glossary, references, index. Walkerana 4: 1–356.

Burch, J. B., 1989. North American freshwater snails: species list, ranges and illustrations. Walkerana 1: 81–365.

Edwards, S. V. & P. Beerli, 2000. Perspective: gene divergence, population divergence, and the variance in coalescence time in phylogeographic studies. Evolution 54: 1839–1854.

Falkner, G., R. A. Bank & T. v. Proschwitz, 2001. Check-list of the marine Molluscan Species-group taxa of the States of Northern, Atlantic and Central Europe (CLECOM I). Heldia 4: 1–76.

Falniowski, A., A. Economou-Amilli & K. Anagnostidis, 1988. *Valvata piscinalis* O.F. Müller (Mollusca, Prosobranchia) and its epizoic diatoms from lake Trichonis, Greece. Internationale Revue der gesamten Hydrobiologie 73: 327–335.

Folmer, O., M. Black, W. Hoeh, R. Lutz & R. Vrijenhoek, 1994. DNA primers for amplification of mitochondrial cytochrome c oxidase subunit I from diverse metazoan invertebrates. Molecular Marine Biology and Biotechnology 3: 294–299.

Frogley, M. R. & R. C. Preece, 2004. A faunistic review of the modern and fossil molluscan fauna from lake Pamvotis, Ioannina, an ancient lake in NW Greece: implications for endemism in the Balkans. In Griffiths, H. I. & B. Kryštufek (eds), Balkan Biodiversity. Kluwer, Dordrecht: 243–260.

Frogley, M. R. & R. C. Preece, 2007. A review of the aquatic Mollusca from Lake Pamvotis, Ioannina, an ancient lake in NW Greece. Journal of Conchology 39: 271–296.

Glöer, P., 2002. Die Süßwassergastropoden Nord- und Mitteleuropas. ConchBooks, Hackenheim.

Glöer, P. & V. Pešić, 2008. The freshwater gastropods of the Skadar Lake with the description of *Valvata montenegrina* n. sp. (Mollusca, Gastropoda, Valvatidae). In Pavić, E. D. & M. Perreau (eds), Advances in the Studies of the Subterranean and Epigean Fauna of the Balkan Peninsula. Volume Dedicated to the Memory of Guido Nonvellier. Monography, Institute for Nature Conservation of Serbia: 341–348.

Gorthner, A., 1992. Bau, Funktion und Evolution komplexer Gastropodenschalen in Langzeit-Seen. Mit einem Beitrag zur Paläobiologie von *Gyraulus "multiformes"* im Steinheimer Becken. Stuttgarter Beiträge zur Naturkunde, Serie B 190: 1–173.

Greenwood, P. H., 1984. What is a species flock? In Echelle, A. A. & I. Kornfield (eds), Evolution of Fish Species Flocks. Orono Press, Maine: 13–19.

Hadžišče, S., 1953. Beitrag zur Kenntnis der Gastropodenfauna des Prespa- und Ohridsees. Periodicum biologorum II/B T. 7: 174–177.

Hadžišče, S., 1956. III. Beitrag zur Kenntnis der Gastropodenfauna des Ohridsees.—Beschreibungen der bis jetzt unbekannten Schnecken und Beispiele der Speciation bei den Gastropoden des Ohridsees. Recueil des Traveaux, Station Hydrobiologique Ohrid 4: 57–107.

Hall, T. A., 1999. BioEdit: a user-friendly biological sequence alignment editor and analysis program for Windows 95/98/NT. Nucleic Acids Symposium Series 41: 95–98.

Haszprunar, G., 1985. The Heterobranchia—a new concept of phylogeny of the higher Gastropoda. Zeitschrift für zoologische Systematik und Evolutionsforschung 23: 15–37.

Huelsenbeck, J. P. & B. Rannala, 1997. Phylogenetic methods come to age: testing hypotheses in an evolutionary context. Science 276: 227–232.

Huelsenbeck, J. P. & F. Ronquist, 2001. MrBayes: Bayesian inference of phylogenetic trees. Bioinformatics 17: 754–755.

Koussouris, T. & M. Pugh-Thomas, 1982. Macrozoobenthic studies in Lake Trichonis—western Greece. Thalassographica 2: 17–25.

Kumar, S., K. Tamura & M. Nei, 2004. Mega3: integrated software for molecular evolutionary genetics analysis and sequence alignment. Briefings in Bioinformatics 5: 150–163.

Maassen, W. J. M., 1980. De zoetwatermollusken van het Ohrid/en Prespameer en hun omgeving. De Kreukel 16: 81–98.

Martens, K., 1997. Speciation in ancient lakes. Trends in Ecology and Evolution 12: 177–182.

Martens, K., G. Coulter & B. Goddeeris, 1994. Speciation in ancient lakes—40 years after Brooks. In Martens, K., B. Goddeeris & G. Coulter (eds), Speciation in Ancient Lakes. Archiv für Hydrobiologie 44: 75–96.

Michel, E., 1994. Why snails radiate: a review of gastropod evolution in long-lived lakes, both recent and fossil. In Martens K., B. Goddeeris & G. Coulter (eds), Speciation in Ancient Lakes. Archiv für Hydrobiologie 44: 285–317.

Miller, M. P., D. E. Weigel, K. E. Mock & B. Roth, 2006. Evidence for an outcrossing reproductive strategy in the hermaphroditic heterobranch gastropod *Valvata utahensis* (Valvatidae), with notes on the genetic differentiation of

V. utahensis and *V. humeralis*. Journal of Molluscan Studies 72: 397–403.

Nishino, M. & N. C. Watanabe, 2000. Evolution and endemism in Lake Biwa, with special reference to its gastropod mollusc fauna. In Rossiter, A. & H. Kawanabe (eds), Ancient Lakes: Biodiversity, Ecology and Evolution—Advances in Ecological Research, Vol. 31. Academic Press, San Diego: 151–180.

Petridis, D. & A. Sinis, 1995. Benthos of Lake Mikri Prespa (North Greece). Hydrobiologia 304: 185–196.

Polinski, V., 1929. Limnološka ispitivanja Balkanskog Poluostrva I. Reliktna fauna gasteropoda Ohridskog Jezera. Glas Srpska Kraljevske Akademije, Belgrade 137: 129–182.

Posada, D. & K. A. Crandall, 1998. Modeltest: testing the model of DNA substitution. Bioinformatics 14: 817–818.

R Development Core Team, 2008. R: A Language and Environment for Statistical Computing. R Foundation for Statistical Computing, Vienna, Austria.

Radoman, P., 1983. Hydrobioidea, a superfamily of Prosobranchia (Gastropoda), I. Systematics. Serbian Academy of Sciences and Arts, Belgrade: 1–256.

Radoman, P., 1985. Hydrobioidea, a superfamily of Prosobranchia (Gastropoda) II. Origin, Zoogeography, Evolution in the Balkans and Asia Minor. Institute of Zoology, Department of Biology, Faculty of Science, Belgrade: 1–173.

Rath, E., 1988. Organization and systematic position of the Valvatidae. Malacological-Review-Supplement 4: 194–204.

Reischütz, A. & P. L. Reischütz, 2003. Hellenika pantoia, 5: Zur Kenntnis der Molluskenfauna des Limni Trichonida und des Limni Lisimachia (Aitolien/Akarnien, Griechenland). Nachrichtenblatt der Ersten Vorarlberger Malakologischen Gesellschaft 11: 28–30.

Salemaa, H., 1994. Lake Ohrid. In Martens, K., B. Goddeeris & G. Coulter (eds), Speciation in Ancient Lakes. Archiv für Hydrobiologie 44: 55–64.

Schütt, H., 1962. Neue Süßwasser-Prosobranchier Griechenlands. Archiv für Molluskenkunde 91: 157–166.

Sitnikova, T. Y., 1994. Recent views on the history and diversity of the Baikalian malacofauna. In Martens, K., B. Goddeeris & G. Coulter (eds), Speciation in Ancient Lakes. Archiv für Hydrobiologie 44: 319–326.

Sitnikova, T. Y., Y. I. Starobogatov, A. A., Shirokaya, I. V. Shibanova, N. V. Korobkova & F. V. Adov, 2004. Bryukhonogie mollyuski (Gastropoda). In Annotirovannyi spisok fauny ozera Baikal i ego vodosbornogo basseina (annotated list of the fauna of the Baikal Lake and its basin), Vol. 1. Nauka, Novosibirsk: 937–1002.

Stanković, S., 1932. Die Fauna des Ohridsees und ihre Herkunft. Archiv für Hydrobiologie 23: 557–617.

Stanković, S., 1960. The Balkan Lake Ohrid and its Living World. Monographiae Biologicae, Vol. IX. Uitgeverij Dr. W. Junk, Den Haag.

Starobogatov, Y. I. & T. Y. Sitnikova, 1998. The Lists of species of animals and plants inhabiting in Baikal. In Kozhova, O. M. & L. R. Izmest'eva (eds), Mollusca. Lake Baikal: Evolution and Biodiversity. Backuys Publishers, Leiden.

Sturany, R., 1894. Zur Molluskenfauna der europäischen Türkei. Annalen des kaiserlich königlichen Naturhistorischen Hofmuseums 9: 369–394.

Swofford, D. L., 1998. Paup*. Phylogenetic Analysis using Parsimony (*and other methods), Version 4.0. Sinauer Associates, Sunderland, MA.

Wilke, T., 2003. *Salenthydrobia* n.gen. (Rissooidea: Hydrobiidae): a potential relict of the Messinian salinity crisis. Zoological Journal of the Linnean Society 137: 319–336.

Wilke, T., C. Albrecht, V. V. Anistratenko, S. Kosal & M. Z. Yildirim, 2007. Testing biogeographical hypotheses in space and time: faunal relationships of the putative ancient Lake Egirdir in Asia Minor. Journal of Biogeography 34: 1807–1821.

Wilke, T., G. M. Davis, Q. Dongchuan & R. C. Spear, 2006. Extreme mitochondrial sequence diversity in the intermediate schistosomiasis host *Oncomelania hupensis robertsoni*: another case of ancestral polymorphism? Malacologia 48: 143–157.

Wilson, A. B., M. Glaubrecht & A. Meyer, 2004. Ancient lakes as evolutionary reservoirs: evidence from the thalassoid gastropods of Lake Tanganyika. Proceedings of the Royal Society B, Biological Sciences 271: 529–536.

Xia, X. & Z. Xie, 2001. Dambe: data analysis in molecular biology and evolution. Journal of Heredity 92: 371–373.

Xia, X., Z. Xie, M. Salemi, L. Chen & Y. Wang, 2003. An index of substitution saturation and its application. Molecular Phylogenetics and Evolution 26: 1–7.

Hydrobiologia (2008) 615:181–199
DOI 10.1007/s10750-008-9568-9

The species flocks of lacustrine gastropods: *Tylomelania* on Sulawesi as models in speciation and adaptive radiation

Matthias Glaubrecht · Thomas von Rintelen

Published online: 11 September 2008

Abstract Endemic radiations provide splendid opportunities for studies in evolutionary biology. Species flocks in ancient lakes, such as in Tanganyika, Malawi or Baikal, have featured prominently in evolutionary biology, viewing these "evolutionary theatres" as hotspots of diversification. However, following a century of neglect, the endemic evolution of limnic cerithioidean gastropods in the two central lake systems on the Indonesian island of Sulawesi (i.e. Lake Poso and the lakes of the Malili system, e.g. Danau Matano, Mahalona and Towuti) also provide instructive model cases for the study of speciation mechanisms, adaptive radiation and annidation (i.e. niche exploitation). We here discuss the evolutionary and taxonomic implications of the lacustrine species flocks in *Tylomelania* from these lakes in Sulawesi as an exceptional endemic assemblage of morphologically distinct viviparous pachychilid gastropods. This first comprehensive compilation of data on both ancient lake systems, Poso and Malili, offers a new perspective on ecological differentiation in this radiation. Presented here within the framework of the theory of evolutionary ecology it provides a research program for acquiring a synthetical perspective that includes morphology, molecular genetics, ecology and biogeography. In this context, it will be possible to compare the species flocks of these truly "Darwinian snails" on Sulawesi with the long enigmatic, so-called thalassoid (i.e. marine-like) gastropod radiation in East African's Lake Tanganyika.

Keywords Freshwater gastropods · Speciation · Adaptive radiation · Molecular phylogeny · Evolutionary ecology

Guest editors: T. Wilke, R. Väinölä & F. Riedel
Patterns and Processes of Speciation in Ancient Lakes: Proceedings of the Fourth Symposium on Speciation in Ancient Lakes, Berlin, Germany, September 4–8, 2006

M. Glaubrecht (✉) · T. von Rintelen
Museum für Naturkunde, Humboldt University Berlin, Invalidenstraße 43, 10115 Berlin, Germany
e-mail: matthias.glaubrecht@museum.hu-berlin.de

Introduction

It has been proposed that in biology concepts play an equally important role as do natural laws in physics and chemistry (e.g. Mayr, 1997, 2004). Given their heuristic significance biological concepts such as, e.g. intralacustrine speciation, adaptive radiation, ecological niche, coevolution, competitive displacement etc. are in need of case studies for testing and illustration of their general validity. Organismic evolutionary biology is largely dependent on suitable model organisms, which were traditionally recruited among vertebrates as, e.g. the well-known African cichlids (most recent review in Salzburger & Meyer, 2004). In this respect, since Brooks' (1950) and Boss' (1978) seminal papers, species flocks in ancient lakes have featured prominently in evolutionary biology, viewing these "evolutionary theatres" (Hutchinson, 1957) as hotspots of diversification. For the most recent general overview, see, e.g. contributions in Rossiter & Kawanabe (2000); some recent detailed studies are, e.g. Herder et al. (2006) for fishes, following the discovery and description of unique telmatherinid fishes from Sulawesi (Kottelat, 1990, 1991; Kottelat et al., 1993), and for gastropods from some prominent ancient lakes, for example von Rintelen et al. (2004, 2007), Wilson et al. (2004), Albrecht et al. (2006) and Glaubrecht (2008).

While invertebrates feature less prominently in this respect, they may have very distinct intrinsic properties, such as widely different dispersal abilities and a less prominent influence of sexual selection. For these reasons, freshwater gastropods are ideally suited for addressing central questions, e.g. in biogeography and biodiversity research, particularly, (i) the question of species numbers in general, (ii) on the nature of species (species concepts and species delineation) and (iii) the mechanisms of speciation (see, e.g. Glaubrecht, 1996, 2000, 2004). In this context, adaptive radiations are well-suited for research looking into the origin of biological diversity in general.

Thus, we use freshwater gastropods as models for the study of general questions in evolutionary biology, e.g. addressing problems in speciation or the causes of (adaptive) radiations. Our research in the last decade has focused on limnic members of the Cerithioidea, a large gastropod group with a worldwide distribution (Glaubrecht, 1996, 1999, 2006; Lydeard et al., 2002). Especially the Thiaridae, Paludomidae and Pachychilidae as tropical taxa with many viviparous taxa have been successfully employed in testing models of intralacustrine radiation and biogeographical hypotheses with a regional focus on East Africa (see, e.g. Glaubrecht & Strong, 2007; Glaubrecht, 2008) and Southeast Asia (e.g. Glaubrecht & von Rintelen, 2003; Glaubrecht & Köhler, 2004; Köhler & Glaubrecht, 2007). The Indonesian island Sulawesi is particularly interesting in this context, due to its location in the centre of the perhaps most complex geological region of the world, where the Asian and Australian plates converge (see, e.g. van Oosterzee, 1997).

Sulawesi's ancient lakes as "evolutionary theatres"

Following a century of neglect, with only cursory mentioning by Woltereck (1931, 1941) and Wesenberg-Lund (1939), the endemic evolution of limnic Cerithioidea gastropods in the two central lake systems on the Indonesian island of Sulawesi (i.e. Lake Poso and the lakes of the Malili system, e.g. Danau Matano, Mahalona and Towuti) provide an excellent opportunity to study speciation mechanisms and adaptive radiation (Fig. 1).

These two ancient lake systems of Sulawesi are situated in the central mountains of the island, with Lake Poso located ca. 80 km NW of the Malili lakes. While Lake Poso (Fig. 1b) is a deep solitary lake (−450 m, 323.2 km^2), the Malili system comprises five lakes sharing a common drainage (Fig. 1c). The three larger lakes of that system are directly connected: Lake Matano (−590 m, 164.0 km^2) flows into Lake Mahalona via Petea River, and Lake Mahalona (−73 m, 24.4 km^2) in turn spills into Lake Towuti via Tominanga River. Lake Towuti (−203 m, 561.1 km^2) is drained by the Larona River into the Gulf of Bone (Teluk Bone). Two smaller satellite lakes, Lake Lontoa (also known as Wawontoa, −3 m, 1.6 km^2) and Lake Masapi (−4 m, 2.2 km^2), are much less directly connected to the system (see Fig. 1c).

Lake Poso and Lake Matano are of tectonic origin, which accounts for their extraordinary depth. Lake Matano is situated in a strike-slip fault, the Matano fault, which was formed in the final juxtaposition process of South-, Southeast- and East Sulawesi since the Pliocene (ca. 4 Ma) to the present day (Wilson & Moss, 1999). The age of Lake Matano has been estimated at 1–2 Ma (G. Hope, pers. comm.), estimates for the other lakes are lacking. The major ancient lakes of Sulawesi are oligotrophic, with a very low nutrient and organic content and a high transparency of up to 22 m in Lake Towuti (Giesen et al., 1991; Giesen, 1994; Haffner et al., 2001).

The fauna of these ancient lakes was initially described by the Swiss naturalists Paul and Fritz Sarasin, who also were the first Europeans to visit the Malili lake system in 1896 (Sarasin & Sarasin, 1905). Among other endemic freshwater molluscs, they described a radiation of 16 species of peculiar gastropods in both lake systems (Sarasin & Sarasin, 1897; 1898). More than a decade later, the Dutch

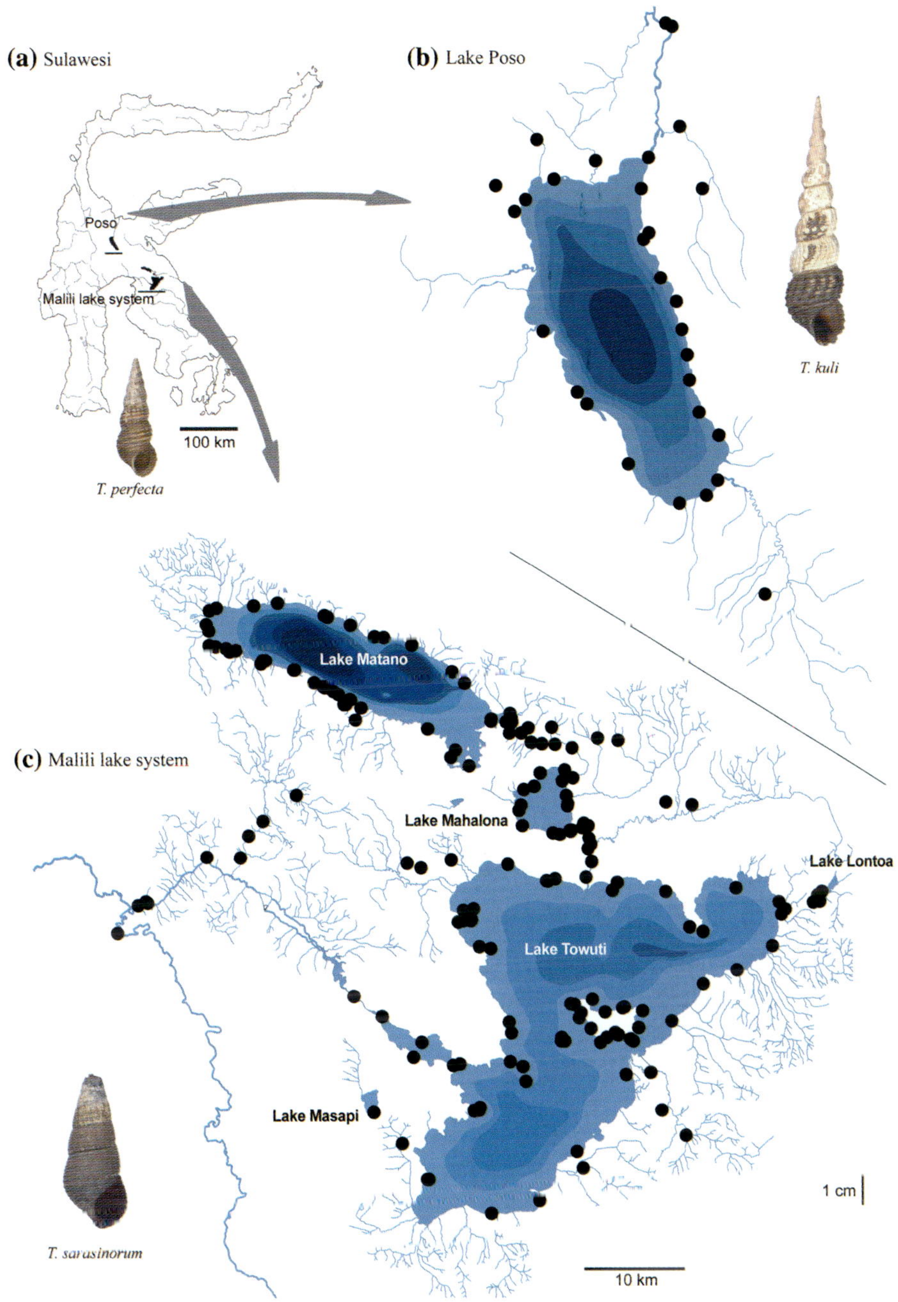

Fig. 1 Sulawesi, Lake Poso and the Malili lake system with all sample sites from the campaigns 1999 to 2005. (**a**) Sulawesi; (**b**) and (**c**) the ancient lakes with sample sites; (**b**) Poso; (**c**) Malili lake system

geologist E. C. Abendanon visited the lakes in 1909/1910, and a subsequent description of his mollusc samples by Kruimel (1913) increased the number of endemic lake taxa by eight to a total of 24 species, with the majority of them ($n = 16$) in the Malili lake system. The species were originally assigned to two different genera, viz *Tylomelania* Sarasin & Sarasin, 1897, which was initially only erected for three endemic Lake Poso species, and the now invalid '*Melania*'. Traditionally assigned to the pantropical freshwater family Thiaridae, it has just recently been recognised that Sulawesi hosts an endemic lineage of

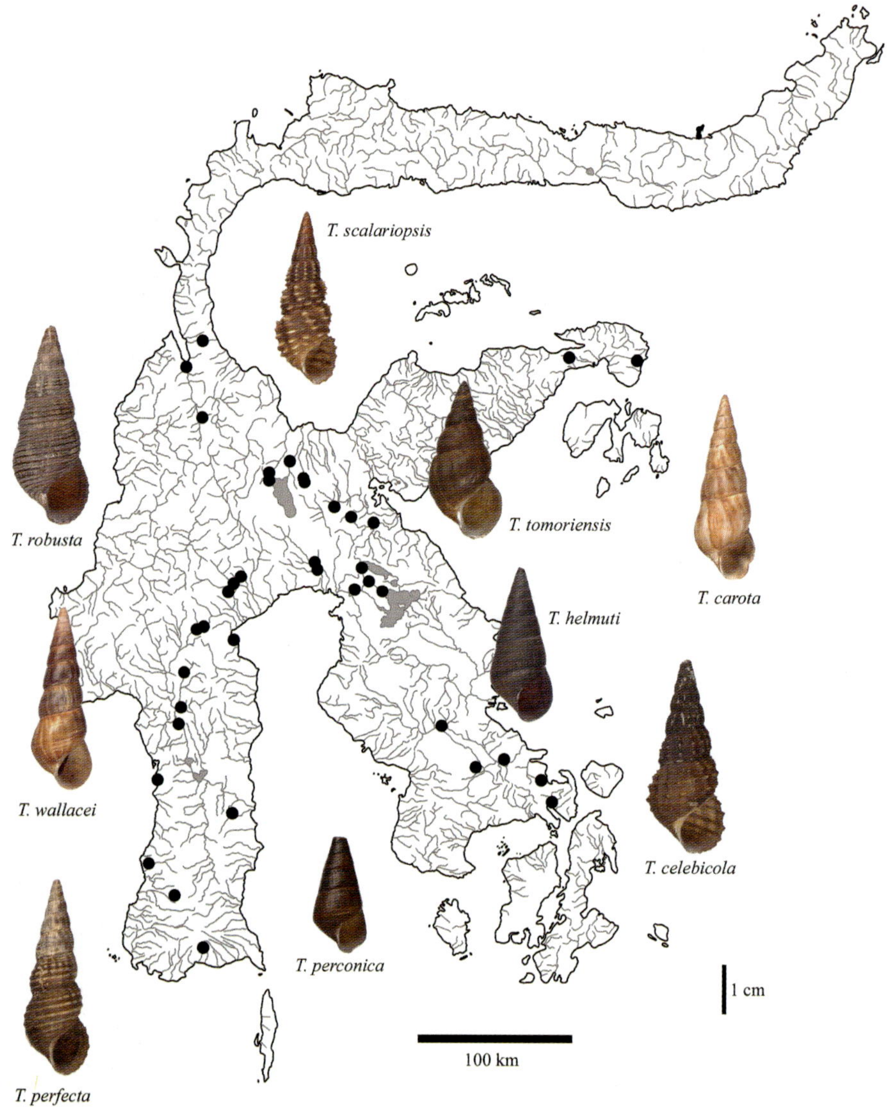

Fig. 2 Sulawesi with riverine sample sites (1999–2005, closely neighboured sample stations have been reduced to just single dots) and species. Scale bar = 1 cm

the Cerithioidean family Pachychilidae (e.g. Lydeard et al., 2002; Glaubrecht & von Rintelen, 2003), and that all pachychilid species from Sulawesi indeed belong to one single endemic genus, i.e. *Tylomelania* (von Rintelen & Glaubrecht, 2005).

The pachychilid freshwater snails have an apparent Gondwana distribution, with representatives occurring in South America (*Doryssa*, *Pachychilus*), Africa (*Potadoma*), Madagascar (*Melanatria*), South and Southeast Asia (*Adamietta*, *Brotia*, *Jagora*, *Paracrostoma*, *Sulcospira*, *Tylomelania*) and the Torres Strait Islands (*Pseudopotamis*) at the Australian margin (Glaubrecht, 2000; Köhler et al., 2004; Köhler & Glaubrecht, 2007). While viviparity in limnic gastropods has been long assumed as an important intrinsic factor in such radiations (see discussion, e.g. Glaubrecht, 1996, 1999, 2006) we found this reproductive strategy to occur in several genera in Asia and in the geographically restricted *Pseudopotamis*, while all other pachychilids are oviparous (Glaubrecht & von Rintelen, 2003; Köhler et al., 2004). *Tylomelania* from Sulawesi shares a unique reproductive anatomy and strategy, i.e. brooding its young in the uterus, with its sister group *Pseudopotamis* (Glaubrecht & von Rintelen, 2003; von Rintelen & Glaubrecht, 2005). On the island of Sulawesi *Tylomelania* is widespread (Fig. 2), with the majority of species being endemic to two ancient lake systems (Fig. 3).

Fig. 3 Shells of the Malili system species of *Tylomelania*. (**a**) *T. insulaesacrae* (LT, NMB 1342a); (**b**) *T. inconspicua* (HT, MNHN); (**c**) *T. kruimeli* (HT, MZB Gst. 11.959); (**d**) *T. palicolarum* (LT, NMB 1331a); (**e**) *T. wesseli* (HT, MZB Gst. 12.104); (**f**) *T. sarasinorum* (LT, ZMA); (**g**) *T. masapensis* (LT, ZMA); (**h**) *T. tominangensis* (LT, ZMA); (**i**) *T. lalemae* (LT, ZMA); (**k**) *T. molesta* (LT, NMB 1340a); (**l**) *T. zeamais* (LT, NMB 1337a); (**m**) *T. abendanoni* (LT, ZMA); (**n**) *T. gemmifera* (LT, NMB 1344a); (**o**) *T. marwotoi* (HT, MZB Gst. 12.319); (**p**) *T. woltereckі* (HT, MZB Gst. 12.281); (**q**) *T. towutica* (LT, ZMA); (**r**) *T. bakara* (HT, MZB Gst. 11.956); (**s**) *T. dubiosa* (HT, MZB Gst. 12.100); (**t**) *T. matannensis* (HT, MZB Gst. 12.309); (**u**) *T. kristinae* (PT, MZB Gst. 12.106); (**v**) *T. towutensis* (PLT, NMB 4790b), (**w**) *T. confusa* (HT, MZB Gst. 12.102); (**x**) *T. mahalonensis* (PLT, ZMA); (**y**) *T. patriarchalis* (LT, NMB 1330a); (**z**) *T. turriformis* (HT, MZB Gst. 12.103). Scale bar = 1 cm. HT, holotype; LT, lectotype; PLT, paralectotype; PT, paratype. Modified from von Rintelen et al. (2007)

Based on phylogenetic analyses using morphological and molecular data on various hierarchical levels, the role of extrinsic and intrinsic factors for speciation and radiation is investigated for the endemic and adaptive radiation of these lacustrine snails. With this study we also explicitly test the fulfilment and, thus, general suitability of the four distinctive criteria of a truly *adaptive* radiation provided by Schluter (2000), viz. (fairly recent) common origin and rapid speciation (as phylogenetic criteria leading to radiation) as well as phenotype–environment correlation and trait utility (as ecological criteria resulting in adaptation). For mollusc radiations, in particular, a rigorous approach along these lines is generally lacking, even though for example Davis (1979, 1982) has implicitly discussed several of the concepts or criteria applied here.

Here, we summarize our current knowledge on the ecological and phylogenetic patterns that have emerged so far from this model system, in particular, from the Malili lakes. We discuss evolutionary hypotheses stemming from these data and we will outline future research perspectives. Also, we suggest priorities in conservation in order to offer future generations the chance to study this unique "submerged Galapagos" or "Wallaces's dreamponds" (Herder et al., 2006) as well.

This and the cited studies by the authors are based on comprehensive sampling in Lake Poso and all five Malili lakes as well as in the connecting and adjacent rivers (Fig. 1), during nine field campaigns. The lakes were visited in 1999 (dry season, MG & TvR), 2000 (wet season, TvR), 2002 (dry season, TvR), 2003 (dry season, TvR & MG), 2004 (three trips, dry and wet season, TvR & MG), 2005 and 2007 (dry season TvR & MG). Samples were obtained by snorkelling and SCUBA diving. In addition, initially a significant amount of material collected in the first systematic expedition to the lakes in 1991 by Philippe Bouchet from the Muséum National d'Histoire Naturelle in Paris was studied (see von Rintelen et al., 2007).

The species flocks in lakes on Sulawesi: species diversity and endemism

Currently, 32 species of *Tylomelania* have been described from both ancient lake systems of Sulawesi (Figs. 3, 4; Table 1). All species are endemic to their respective limnic system. While 25 species are recognised in the Malili system, according to our recent revision (Fig. 3; von Rintelen et al., 2007), only seven taxa have so far been described from Lake Poso and Poso River (Fig. 4). Diversity in Lake Poso is expected to be much higher, though, with estimates ranging as high as 24 species (von Rintelen et al., unpubl. data; Marwoto, pers. comm.). The Malili lakes probably also harbour between 1 and 5, yet undescribed taxa, depending on the outcome of ongoing research on the fine-scale differentiation of genetics and morphology in that system (von Rintelen et al., unpubl. data).

In the Malili system, endemism is not restricted to the lake system as a whole, however, but 20 species (80%) are single lake endemics (von Rintelen et al., 2007). Lake Towuti is most species-rich with 10 species, six of which are endemic, while Lake Mahalona harbours eight species (five endemics) and Lake Matano seven (six endemics). Tominanga River between Lake Mahalona and Towuti hosts four species, only one of which is endemic, though. The smaller satellite lakes only harbour one (Lake Masapi) or two (Lake Lontoa) endemic species, respectively.

The high diversity in the lakes contrasts with a low number of described species in the rivers of Sulawesi in general. Only nine riverine taxa (Fig. 2) have been described so far, five of which also occur in the vicinity of the ancient lakes. However, we anticipate that species numbers are bound to increase considerably here as well. A preliminary assessment of riverine diversity around the Malili lakes has revealed the occurrence of nine or 10 species in this area, only one of which had been described when the study was started and one more by the authors (von Rintelen & Glaubrecht, 2003).

The high amount of endemism in each lake or river is a striking feature, which suggests a strong influence of geographic factors in species divergence, i.e. allopatric speciation, which is perhaps not entirely unexpected given the spatial structure of the system (cf. Fig. 1c). A few taxa also show a highly localized occurrence within single lakes, such as, e.g. in Lake Towuti *T. bakara*, which is apparently confined to one cape (cf. Fig. 12), or *T. kristinae* at Loeha Island. The available data do not offer an easy explanation for these restricted distribution ranges, since these taxa are neither by any means specialized to

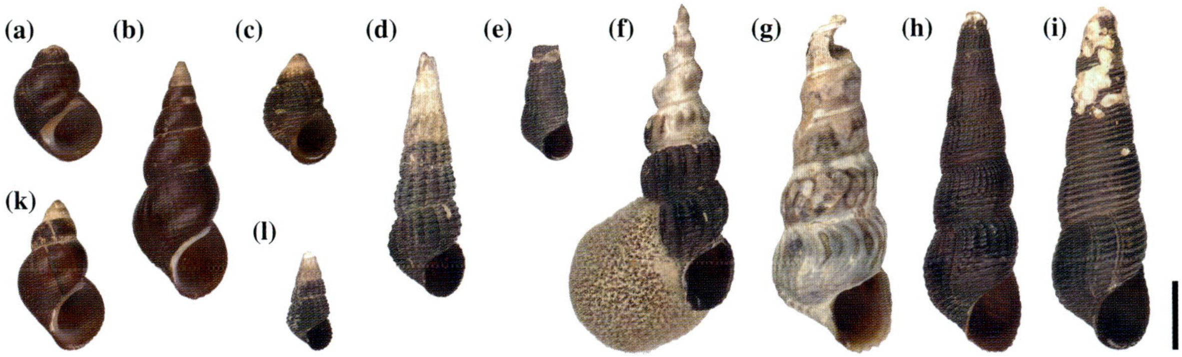

Fig. 4 Lake Poso species of *Tylomelania*. (**a**) *T. neritiformis*, LT, NMB 1349a; (**b**) *T. porcellanica,* LT, NMB 1347a; (**c**) *T. carbo*, LT, NMB 1343a; (**d**) *T.* spec Poso C, ZMB 190201; (**e**) *T. spec Poso D*, ZMB 190195; (**f**) *T. centaurus*, HT, NMB 1339a.; (**g**) *T. kuli*, LT, NMB 1329a; (**h**) *T. toradjarum*, LT, NMB 1328a; (**i**) *T. spec Poso A*, ZMB 190205; (**k**) *T. connectens*, HT, NMB 1335a; (**l**) *T. spec Poso B*, ZMB 190198. Scale bar = 1 cm. HT, holotype; LT, lectotype; ST, syntype

substrates that only occur locally nor are there any obvious obstacles to dispersal.

Lake colonization

The molecular phylogeny based on two mitochondrial gene fragments (COI and 16S) of the mitochondrial DNA, with a total of 1,535 bp reveals four strongly supported clades within the lakes, i.e. one clade in Lake Poso and three clades in the Malili system (Fig. 5; von Rintelen et al., 2004). Riverine taxa are sister groups to three lacustrine clades, the Poso and two of the Malili clades. Evidence from mitochondrial data suggests four independent colonization events in the lakes, three of these in the Malili lakes alone (von Rintelen et al., 2004). A separate invasion of Lake Poso and the Malili lakes is an expected result given that the two lake systems were never connected. However, it is surprising that colonization took place independently in different ancestral lineages in the three major lakes of the Malili system, which are directly connected by rivers.

As outlined below, each colonization was followed by diversification into an array of morphologically distinct and ecologically specialized species, thus fulfilling criteria of a truly adaptive radiation sensu Schluter (2000).

Coevolution with crabs

Dramatic changes in shell morphology are associated with lake colonization (von Rintelen et al., 2004). With one exception (clade Malili 1) the species in each lacustrine clade share characteristic shell features such as the presence of axial ribs, providing additional support for the independent colonization particularly of the Malili lake system. Species can be distinguished by their characteristic shells, though intraspecific variability is rather high (von Rintelen & Glaubrecht, 2003; von Rintelen et al., 2007). In each lacustrine clade convergent evolution of thicker shells relative to riverine species occurred in almost all cases (Fig. 6a). Shell thickness can be regarded as indicative of shell strength (Fig. 6d) and is used here as an estimator of resistance to crab predation. These findings coincide with the occurrence of one species of molluscivorous crabs of the Sundatelphusidae and Parathelphusidae in each of the lakes (Fig. 6b; unpubl. data), which possess pronounced dentition on their chelae enabling them to crack shells. These data on lacustrine gastropod shell strength, structure and also the frequent occurrence of shell repair in lacustrine *Tylomelania* (Fig. 6c), in combination with the occurrence of large molluscivorous crabs, suggests that evolution in the face of crab predation is a driving factor in initial shell divergence upon colonization of the lakes (von Rintelen et al., 2004). The exclusive presence of massive and dentitioned chelae in just the molluscivorous crab species makes it very likely that there is true coevolution, i.e. an evolutionary response also on the crab side to the development of stronger gastropod shells.

Table 1 *Tylomelania* species in rivers and the ancient lakes on Sulawesi

Species	Distribution
Malili lake system	
T. abendanoni Kruimel, 1913	Lake Lontoa
T. bakara von Rintelen & Glaubrecht, 2003	Lake Towuti
T. confusa von Rintelen et al., 2007	Lake Mahalona
T. amphiderita von Rintelen et al., 2007	Lake Towuti
T. gemmifera Sarasin & Sarasin, 1897	Lake Matano
T. inconspicua von Rintelen et al., 2007	Lake Mahalona
T. insulaesacrae Sarasin & Sarasin, 1897	Lake Mahalona, Towuti
T. kristinae von Rintelen et al., 2007	Lake Towuti
T. kruimeli von Rintelen & Glaubrecht, 2003	Lake Mahalona
T. lalemae Kruimel, 1913	Lake Towuti
T. mahalonensis Kruimel, 1913	Lake Mahalona
T. marwotae von Rintelen et al., 2007	Lake Mahalona
T. masapensis Kruimel, 1913	Lake Masapi
T. matannensis von Rintelen et al., 2007	Lake Matano
T. molesta Sarasin & Sarasin, 1897	Lake Matano
T. palicolarum Sarasin & Sarasin, 1897	Lake Matano, Mahalona, Towuti
T. patriarchalis Sarasin & Sarasin, 1897	Lake Matano
T. sarasinorum Kruimel, 1913	Lake Towuti
T. tominangensis Kruimel, 1913	Tominanga River, Lake Lontoa
T. towutensis Sarasin & Sarasin, 1897	Lake Towuti, Tominanga River
T. towutica Kruimel, 1913	Lake Towuti, Tominanga River
T. turriformis von Rintelen et al., 2007	Lake Matano
T. wesseli von Rintelen et al., 2007	Tominanga River
T. woltereckí von Rintelen et al., 2007	Lake Mahalona
T. zeamais Sarasin & Sarasin, 1897	Lake Matano
Lake Poso	
T. carbo Sarasin & Sarasin, 1897	Lake Poso
T. connectens Sarasin & Sarasin, 1898	Poso River
T. centaurus Sarasin & Sarasin, 1898	Lake Poso
T. kuli Sarasin & Sarasin, 1898	Lake Poso
T. neritiformis Sarasin & Sarasin, 1897	Poso River
T. porcellanica Sarasin & Sarasin, 1897	Poso River
T. toradjarum Sarasin & Sarasin, 1897	Lake Poso
Rivers	
T. carota Sarasin & Sarasin, 1897	Kalaena catchment
T. perfecta Sarasin & Sarasin, 1898	Southern Southwest Sulawesi
T. robusta Sarasin & Sarasin, 1898	Northern Southwest Sulawesi
T. scalariopsis Sarasin & Sarasin, 1898	Lake Poso catchment
T. wallacei Sarasin & Sarasin, 1897	Southern Southwest Sulawesi
T. tomoriensis Sarasin & Sarasin, 1897	Tomori region
T. celebicola Sarasin & Sarasin, 1897	Tomori region
T. helmuti von Rintelen & Glaubrecht, 2003	Malili catchment

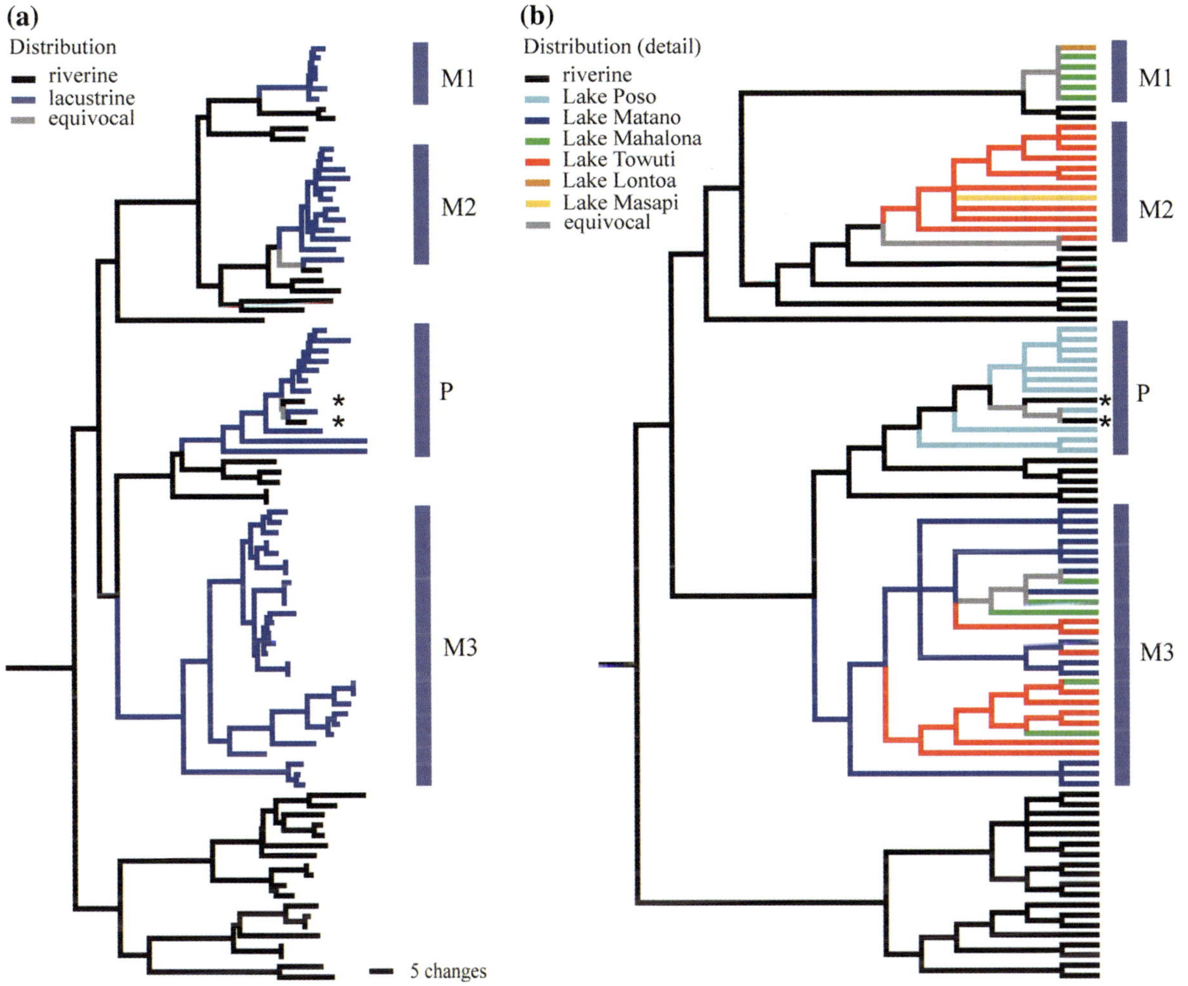

Fig. 5 Molecular phylogeny of *Tylomelania* based on 1,535 bp of mtDNA (16S & COI). The four blue bars on the right mark the four lake clades representing four independent lake colonization events. (**a**) Bayesian inference phylogram. Lacustrine taxa are highlighted in blue and riverine species in black. (**b**) Maximum parsimony cladogram with details of lake colonization. Branch colours correspond to single lakes as given in the legend. Modified from von Rintelen et al. (2004)

Trophic (radula) specialization and ecology

A striking pattern found in all lacustrine lineages is the high variety of radula morphology (Figs. 7–10), which contrasts sharply with the situation among the riverine species, where the vast majority of taxa possess almost identical radulae (von Rintelen et al., 2004). The radula is a pivotal part of the trophic system in gastropods and radular morphological differences are indicative of food and substrate preferences (Hawkins et al., 1989). The molluscan radula is generally considered a conservative character with little variation at the species level, but see Padilla (1998), Reid and Mak (1999) and Reid (2000). In contrast, at least nine different species-specific radula morphologies distinguished by the shape and relative size of their denticulation are present within the ancient lake clades, with 3–6 phenotypes found in each clade (von Rintelen et al., 2004). Preliminary results from more extensive studies on radulae, in particular, of species from Lake Poso indicate that the variety of radula forms may be even twice as high.

Trophic morphology and substrate were supposed to be highly correlated in all clades (Fig. 10; von Rintelen et al., 2004). All species in the major lakes of the Malili system for which we have data on substrate preferences available are specialized on

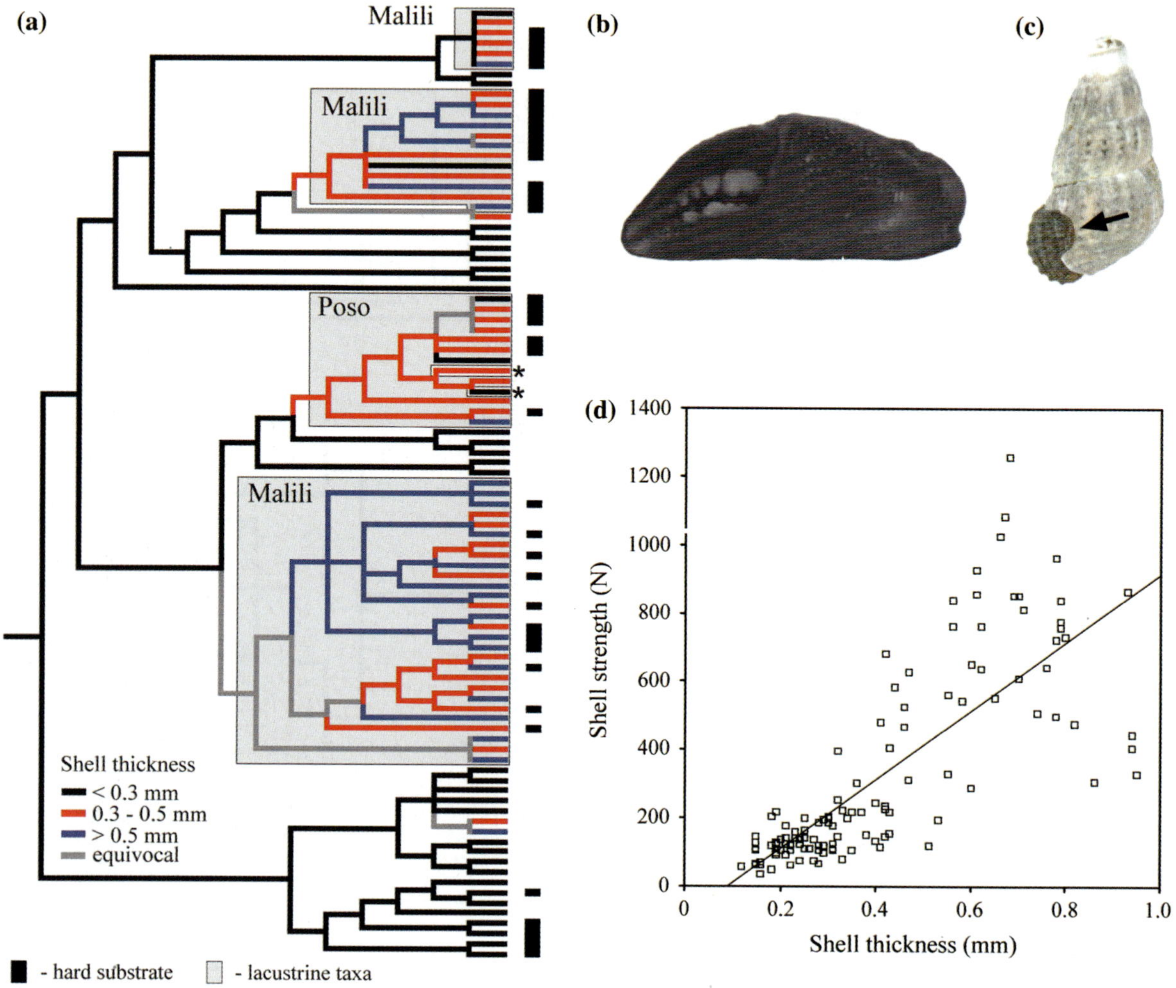

Fig. 6 Shell strength of lacustrine *Tylomelania* species and crab predation. (**a**) Maximum parsimony tree based on 1,535 bp of mtDNA (16S & COI) with colours corresponding to shell thickness; (**b**) Left claw of molluscivorous lacustrine crab species (*Syntripsa matannensis* (Schenkel, 1902)); (**c**) shell with repair scar (*T. zeamais*, Lake Matano). Scale bar = 1 cm; (**d**) scatter plot showing correlation between shell thickness and shell strength. Modified from von Rintelen et al. (2004)

either soft (mud, sand) or hard (rock, sunken wood) substrates, with about 50% of species occurring on either substrate category (von Rintelen et al., in press). Soft substrate species (Fig. 7a, b) mostly have identical or very similar radulae to those found in riverine taxa (Fig. 7b), while hard substrate taxa particularly from rocks often have strongly enlarged teeth in a variety of shapes (Fig. 7c–f). This tight correlation between enlargement of radula denticles and hard substrate described by von Rintelen et al. (2004) is generally supported. However, more detailed radula descriptions have revealed a rather complex pattern, as it is evident from cases of extensive radula polymorphism in many species both from hard and soft substrates (Fig. 8; von Rintelen et al., in press). The role of radula (i.e. trophic) specialization in the ecological diversification of the lake species as stressed by von Rintelen et al. (2004) remains unchallenged, though. It is further supported by the parallel occurrence of accordingly modified radulae in soft and hard substrate dwellers in both ancient lake systems on Sulawesi (Fig. 9) and in the only constantly rock-dwelling riverine taxa found so far from the Maros area in Southwest Sulawesi (von Rintelen et al., unpubl. data). These observations suggest a functional role for the differences found,

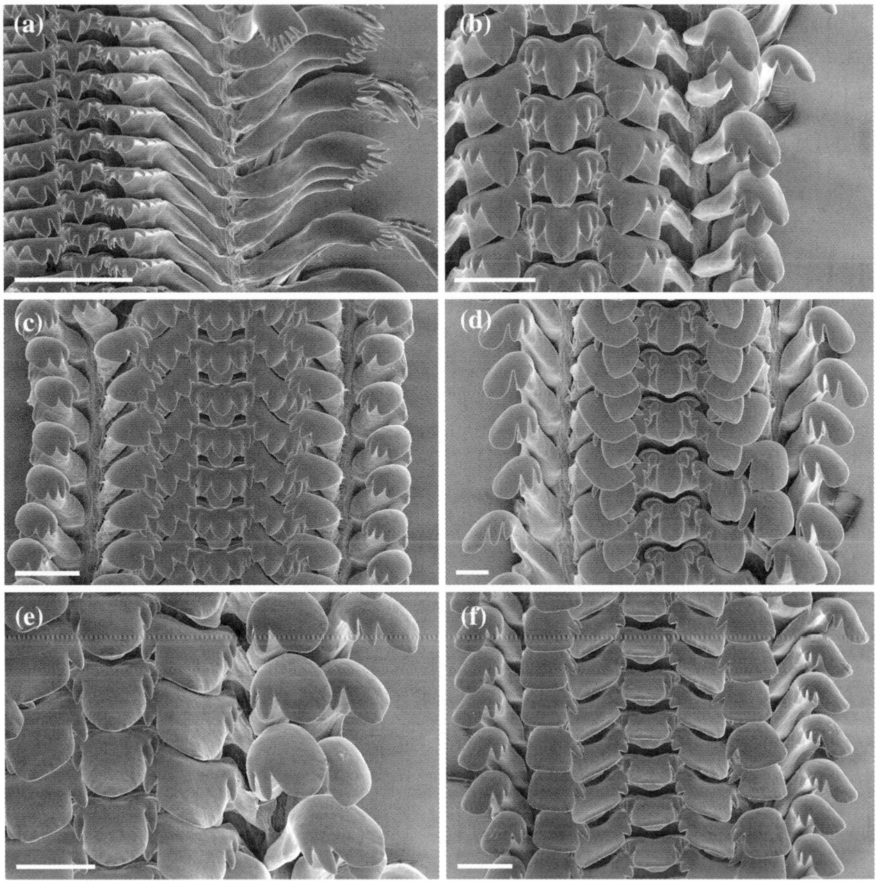

Fig. 7 Characteristic radulae of Malili system *Tylomelania*. (**a**) and (**b**) Soft substrate dwellers. (**a**) *T. gemmifera* (ZMB 190095, Lake Matano); (**b**) *T. patriarchalis* (ZMB 190061, Lake Matano); (**c–f**) hard substrate dwellers. (**c**) *T. kruimeli* (ZMB 190152, Lake Mahalona); (**d**) *T. insulaesacrae* (ZMB 190180, Lake Towuti); (**e**) *T. matannensis* (ZMB 190094, Lake Matano); (**f**) *T. bakara* (ZMB 190131, Lake Towuti). Scale bar = 100 μm

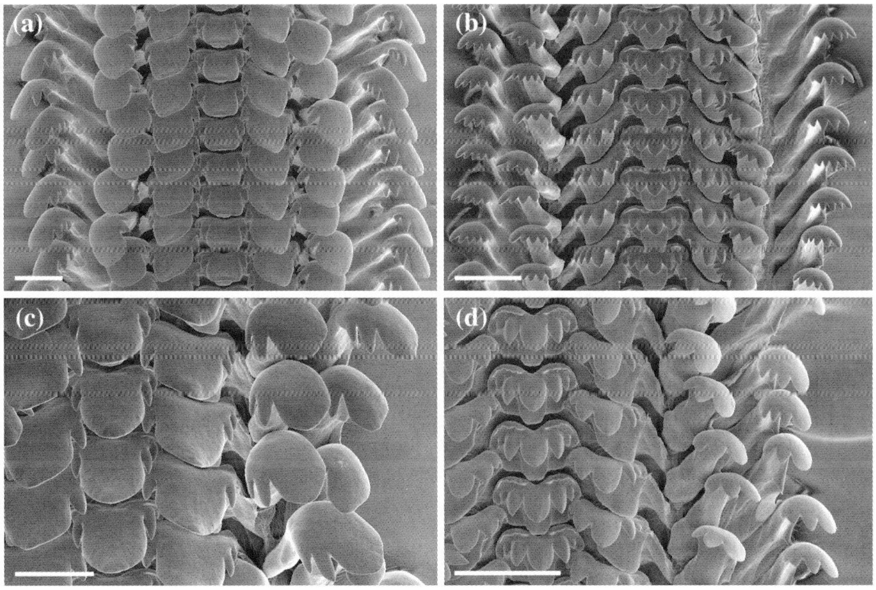

Fig. 8 Radula polymorphism in lacustrine *Tylomelania* species. (**a**) and (**b**) *T. sarasinorum* (ZMB 190123, Lake Towuti); (**a**) rock; (**b**) wood. (**c**) and (**d**) *T. matannensis* (Lake Matano); (**c**) ZMB 190094, rock; (**d**) ZMB 190100, wood? Scale bar = 100 μm

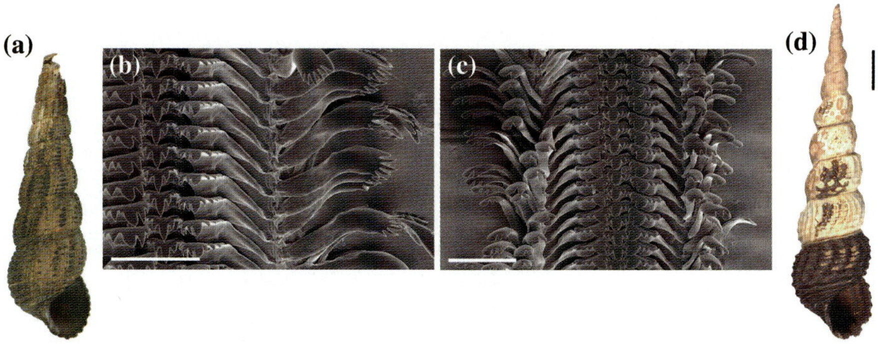

Fig. 9 Radula and shell convergence in Lake Poso and the Malili lakes. (**a**) and (**b**) *T. gemmifera*, ZMB 190094, Lake Matano, Malili system; (**a**) shell; (**b**) radula. (**c**) and (**d**) *T. Kuli*, ZMB 190011, Lake Poso; (**c**) radula; (**d**) shell. Scale bar (shells) = 1 cm, (radula) = 100 μm

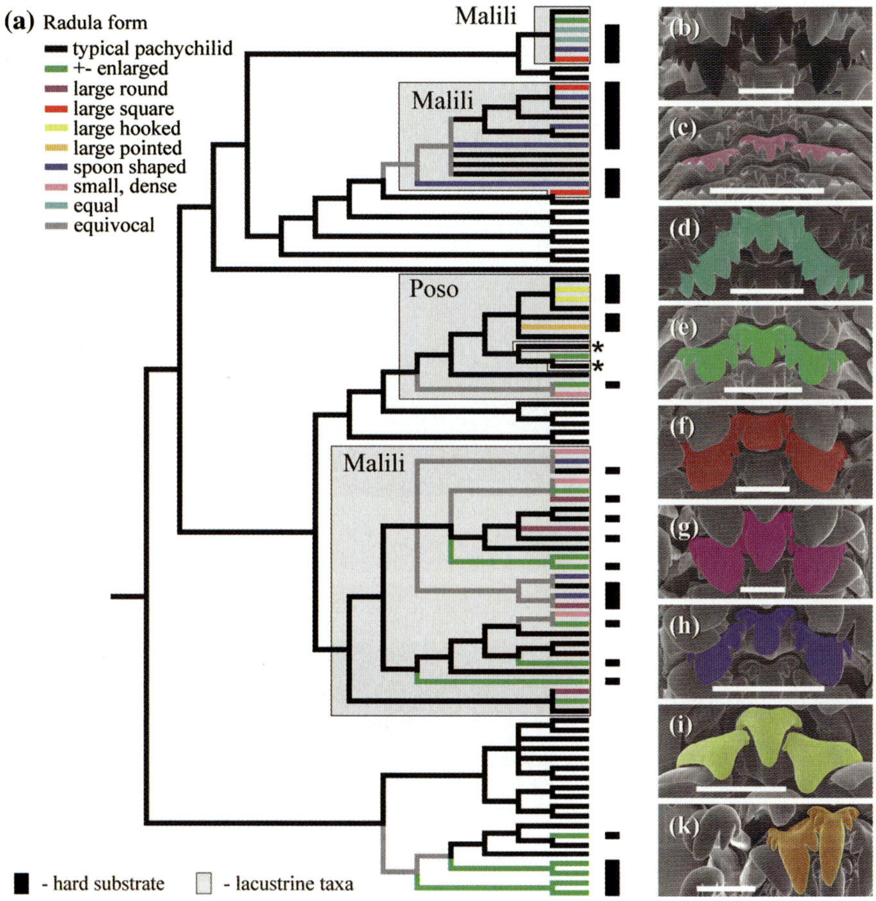

Fig. 10 Radula diversification in *Tylomelania* species. (**a**) Maximum parsimony tree based on 1,535 bp of mtDNA (16S & COI) with colours corresponding to the radula form shown in (**b–k**); (**b–k**) major radula types from Malili and Lake Poso Tylomelania; (**b**) typical pachychilid (riverine), *T. perfecta*; (**c**) short with dense dentation, *T. gemmifera*; (**d**) equal sized denticles, *T. kruimeli*; (**e**) enlarged major denticles. *T. towutica*; (**f**) large, square major denticles, *T. sarasinorum*; (**g**) large, round major denticles, *T. matannensis*; (**h**) small, spoon-shaped major denticles, *T. insulaesacrae*; (**i**) single, hooked denticles, *T. carbo*; (**k**) long, pointed major denticles, *T. spec Poso 2*. Scale bar = 100 μm

although a detailed understanding of the underlying mechanisms requires further investigation.

Habitat specialization (substrate and to a lesser degree depth preferences) and radula differentiation in ancient lake *Tylomelania* enables more than five, possibly up to seven species to coexist at localities with sufficiently structured habitats (Fig. 11; von Rintelen et al., 2007, and unpubl. data). This level of single-site diversity was only observed in taxa from Lake Poso and within the Malili system in Lake Mahalona and Lake Towuti. The absolute numbers of syntopic species in the lakes may not be impressive for themselves, but they become more striking when contrasted with the situation in rivers, where only a single species is found in almost all cases.

Thus, a strong role for ecological factors in diversification and possibly even speciation itself is suggested by the marked substrate preferences of

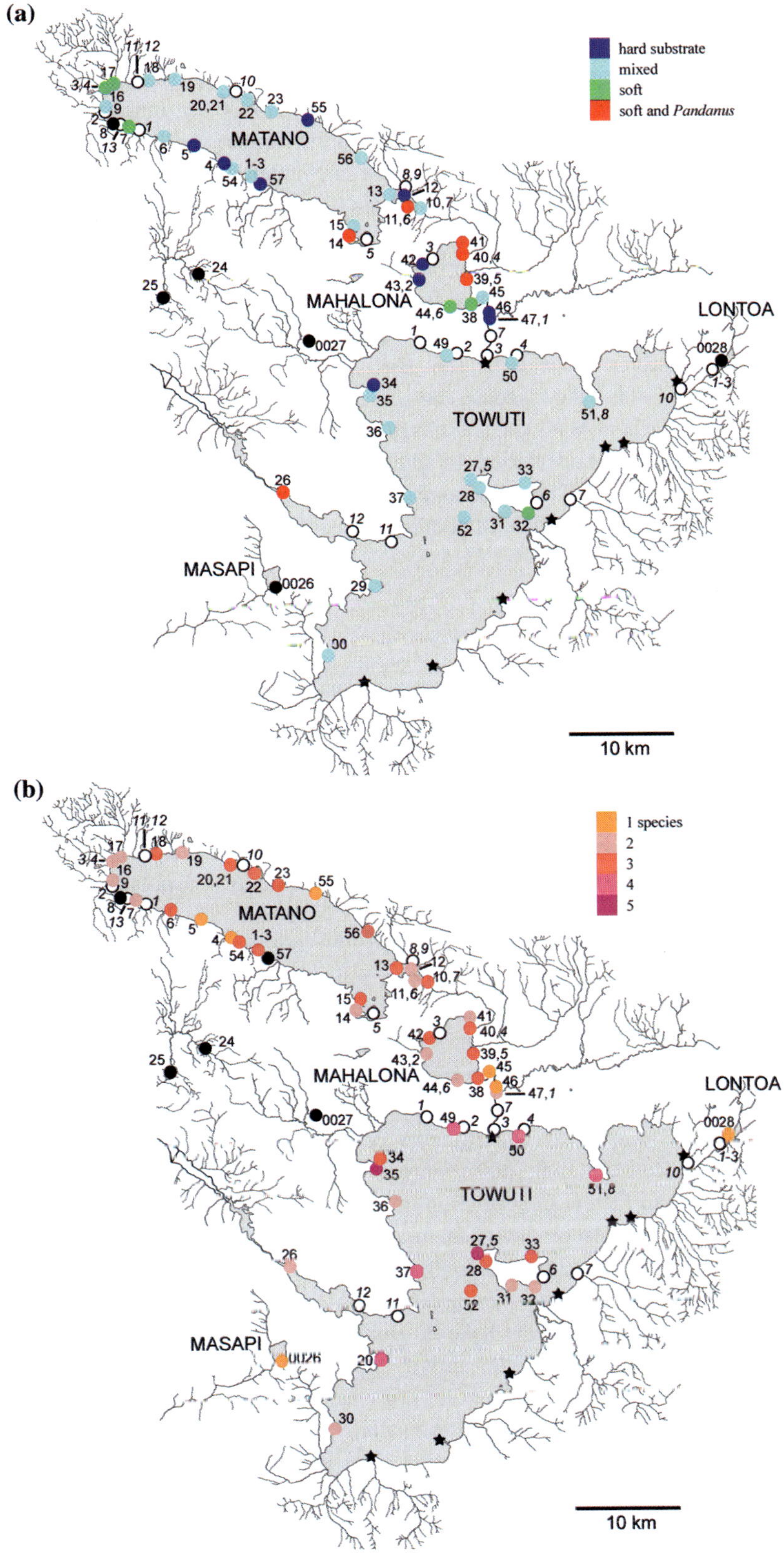

Fig. 11 Substrate and species number at the Malili lake system sample sites. (**a**) Substrate; (**b**) number of *Tylomelania* species. Only sites sampled in 1999 and 2000 have been considered, the numbers are the original field codes. No information has been available for sites that have not been colour-coded. Asterisks: no snails found

most *Tylomelania* species, coupled with a correlated species-specific radula (trophic) morphology and genetic clustering in some taxa. However, our detail studies on selected species and species complexes so far reveal a rather complicated situation in the limnic systems on Sulawesi. The phenomena encountered range from species with highly characteristic specific radulae irrespective of their occurrence on various substrates to species where two fundamentally different radula morphologies are found in the same population on the same substrate. All other morphological features, especially the shell, are highly variable in the lake species, but so far we lack evidence that conchological features play a crucial role in the causation of speciation. In addition, the molecular phylogeny based on mitochondrial genes does not correspond to morphologically delimited species.

Speciation patterns

In a broader perspective, these results from *Tylomelania* suggest a wider applicability of a radiation model proposed for vertebrates by Streelman & Danley (2003), which assumes that trophic specialization and habitat specialization drive the initial stages of adaptive radiation. Similar, albeit somewhat less clearly phrased observations suggest that among molluscs this may also apply to the riverine pomatopsid radiation in the Mekong (Davis, 1979), but the lack of a rigorous phylogeny does not allow a detailed comparison with the Sulawesi snails.

The finding of largely species-specific radula types and the correlation between trophic morphology and substrate in *Tylomelania* as well as marked niche differences in species (von Rintelen & Glaubrecht, 2003; von Rintelen et al., 2007) suggest a strong role for ecological factors in speciation in these snails. In gastropods, studies on intertidal marine species of *Littorina* have indicated differentiation across ecotones and assortative mating between morphs (Johannesson et al., 1995; Kyle & Boulding, 1998; Wilding et al., 2001). The limnic gastropods of Sulawesi offer excellent opportunities to test the importance of these factors in freshwater.

Nevertheless, allopatric speciation (i.e. speciation in geographic separation) is probably the predominant mode of speciation at least in the Malili lakes. Here, different species of single putative species-groups reveal an allopatric distribution pattern in the three major lakes (von Rintelen et al., 2007) and single lake endemism is high (see above). However, independent parallel evolution of hard and soft substrate dwellers with associated trophic traits (and size differences) occurred in all three Malili lineages, suggesting an important role for ecology-driven differentiation, even if the initial phase of allopatric speciation was not associated with ecological divergence. Looking beyond the Malili lakes, a particularly striking case of parallel evolution is represented by two shallow water mud-dwellers in Lake Poso (*T. kuli*) and the Malili system (*T. gemmifera*) (Fig. 9). Similar patterns might be suspected for several other species among the Malili lake clades, but the failure of a simple DNA-taxonomy approach in our system (see below) prohibits a more detailed discussion of convergent evolution for the time being.

Molecular taxonomy and introgression

While the molecular phylogeny provides a rather clear-cut picture at a higher level, there is virtually no resolution at the species level for *Tylomelania*. All lacustrine morphospecies of which more than one specimen or population has been sequenced appear polyphyletic in the molecular phylogeny (Fig. 12). For a more general discussion of the DNA taxonomy approach we refer to Funk & Omland (2003), Sites & Marshall (2004), Will & Rubinoff (2004) and Vogler & Monaghan (2006). This lack of resolution is even more remarkable since there is no lack of genetic structure per se in the data.

Several phenomena might account for this problem, such as incomplete lineage sorting, i.e. the maintenance of ancestral genetic polymorphism across species boundaries, or introgressive hybridization. We particularly suspect a pivotal role of hybridization here, as concluded from several significant mismatches between observed morphological features of species and their placement in the molecular phylogeny, which cannot be parsimoniously explained by other hypotheses. For example, a clade comprising smooth-shelled species mostly from Lake Towuti comprises some taxa with axial ribs (Fig. 12a). The presence or absence of axial ribs is otherwise considered a distinguishing feature of the major clades of *Tylomelania* (von Rintelen et al.,

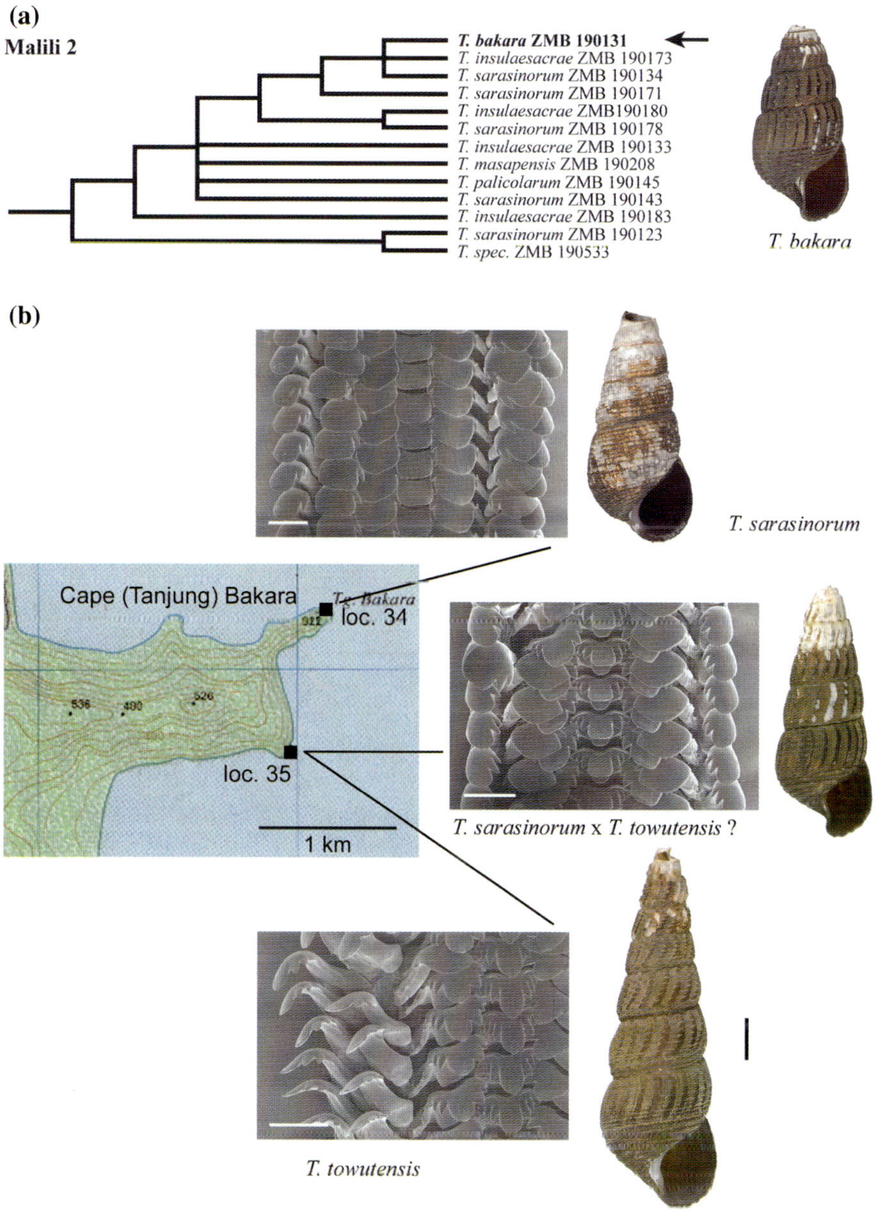

Fig. 12 Putative hybridization between *T. sarasinorum* and *T. towutensis* at Cape Bakara, Lake Towuti. (**a**) Maximum parsimony cladogram showing Malili clade 2 enlarged from Fig. 5. The arrow highlights *T. bakara*, the only axial ribbed species in this clade. (**b**) Map of Cape Bakara and the species suspected to be involved in hybridization. Morphs intermediate between *T. sarasinorum* and *T. towutensis* occur at loc. 35

2007). The ‘misplaced’ species, *T. bakara* and a similar form from the same area (Fig. 12b), are suspected of being of hybrid origin. In addition, in the lacustrine clade from Lake Poso also riverine species appear in terminal position (Fig. 4), suggesting the occurrence of hybridization between lacustrine and riverine species. While this hypothesis needs to be tested using nuclear markers, it might provide the second case of widespread introgression of fluviatile haplotypes in lake populations or vice versa in the ancient lakes of Sulawesi, which has been shown already to be a major factor in Malili telmatherinid fishes (Herder et al., 2006).

The framework of evolutionary ecology

Utilizing several biological concepts, e.g. the concept of species, ecological niche, allopatric speciation and annidation, Glaubrecht (1996) has introduced and described an evolutionary ecology approach to the study of limnic gastropods and, in particular, in ancient lake radiations. He defined evolutionary ecology in this context as "aiming to reconstruct the origin and alteration of the ecological interrelations of organisms and their respective environments in the course of evolution", an idea which has recently been taken up also by Sudhaus (2004). This approach provides a research program for acquiring a synthetical perspective that includes morphology, molecular genetics, ecology and biogeography. Within this framework of evolutionary ecology, and based on a revised Cerithioidean phylogeny (Glaubrecht, 1996; Glaubrecht et al., unpubl. data; Lydeard et al., 2002), it will now be possible to compare the Sulawesi species flocks to the long enigmatic and so-called thalassoid (i.e. marine-like) gastropod radiation in East African's Lake Tanganyika, for which morphological data (Glaubrecht & Strong, 2007; Glaubrecht, 2008) and molecular data (e.g. Wilson et al., 2004) are available now, too. Discussing the evolutionary and taxonomic implications of the species flocks in *Tylomelania* we have provided also the groundwork to compare this lacustrine radiation with an exceptional endemic assemblage of morphologically distinct viviparous pachychilid gastropods found within the Kaek River in Thailand (Glaubrecht & Köhler, 2004). Finally, it will be possible to decipher in how far different lacustrine model systems and their respective faunal elements are distinctive with respect to the influence of intrinsic factors (e.g. crucial biological features such as viviparity, dispersal and trophic specialization) versus extrinsic factors (e.g. palaeohydrology, habitat fragmentation), as was initially outlined for limnic Cerithioidea by Glaubrecht (1996). It will then become evident why and how earlier attempts to evaluate these factors were essentially suffering from lacking insight into the systematics and morphology (i.e. the reproductive biology) of the limnic gastropods involved, and how crucial differences in the geological setting were, for example, rifting in the ca. 12-myr-old Lake Tanganyika versus Sulawesi's composite terrane origin with ca. 1- to 2-myr-old lakes.

Conclusion

The two ancient lake systems on the Indonesian island Sulawesi, Lake Poso and the five lakes of the Malili system, harbour endemic species flocks of at least 32 and possibly more than 50 taxa of viviparous pachychilid gastropods whose origin and that of *Tylomelania* on Sulawesi, in general, can be linked to the tectonic events shaping the island in the Tertiary. Evidence from molecular data has revealed four independent colonization events in the lakes, three of which occurred in the Malili system. Subsequent intralacustrine radiation in all clades in both lake systems has produced several species flocks. Each colonization was followed by a radiation into an array of morphologically distinct and ecologically diverse species, thus fulfilling Schluter's (2000) criteria for defining adaptive radiation. These multiple radiations of *Tylomelania* in the individual lake systems of Poso and Malili include a fair amount of parallel (convergent) evolution and, therefore, provide a unique opportunity to study, e.g. key factors in the origin of species.

However, the details of the patterns of speciation and diversification in the lakes are rather complex. While a geographic component is clearly recognisable, both on the level of clades as well as within closely related species groups, the occurrence of presumable hybridization with riverine taxa in both lake systems renders a simple interpretation of the mitochondrial data doubtful. Indeed, the widespread mismatch between the morphological and molecular data suggests that introgressive hybridization is a major factor in the evolution of the lacustrine species flocks, which might provide good model systems to test, among others, the syngameon hypothesis of adaptive radiation by Seehausen (2004).

While not refuting the possibility of ecological speciation (see Schluter, 2000), the contradictory patterns found particularly in the trophic specialisation in *Tylomelania* as described above caution against simple explanatory models for the evolution of diversity in the Sulawesi lakes' radiations. Therefore, major aims of our future research are (i) to obtain a fully resolves species phylogeny based also on nuclear markers and (ii) a closer look into the adaptation of species to their environment (substrate) and the causes of the large intraspecific variation found.

Outlook

The results presented here indicate the potential of the *Tylomelania* (and other gastropod) species flock in the ancient lakes on Sulawesi to serve as another model system for the study of evolutionary factors in speciation, adaptive radiation and annidation (i.e. niche partitioning). The existence of three separate adaptive radiations in the Malili lakes offers, in combination with the Lake Poso flock, an outstanding opportunity to study *in parallel* evolutionary processes such as the origin of specialization. The indicated involvement of coevolution at a basal level in this system is unique among radiations in ancient lakes. The species diversity in the Sulawesi lakes is comparable to species numbers observed in other well-studied adaptive radiations, such as, e.g. the Galapagos finches or Caribbean *Anolis*, and thus render them indeed truly "Darwinian snails".

We believe that beyond the singular possibility to study coevolution in the context of radiation, the relative simplicity of the Sulawesi system as compared, e.g. to the Great African lakes will especially favour tests of ecological speciation by comparing trait variance under different natural conditions. Recently started work using highly variable nuclear markers (ITS and AFLPs) is expected to yield a highly resolved phylogeny at the population and species level, and should permit, in conjunction with quantitative ecological data, an elaboration of our hypotheses on ecological speciation in the Sulawesi snails. The suspected importance of introgressive hybridization in the system will be another research focus. In this respect, we are particularly interested in how it contributes to speciation and how species boundaries are maintained despite hybridization.

Conservation

The Malili lakes are a magnificent and extraordinary limnic model system, both in a scientific respect, as amply described above, but also for the value of their sheer diversity in living forms. For these reasons their conservation should have a high priority.

Unfortunately, the lakes' environment is threatened by several factors. The activities of P.T. INCO are a critical issue. While appreciable efforts have been made by this company to preserve the water quality in the lakes, less attention is payed to the importance of protecting the rivers connecting the lakes and draining the area. Petea River, for example, has been fitted with a sluice to regulate water flow, and Larona River is fitted with yet another dam. Any further action especially on Petea and also on the Tominanga River in the future is going to have direct effects on the whole system. Besides the risk that the endemic species in these rivers would be lost, these waterways function as '*species valves*' between the lakes, which apparently in the past have played an important role in speciation processes in the lakes. The Malili lakes would be lost as a model system for evolutionary biology if the rivers were to be canalized. This would put another organismic and evolutionary 'hotspot' of Southeast Asian biodiversity at risk.

Further dangers stem from the continued activities of illegal loggers in the area, which threaten to damage lake habitats by erosion effects; some areas along the Lake Towuti shoreline are already degraded and devoid of snails. The growth of the local communities is another point of concern, since it seems inevitably coupled with pollution and direct habitat destruction. While these issues are notoriously difficult to tackle, at least efforts should be made to increase community awareness about the problems, which have the potential to threaten the future of the local people as well.

Acknowledgements We thank the editors for inviting us to present our findings not only during the 2006 SIAL meeting held in Berlin, but also for contributing this review to the special issue. Christian Albrecht and George Davis as reviewers and Tom Wilke provided constructive comments helping to improve the manuscript. We are most grateful to Kristina von Rintelen (Berlin) for her immense help both during the recent field campaigns and in analyzing data. Ristiyanti Marwoto (Zoological Museum Bogor) was very helpful in the preparation of the fieldwork in Indonesia and A. Munandar provided assistance in the field. We thank LIPI for permits to conduct research in Indonesia. We are most indebted to INCO for its invaluable logistic support at the lakes during our many visits. This work was funded by research grants GL 297/1 and GL 297/7 awarded to M. G. from the Deutsche Forschungsgemeinschaft.

References

Albrecht, C., S. Trajanovski, K. Kuhn, B. Streit & T. Wilke, 2006. Rapid evolution of an ancient lake species flock: freshwater limpets (Gastropoda: Ancylidae) in the Balkan Lake Ohrid. Organisms, Diversity & Evolution 6: 294–307.

Boss, K. J., 1978. On the evolution of gastropods in ancient lakes. In Fretter, V. & J. F. Peake (eds), Systematics, Evolution and Ecology, Vol. 2A. Academic Press, London: 385–428.

Brooks, J. L., 1950. Speciation in ancient lakes. Quarterly Review of Biology 25: 30–60, 131–176.

Davis, G. M., 1979. The origin and evolution of the gastropod family Pomatiopsidae with emphasis on the Mekong River Triculinae. Monographs of the Academy of Natural Sciences of Philadelphia 20: 1–120.

Davis, G. M., 1982. Historical and ecological factors in the evolution, adaptive radiation, and biogeography of freshwater mollusks. American Zoologist 22: 375–395.

Funk, D. J. & K. E. Omland, 2003. Species-level paraphyly and polyphyly: frequency, causes, and consequences, with Insights from animal mitochondrial DNA. Annual Review of Ecology and Systematics 34: 397–423.

Giesen, W., 1994. Indonesia's major freshwater lakes: a review of current knowledge, development processes and threats. Mitteilungen der Internationalen Vereinigung für Theoretische und Angewandte Limnologie 24: 115–128.

Giesen, W., M. Baltzer & R. Baruadi, 1991. Integrating Conservation with Land-use Development in Wetlands of South Sulawesi. Directorate General of Forest Protection and Nature Conservation, Bogor.

Glaubrecht, M., 1996. Evolutionsökologie und Systematik am Beispiel von Süß- und Brackwasserschnecken (Mollusca: Caenogastropoda: Cerithioidea): Ontogenese-Strategien, paläontologische Befunde und historische Zoogeographie. Backhuys, Leiden: 499 pp.

Glaubrecht, M., 1999. Systematics and the evolution of viviparity in tropical freshwater gastropods (Cerithioidea: Thiaridae sensu lato)—an overview. Courier Forschungs-Institut Senckenberg 203: 91–96.

Glaubrecht, M., 2000. A look back in time—toward an historical biogeography as a synthesis of systematic and geologic patterns outlined with limnic gastropods. Zoology 102: 127–147.

Glaubrecht, M., 2004. Leopold von Buch's legacy: treating species as dynamic natural entities, or why geography matters. American Malacological Bulletin 19: 111–134.

Glaubrecht, M., 2006. Independent evolution of reproductive modes in viviparous freshwater Cerithioidea (Gastropoda, Sorbeoconcha)—a brief review. Basteria 69 (Suppl. 3): 32–38.

Glaubrecht, M., 2008. Adaptive radiation of thalassoid Cerithioidean gastropods in Lake Tanganyika, East Africa: morphology and systematization of a paludomid species flock in an ancient lake. Zoosystematics and Evolution 84: 69–120.

Glaubrecht, M. & F. Köhler, 2004. Radiating in a river: systematics, molecular genetics and morphological differentiation of viviparous freshwater gastropods endemic to the Kaek River, central Thailand (Cerithioidea, Pachychilidae). Biological Journal of the Linnean Society 82: 275–311.

Glaubrecht, M. & E. E. Strong, 2007. Ancestry to an endemic radiation in Lake Tanganyika? Evolution of the viviparous gastropod *Potadomoides* Leloup, 1953 in the Congo River system (Cerithioidea, Paludomidae). Biological Journal of the Linnean Society 92: 367–401.

Glaubrecht, M. & T. von Rintelen, 2003. Systematics and zoogeography of the pachychilid gastropod *Pseudopotamis* Martens, 1894 (Mollusca: Gastropoda: Cerithioidea): a limnic relict on the Torres Strait Islands, Australia? Zoologica Scripta 32: 415–435.

Haffner, G. D., P. E. Hehanussa & D. Hartoto, 2001. The biology and physical processes of large lakes of Indonesia: Lakes Matano and Towuti. In Munawar, M. & R. E. Hecky (eds), The Great Lakes of the World (GLOW): Food-Web Health and Integrity. Backhuys Publishers, Leiden: 183–192.

Hawkins, S. J., D. C. Watson, A. S. Hill, P. Harding, M. A. Kyriakides, S. Hutchinson & T. A. Norton, 1989. A comparison of feeding mechanisms in microphageous, herbivorous, intertidal prosobranchs in relation to resource partitioning. Journal of Molluscan Studies 55: 151–165.

Herder, F., A. W. Nolte, J. Pfänder, J. Schwarzer, R. K. Hadiaty & U. K. Schliewen, 2006. Adaptive radiation and hybridization in Wallace's dreamponds: evidence from sailfin silversides in the Malili lakes of Sulawesi. Proceedings of the Royal Society London B 273: 2209–2217.

Hutchinson, G. E., 1957. A Treatise on Limnology. Wiley, New York.

Johannesson, K., E. Rolan-Alvarez & A. Ekendahl, 1995. Incipient reproductive isolation between two sympatric morphs of the intertidal snail *Littorina saxatilis*. Evolution 49: 1180–1190.

Köhler, F. & M. Glaubrecht, 2007. Out of Asia and into India: on the molecular phylogeny and biogeography of the endemic freshwater pachychilid gastropod *Paracrostoma* Cossmann, 1900 (Caenogastropoda: Pachychilidae). Biological Journal of the Linnean Society 91: 627–651.

Köhler, F., T. von Rintelen, A. Meyer & M. Glaubrecht, 2004. Multiple origin of viviparity in Southeast Asian gastropods (Cerithioidea: Pachychilidae) and its evolutionary implications. Evolution 58: 2215–2226.

Kottelat, M., 1990. Sailfin silversides (Pisces: Telmatherinidae) of Lakes Towuti, Mahalona and Wawontoa (Sulawesi, Indonesia) with descriptions of two new genera and two new species. Ichthyological Exploration of Freshwaters 1: 227–246.

Kottelat, M., 1991. Sailfin silversides (Pisces: Telmatherinidae) of Lake Matano, Sulawesi, Indonesia, with descriptions of six new species. Ichthyological Exploration of Freshwaters 1: 321–344.

Kottelat, M., A. J. Whitten, S. N. Kartikasari & S. Wirjoatmodjo, 1993. Freshwater Fishes of Western Indonesia and Sulawesi. Periplus, Jakarta.

Kruimel, J. H., 1913. Verzeichnis der von Herrn E. C. Abendanon in Celebes gesammelten Süsswasser-Mollusken. Bijdragen tot de Dierkunde 19: 217–235.

Kyle, C. J. & E. G. Boulding, 1998. Molecular genetic evidence for parallel evolution in a marine gastropod, *Littorina subrotundata*. Proceedings of the Royal Society London B 265: 303–308.

Lydeard, C., W. E. Holznagel, M. Glaubrecht & W. F. Ponder, 2002. Molecular phylogeny of a circum-global, diverse gastropod superfamily (Cerithioidea: Mollusca: Caenogastropoda): pushing the deepest phylogenetic limits of mitochondrial LSU rDNA sequences. Molecular Phylogenetics and Evolution 22: 399–406.

Mayr, E., 1997. This is Biology. Belknap Press, Cambridge, MA.

Mayr, E., 2004. What makes Biology Unique? Considerations on the Autonomy of a Scientific Discipline. Cambridge University Press, Cambridge, UK.

Padilla, D. K., 1998. Inducible phenotypic plasticity of the radula in *Lacuna* (Gastropoda: Littorinidae). Veliger 4: 201–204.

Reid, D. G., 2000. The use of the radula in the taxonomy and phylogeny of gastropods: cautionary cases of convergence, intraspecific variation and plasticity. Phuket Marine Biological Center Special Publication 21: 329–345.

Reid, D. G. & Y. M. Mak, 1999. Indirect evidence for ecophenotypic plasticity in radular dentition of *Littoraria* species (Gastropoda: Littorinidae). Journal of Molluscan Studies 65: 355–370.

Rossiter, A. & H. Kawanabe (eds), 2000. Ancient Lakes: Biodiversity, Ecology and Evolution. Academic Press, San Diego.

Salzburger, W. & A. Meyer, 2004. The species flocks of East African cichlid fishes: recent advances in molecular phylogenetics and population genetics. Naturwissenschaften 91: 277–290.

Sarasin, P. & F. Sarasin, 1897. Über die Molluskenfauna der großen Süßwasser-Seen von Central-Celebes. Zoologischer Anzeiger 539(540): 308–320.

Sarasin, P. & F. Sarasin, 1898. Die Süßwassermollusken von Celebes. Kreidel, Wiesbaden.

Sarasin, P. & F. Sarasin, 1905. Reisen in Celebes ausgeführt in den Jahren 1893–1896 und 1902–1903. Kreidel, Wiesbaden.

Schluter, D., 2000. The Ecology of Adaptive Radiation. Oxford University Press, Oxford.

Seehausen, O., 2004. Hybridization and adaptive radiation. Trends in Ecology and Evolution 19: 198–207.

Sites, J. W. & J. C. Marshall, 2004. Operational criteria for delimiting species. Annual Review of Ecology and Systematics 35: 199–227.

Streelman, J. T. & P. D. Danley, 2003. The stages of vertebrate evolutionary radiation. Trends in Ecology and Evolution 18: 126–131.

Sudhaus, W., 2004. Radiation within the framework of evolutionary ecology. Organisms, Diversity & Evolution 4: 127–134.

van Oosterzee, P., 1997. Where Worlds Collide. The Wallace line. Cornell University Press, Ithaca.

von Rintelen, T., P. Bouchet & M. Glaubrecht, 2007. Ancient lakes as hotspots of diversity: a morphological review of an endemic species flock of *Tylomelania* (Gastropoda: Cerithioidea: Pachychilidae) in the Malili lake system on Sulawesi, Indonesia. Hydrobiologia 592: 11–94.

von Rintelen, T. & M. Glaubrecht, 2003. New discoveries in old lakes: three new species of *Tylomelania* Sarasin & Sarasin, 1897 (Gastropoda: Cerithioidea: Pachychilidae) from the Malili lake system on Sulawesi, Indonesia. Journal of Molluscan Studies 69: 3–17.

von Rintelen, T. & M. Glaubrecht, 2005. Anatomy of an adaptive radiation: a unique reproductive strategy in the endemic freshwater gastropod *Tylomelania* (Cerithioidea: Pachychilidae) on Sulawesi, Indonesia, and its biogeographic implications. Biological Journal of the Linnean Society 85: 513–542.

von Rintelen, T., A. B. Wilson, A. Meyer & M. Glaubrecht, 2004. Escalation and trophic specialization drive adaptive radiation of viviparous freshwater gastropods in the ancient lakes on Sulawesi, Indonesia. Proceedings of the Royal Society London B 271: 2541–2549.

Vogler, A. P. & M. T. Monaghan, 2006. Recent advances in DNA taxonomy. Journal of Zoological Systematics and Evolutionary Research 45: 1–10.

Wesenberg-Lund, C., 1939. Biologie der Süsswassertiere. Wirbellose Tiere. Julius Springer, Wien.

Wilding, C. S., R. K. Butlin & J. Grahame, 2001. Differential gene-exchange between morphs of *Littorina saxatilis* detected using AFLP markers. Journal of Evolutionary Biology 14: 611–619.

Will, K. W. & D. Rubinoff, 2004. Myth of the molecule: DNA barcodes for species cannot replace morphology for identification and classification. Cladistics 20: 47–55.

Wilson, A. B., M. Glaubrecht & A. Meyer, 2004. Ancient lakes as evolutionary reservoirs: evidence from the thalassoid gastropods of Lake Tanganyika. Proceedings of the Royal Society London B 271: 529–536.

Wilson, M. E. J. & S. J. Moss, 1999. Cenozoic palaeogeographic evolution of Sulawesi and Borneo. Palaeogeography, Palaeoclimatology, Palaeoecology 145: 303–337.

Woltereck, R., 1931. Beobachtungen und Versuche zum Fragenkomplex der Artbildung. I. Wie entsteht eine endemische Rasse oder Art? Biologisches Zentralblatt 51: 231–253.

Woltereck, R., 1941. Die Seen und Inseln der "Wallacea"-Zwischenregion und ihre endemische Tierwelt. Erster Teil: Vorgeschichte und Aufgabe der Forschungsreise. Internationale Revue der gesamten Hydrobiologie und Hydrographie 41: 1–36.

Hydrobiologia (2008) 615:201–213
DOI 10.1007/s10750-008-9561-3

SPECIATION IN ANCIENT LAKES

Gene flow at the margin of Lake Matano's adaptive sailfin silverside radiation: Telmatherinidae of River Petea in Sulawesi

Julia Schwarzer · Fabian Herder · Bernhard Misof · Renny K. Hadiaty · Ulrich K. Schliewen

Published online: 13 September 2008

Abstract Classical speciation concepts focus almost exclusively on the evolution of strict reproductive isolation as a prerequisite for speciation. However, there is a growing body of evidence indicating that speciation is possible despite or even triggered by gene flow among populations or species. Previous findings indicate that introgressive hybridization is a dominant phenomenon in the adaptive radiation of sailfin silversides (Telmatherinidae) endemic to Lake Matano (Sulawesi). In this study, we investigate patterns of genotypic and phenotypic variation of "sharpfin" sailfin silversides in the outlet area of L. Matano and six locations along River Petea, which is the only connection between L. Matano and other lakes and streams of the Malili Lakes system. Fieldwork revealed no hints for a previously cited major waterfall in River Petea, which was thought to separate L. Matano's sailfin silverside radiation from the diversity of the downstream lake drainages. Likewise, genomic (AFLP) and morphometric data suggest high levels of gene flow between upper and lower stretches of this river, as well as between riverine Petea and lacustrine Matano populations. Increasing levels of genotypic and phenotypic dissimilarity are correlated with distance over a remarkably short geographic range.

Keywords Malili Lakes · Telmatherinidae · Introgression · Clinal differentiation · Isolation by distance · AFLP · Geometric morphometrics

Guest editors: T. Wilke, R. Väinölä & F. Riedel
Patterns and Processes of Speciation in Ancient Lakes: Proceedings of the Fourth Symposium on Speciation in Ancient Lakes, Berlin, Germany, September 4–8, 2006

J. Schwarzer (✉) · F. Herder · B. Misof
Zoologisches Forschungsmuseum Alexander Koenig, Adenauerallee 160, 53111 Bonn, Germany
e-mail: j.schwarzer.zfmk@uni-bonn.de

J. Schwarzer · F. Herder · U. K. Schliewen
Zoologische Staatssammlung München, Münchhausenstr. 21, 81247 München, Germany

R. K. Hadiaty
Ichthyology Laboratory, Division of Zoology, Research Center for Biology, Indonesian Institute of Sciences (LIPI), JL. Raya Bogor Km 46, Cibinong 16911, Indonesia

Introduction

Allopatric speciation through the complete cessation of gene flow in geographically fully separated areas has been the dominant view of speciation over the past six decades (Turelli et al., 2001). Recently, concepts of sympatric or parapatric speciation in the presence of gene flow have gained more consideration (Coyne & Orr, 2004). This discussion is accompanied by the idea that introgression and gene flow among recently diverged populations could affect single genes or genomic regions rather than the whole genome (Templeton, 1981; Feder, 1998). This, in turn, suggests that adaptive speciation might be restricted to those traits directly affected by distinct selective regimes, without the prerequisite of strict reproductive isolation (Wu, 2001; Seehausen, 2004; Mallet, 2007). The idea of "porous genomes" (Shaw & Danley, 2003) meets mounting support from field studies suggesting that speciation has proceeded despite ongoing gene flow or even triggered by reticulate processes (e.g. Shaw et al., 2000; Salzburger et al., 2002; Schliewen & Klee, 2004; Nolte & Sheets, 2005; Schelly et al., 2006).

Introgressive hybridization has been demonstrated to be a major phenomenon in the species flock of sailfin silversides (Teleostei: Atheriniformes: Telmatherinidae) endemic to the ancient (Brooks, 1950) Malili Lakes in Central Sulawesi, Indonesia (Herder et al., 2006a, Fig. 1). The three major lakes of the system, Lakes Matano, Mahalona and Towuti, are deep tropical lakes of tectonic origin (Abendanon, 1915; Haffner et al., 2001). They form a cascade interconnected by rivers (Fig. 1), inhabited by endemic radiations of various freshwater organisms, including gastropods (von Rintelen et al., 2004; von Rintelen & Glaubrecht, 2005), bivalves (von Rintelen & Glaubrecht, 2006), crustaceans (Fernandez-Leborans et al., 2006; Zitzler & Cai, 2006) and fish (Kottelat, 1990a, b, 1991; Whitten et al., 2002; Roy et al., 2004; Gray & McKinnon, 2006; Herder et al., 2006a, b).

The lacustrine ichthyofauna of the Malili Lakes is characterized by local endemism either to L. Matano (Ahmad, 1977; Haffner et al., 2001) or to Lakes Mahalona/Towuti (Kottelat, 1990a, b, 1991; Herder et al., 2006b). L. Matano, the uppermost and with approx. 590 m the deepest lake of this drainage (Ahmad, 1977; Haffner et al., 2001), harbours an endemic adaptive radiation of Telmatherinidae (Herder et al., 2006a, in press). According to nuclear markers and consistent to morphological concepts (Kottelat, 1991; Herder et al., 2006b), two major phylogenetic clades are well supported within the L. Matano radiation, "roundfins" and "sharpfins" (Herder et al., 2006a). In contrast, analyses of mitochondrial markers reveal three likewise well-supported clades, dividing the clade of "sharpfins" into two (Roy et al., 2004, 2007a, b; Herder et al., 2006a). Interestingly, inclusion of stream populations inhabiting most permanently flowing waters of the

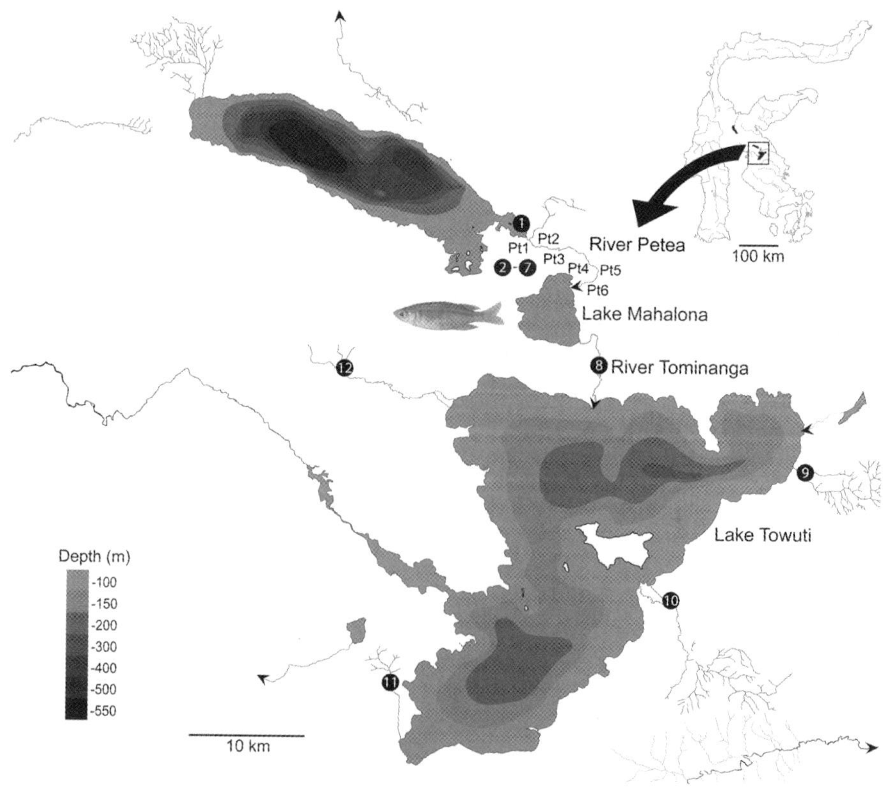

Fig. 1 The Malili Lakes system with sample sites included in this study. L. Matano is a very deep (590 m), graben lake, drained by the steep R. Petea to L. Mahalona, from where the flow continues to L. Towuti, which drained to the sea. Sampling locations are indicated as dots, with numbers referring to the following location names: 1, L. Matano Outlet; 2–7, R. Petea sample sites Pt1 to Pt6; 8, R. Tominanga; 9–12, L. Towuti streams. Map by T. & K. von Rintelen, modified with permission

Malili Lakes system (Herder et al., 2006a) revealed that parts of the stream individuals share the lacustrine "sharpfin" haplotype, strongly suggesting introgressive hybridization from stream to "sharpfin" populations (Herder et al., 2006a). R. Petea is the only connection of L. Matano with the remaining lakes of the Malili Lakes system. Its potential as a barrier for the dispersal of freshwater organisms has been the object of discussions and speculations in the recent past (Roy et al., 2004, 2006, 2007a, b; Herder et al., 2006a, b). This short river drops 72 m of elevation along its only approx. 9 km length, initiating speculations about a major waterfall (Roy et al., 2004). Conspicuously, stream-inhabiting individuals from R. Petea are the only known river-dwelling "sharpfins", sharing a clade with lacustrine "sharpfins" in the nuclear multilocus dataset (see Herder et al., 2006a).

In the present study, we focus on sailfin silversides inhabiting R. Petea. We use individual-based population assignment tests based on Amplified Fragment Length Polymorphism (AFLP) data (Duchesne & Bernatchez, 2002; Campbell et al., 2003) to test for potential gene flow between lacustrine and river-dwelling "sharpfins", based on individuals inhabiting different sections of the river itself and a representative "sharpfin" sample of the lake's outlet area. We hypothesize that genetic contact between river and lake populations is a strong or even ongoing phenomenon, which should be detectable within a multilocus nuclear dataset. For this purpose, we collected individuals on a fine scale, based on six sampling locations along R. Petea and the river's transition area to L. Matano. Consequently, we estimate the level of genetic and morphometric differentiation between lacustrine and riverine "sharpfins" and test for isolation by distance.

Materials and methods

Study site and sampling

Samples were collected between February and April 2004 at the Malili Lakes system in Central Sulawesi (Indonesia) at six sampling sites along R. Petea, named downstream Pt1 to Pt6, with 19 to 23 individuals taken from each site (total $N = 123$, Fig. 2). In addition, 450 "sharpfin" specimens were collected at the transition area of L. Matano and R. Petea, from which a pooled sample of 133 specimens was randomly taken ("L. Matano Outlet"). The random sampling approach was applied because phenotypic variation present in the outlet area did not allow unequivocal assignment of all individuals to "sharpfins" known from L. Matano (see Kottelat, 1990a, b, 1991; Herder et al., 2006b). Finally, riverine specimens from rivers Beau, Lemolemo, Wawondula and Tominanga as well as from the southern part of L. Towuti ($N = 16$, see Fig. 1) were included for a genetic comparison as these populations partly share mtDNA-haplotypes with "sharpfins" of R. Petea and L. Matano (rivers Wawondula and Tominanga; Herder et al., 2006a). In contrast, lacustrine species from L. Mahalona and L. Towuti were not included, because previous phylogenetic analyses (Herder et al., 2006a) as well as high F_{ST} values (between 0.2 and 0.3, JS unpublished data) suggest that they are genetically highly distinct from R. Petea and L. Matano "sharpfins". In total, 272 specimens were used for AFLP analysis and 103 specimens were used for morphometric analyses (101 for landmarks, 103 for head width parameters).

Genetic analysis

DNA extraction was performed using the DNeasy Tissue Kit (Qiagen) according to the manufacturer's protocol. Genomic DNA was extracted from the right pectoral fin, including muscle tissue. AFLP analysis was carried out following the slightly modified protocol of the AFLP method (Vos et al., 1995), described in Schliewen & Klee (2004). Restriction and ligation were carried out in a single step in a thermocycler (2 h at 37°C, 8 h at 16°C). Genomic DNA was digested with restriction enzymes *Mse*I (1 unit) and *Eco*RI (5 units). Polymerase chain reaction (PCR) adaptors specific to the cutting sites were ligated using DNA-Ligase (1 unit; all enzymes from New England Biolabs). Preselective PCR was performed with one selective base on each primer (*Mse*I-C and *Eco*RI-A). In selective amplifications, two additional bases were added to each 3′end of six primer pairs (*Mse*I/*Eco*RI^{DYE}): AA-CT^{FAM}, AA-GG^{JOE}, AT-CT^{FAM}, AT-GG^{JOE}, TA-CT^{FAM} and TA-CT^{JOE}. DNA quality and concentration was checked before restriction-ligation, PCR product quality after preselective amplification. The fragments were separated with 6% LongRanger polyacrylamid gels (FMC Bioproducts) on an ABI PRISMTM 377 automated sequencer, adding internal size standard to each lane (GS-500 ROX; Applied Biosystems). Signal detection was carried out by GeneScan version 3.1 software (Applied Biosystems).

Fragment analysis and subsequent quality check of the AFLP profiles were performed as specified in Herder et al. (2006a); fragment categories were created according to fragment distribution. Scoring for the presence/absence of the peaks was conducted between 100 and 499 bases with threshold set to 100 normalized units. Bin (category) width set to 0.6 bases reduces repeated scoring of the same fragment. Using spreadsheet routines bins differing by less than 0.35 bases were excluded, corresponding to the double standard deviation of the sequencer (Lazaruk et al., 1998). Repeated analysis of four samples was used to exclude bins without full reproducibility (<5%), resulting in the recognition of 446 polymorphic AFLP loci. AFLP data were analysed using AMOVA (Analysis of Molecular Variance) implemented in the Arlequin software package version 2.0 (Schneider et al., 2000). This procedure partitions the genotypic variance, resulting in a F_{ST} that refers to the correlation of genotypes rather than to the correlation of genes as in the classical F_{ST} (Excoffier et al., 1992). Assuming that the mating pattern is identical in all studied populations, based on prior knowledge that *Telmatherina* show a very conserved mating behaviour (Gray & McKinnon, 2006), this measure is informative with respect to the differentiation between populations (Schneider et al., 2000). Pairwise F_{ST} values for each combination of populations were determined and the significance of F_{ST} values was calculated from 1,000 permutations using the non-parametric permutation approach of Excoffier et al. (1992). Afterwards *P*-values were Bonferroni corrected based on the number of pairwise comparisons. Simple Mantel tests for "isolation by distance" (Wright, 1943, 1946) were performed to test for correlations between distance matrices. The software zt (Bonnet & Van de Peer, 2002) was used to test for pairwise relationships between the natural log of geographic distance (km) and genetic differentiation ($F_{ST}/1 - F_{ST}$). Shoreline distances were measured on a detailed geographical map (1:50,000, Lembar 2113-34 und 2213-13, 1:50,000, Peta Rupabumi Indonesia), verified by own ground checks. These units for distance and genetic differentiation were chosen as recommended in Rousset (1997) for a two-dimensional system. Each Mantel test was based on 10^7 permutations.

AFLP-based assignment tests

The probability of a composite AFLP genotype belonging to either the population (=sampling site) of its origin or one of the other populations was calculated for each individual from R. Petea, L. Matano Outlet and L. Towuti streams according to Paetkau et al. (1995), adjusted for dominant markers as implemented in AFLPOP version 1.0 (Duchesne & Bernatchez, 2002). Specimens were reassigned on the basis of log-likelihood values, which were computed for each population. By this means, differences in reassignment probability, represented through log-likelihood values, could be visualized on an individual level for each population. Differences between these values provided an estimate of the power of the assignments. Low log-likelihood differences (LD) may represent inaccurate assignments, as individuals could not be clearly allocated to one population. In this study, individuals were reassigned to a given population on the basis of the highest likelihood of occurrence (Potvin & Bernatchez, 2001). No threshold for reassignment was chosen, based on a series of simulations run with different values for the minimum likelihood difference for assignment (MLD, Duchesne & Bernatchez, 2002), resulting in MLD $= 0$ as optimal value. The log-likelihood differences and the proportion of reassignment were used as an estimator for genetic identity of sampled populations.

Assuming that only non-introgressed genotypes can be clearly assigned (meaning with a high log-likelihood difference to the second likely population) to their original population whereas mixed genotypes are not clearly assignable in this approach (Manel et al., 2005) as they represent a composite of at least two populations (Arnold, 1997), the minimum likelihood difference for AFLPOP assignment (MLD) was increased from $z = 0$ to $z = 1$ in a second assignment run. Thereby, only specimens with genotypes 10 times more likely allocated to one population than to the second likely one were assigned (Duchesne & Bernatchez, 2002). We use this approach as simple estimate of the amount of "non-introgressed"

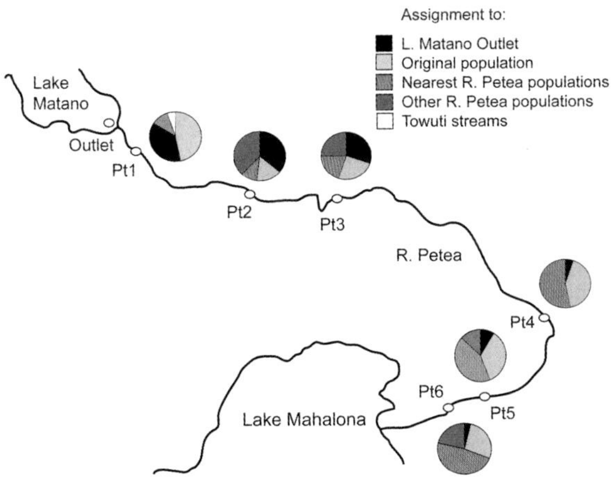

Fig. 2 Clinal differentiation in genetic traits along R. Petea "sharpfins". Riverine specimens collected closer to L. Matano (Pt1 ($N = 19$), Pt2 ($N = 19$) to Pt3 ($N = 20$)), were more similar to lacustrine specimens than those in greater distance to L. Mahalona (Pt4 ($N = 19$), Pt5 ($N = 23$) to Pt6 ($N = 23$)). The pie charts above R. Petea sample locations represent the assignment patterns of each local population

genotypes in different sections of R. Petea and in L. Matano Outlet.

Anticipating low local reassignment to each sample site, and to evaluate more general patterns, populations Pt1, Pt2 and Pt3 were in a following step merged into "upper R. Petea" and Pt4, Pt5 and Pt6 into "lower R. Petea" (see also Fig. 2). By this means, each specimen was assigned either to "Lower R. Petea", "Upper R. Petea", "L. Matano Outlet", "Towuti streams" or to "None". We used the proportion of clearly assigned specimens per population as an estimator for the proportion of non-introgressed lacustrine or riverine specimens in our dataset.

Morphometric analyses

To characterize potential differences in body shape, lacustrine specimens from L. Matano Outlet and four riverine R. Petea populations (Pt2, Pt3, Pt5 and Pt6) were morphometrically analysed. As head width has previously been used as taxonomical character to distinguish stream-dwelling sailfin silversides (Kottelat, 1991), this parameter was included as an addition to the two-dimensional shape analyses. Because of a strong sexual dimorphism present in telmatherinids (Kottelat, 1990a, 1991; Herder et al., 2006b) only adult male specimens were used in the morphometric approach. However, this resulted in the exclusion of sample sites Pt1 and Pt4, as few adult males were available from these sites. Ultimately, the following samples were included in the analysis: L. Matano Outlet ($N_{\text{Landmarks}} = 73$, $N_{\text{Head}} = 74$), R. Petea populations Pt2 + Pt3 ($N_{\text{Head}} = 20$), Pt2 ($N_{\text{Landmarks}} = 4$), Pt3 ($N_{\text{Landmarks}} = 15$) and Pt5 + Pt6 ($N_{\text{Landmarks/Head}} = 9$).

Pre-analyses of the trait "interorbital width" revealed no significant differences between Pt2 and Pt3 (Mann–Whitney U test, $N_1 = 16$, $N_2 = 4$, $Z = -1.134$, $P = 0.257$) and between Pt5 and Pt6 (Mann–Whitney U test, $N_1 = 6$, $N_2 = 4$, $Z = -0.640$, $P = 0.522$). On the basis of geographical proximity and small sample size, specimens from Pt5 and Pt6 and Pt2 and Pt3 were combined in the classical morphometric analysis. Goodall's F test based on 2,500 bootstrap replicates also revealed no significant difference in body shape between the above-mentioned groups (2,500 repeats, Pt5/6: $F = 0.97$, $P = 0.43$; Pt2/3: $F = 1.91$, $P = 0.06$). However, due to the marginally significant p-value between specimens from Pt2 and Pt3, we chose a conservative approach and treated them separately in the geometric morphometric approach.

A landmark-based geometric morphometric (GM) approach (Rohlf & Marcus, 1993) based on the x, y coordinates of 20 homologous landmarks was used to quantify differences in body shape between lacustrine and riverine populations (Fig. 3). Data acquisition was conducted using the software packages Tpsutil and Tpsdig (Rohlf, 2004a, b). All subsequent morphometric analyses were employed using the IMP software package (Sheets, 2002). To remove non-shape variation, a Generalized Procrustes Analysis (Rohlf, 1990; Rohlf & Slice, 1990) was performed which translates all specimens to a

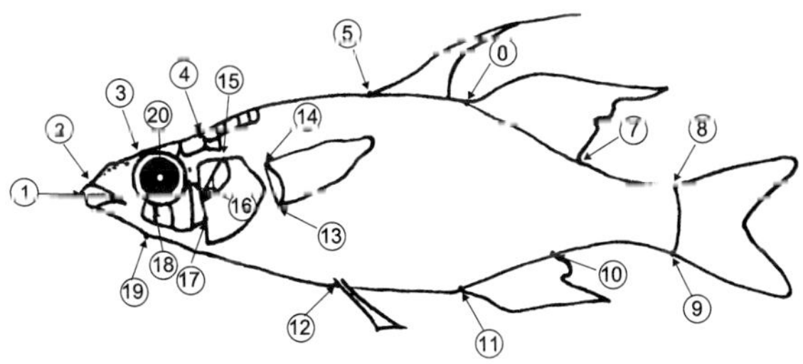

Fig. 3 Position of the landmarks used in the geometric morphometric approach. Twenty landmarks were chosen to analyse variability in body shape

Table 1 Characterization of all calliper-based measurements to determine the head and body width

Measurement	Description
Interorbital width	Distance of the dorsal between bony margin on the left and right orbit
Body width	Distance between the upper margins of pectoral fins
Head width	Distance between the dorsal end of opercular openings
Snout width	Distance from the right to the left corner of the mouth
Standard length	Distance from upper jaw symphysis to caudal margin of hypuralia

The measured sections were chosen following Kottelat (1990a). All measurements are highly correlated and reflect the same differentiation pattern between the tested populations

common location, scales them to a unit size, and rotates them until corresponding landmarks line up as closely as possible. After superimposition, the data were translated into Partial Warps using the thin-plate spline model (Bookstein, 1989). These linearly transformed variables are suitable for use with standard multivariate statistical methods, since they carry the same number of variables as degrees of freedom (Zelditch et al., 2004). PCA reduction was conducted to scale down the number of variables in the dataset using the program CVA-Gen6o. Consequently, instead of using landmark coordinates or partial warp scores in the subsequent analysis, PCA axis scores of the specimens were used (Sheets, 2002). A Canonical Variates Analysis (CVA) based on the reduced dataset (consisting of eight PC Axes) was accomplished. The CVA was used to describe the potential morphological discrimination along predefined groups, displaying the greatest variance between groups relative to the variance within groups.

As a two-dimensional landmark approach was conducted, we additionally took five calliper-based measurements characterizing head and body width of each specimen (interorbital width, head width, body width, snout width and standard length, Table 1) using a digital calliper to the nearest of 0.1 mm. As the values for all measurement sections were highly correlated (Pearson correlation, $P < 0.001$, coefficient of correlation: all $r > 0.702$), only the results for interorbital width are presented. However, pre-analyses revealed that all measured sections reflect the same result. Residuals were calculated by regression analysis with "standard length" as the independent and "interorbital width" as the dependent variable. Differences in residuals between riverine and lacustrine groups were analysed using one-way ANOVA and Post hoc tests.

Results

Genetic differentiation

F_{ST} values ranged from 0.01 to 0.09 between L. Matano's and R. Petea's "sharpfins" and from 0 to 0.08 between individual R. Petea populations (Table 2). The riverine populations from streams near L. Towuti were differentiated from R. Petea and L. Matano "sharpfins" with F_{ST} values between 0.14 and 0.21, showing the highest differentiation from lacustrine sharpfins (Table 2). The Mantel test revealed significant correlation between geographic and genetic distances of populations from L. Matano Outlet and R. Petea sample sites ($N = 7$, $r = 0.55$, $P = 0.02$), indicating that genetic differentiation of the riverine Petea "sharpfin" populations to L. Matano Outlet "sharpfins" increases with higher geographical distance from L. Matano.

The likelihood-based assignment revealed that 71% of L. Matano Outlet "sharpfins" and 87.5% of riverine populations from streams surrounding L. Towuti were reassigned to their original population. Reassignment success of R. Petea populations (Pt1 to Pt6) was comparatively low, ranging from 16% (Pt2) to 42% (Pt4) (Table 3). Log-likelihood distance to the next likely population was overall comparatively low (Table 3).

Assignment of R. Petea specimens to L. Matano Outlet ranged from 4.5% (Pt6) to 47% (Pt1, Fig. 2). Twenty-seven percent of lacustrine Matano "sharpfins" were assigned to R. Petea sample sites. The assignment to the lacustrine Matano population differed significantly between "upper R. Petea" populations (Pt1, Pt2 and Pt3, assignments between 32% and 47%) and "lower R. Petea" populations (Pt4, Pt5 and Pt6, assignments between 4% and 8%; χ^2-test, $\chi^2 = 22.34$, $P < 0.001$; Fig. 2).

Table 2 Pairwise F_{ST} estimates across all loci (below diagonal in bold letters)

Sample location	L. Matano Outlet	R. Petea						L. Towuti streams
		Pt1	Pt2	Pt3	Pt4	Pt5	Pt6	
L. Matano Outlet	–	0.36	1.80	3.30	6.30	7.90	8.20	19.50
Pt1	**0.01**$^{n.s.}$	–	1.50	3.00	6.00	7.60	7.90	19.20
Pt2	**0.03***	**0.03***	–	1.50	4.50	6.10	6.40	17.70
Pt3	**0.04***	**0.03***	**0.00**$^{n.s.}$	–	3.00	4.60	4.90	16.20
Pt4	**0.09***	**0.08***	**0.06***	**0.06***	–	1.60	1.90	13.20
Pt5	**0.02***	**0.02**$^{n.s.}$	**0.02***	**0.02**$^{n.s.}$	**0.04***	–	0.30	11.60
Pt6	**0.05***	**0.04***	**0.02**$^{n.s.}$	**0.04***	**0.01**$^{n.s.}$	**0.01**$^{n.s.}$	–	11.30
L. Towuti streams	**0.21***	**0.16***	**0.15***	**0.13***	**0.15***	**0.14***	**0.14***	–

Significance was Bonferroni corrected resulting in: $\alpha < 0.002$, *, significant; n.s., not significant

Geographic distances between sample locations (km) are shown above diagonal

Table 3 Assignment and log-likelihood differences with MLD $z = 0$

Sample location	L. Matano Outlet	R. Petea						L. Towuti streams
		Pt1	Pt2	Pt3	Pt4	Pt5	Pt6	
L. Matano Outlet	97 (72%)	9 (47%)	7 (37%)	6 (30%)	1 (5%)	2 (9%)	1 (4.5%)	1 (10%)
Pt1	5 (4%)	7 (37%)	1 (5%)	1 (5%)	0	2 (9%)	1 (4.5%)	0
Pt2	8 (6%)	0	3 (16%)	3 (15%)	0	0	3 (13%)	0
Pt3	2 (1.5%)	0	2 (10.5%)	5 (25%)	0	1 (4.5%)	1 (4.5%)	0
Pt4	3 (2.3%)	0	1 (5%)	0	8 (42%)	4 (17%)	5 (22%)	0
Pt5	17 (13%)	3 (16%)	2 (10.5%)	5 (25%)	3 (16%)	8 (35%)	6 (26%)	0
Pt6	0	0	3 (16%)	0	7 (37%)	6 (26%)	6 (26%)	0
Towuti streams	0	0	0	0	0	0	0	9 (90%)
Log-likelihood difference (median)	1.69	0.8	1.12	0.78	0.85	0.96	1.23	4.6
± s.d.	2.07	0.85	0.96	0.70	0.72	1.34	1.06	3.14

Total number of assigned specimens, percentage values of assignment (brackets) and median MLD for the assignment of each population are shown

The rejection level for assignment was increased from the lowest MLD $z = 0$ to the next higher level $z = 1$, which forces assignment of specimens to a population only when there is a 10 or more times difference between the most likely and the second most likely population. This had a pronounced impact on the number of non-classified individuals (Table 4). The reassignment success differed between L. Matano and R. Petea "sharpfins", with 49.2% of the specimens correctly reallocated to L. Matano Outlet, 57% to lower R. Petea sample sites (Pt4, Pt5 and Pt6) but only 7% to "upper R. Petea" sample sites (Pt1, Pt2 and Pt3).

Forty-six percent of the specimens from L. Matano Outlet and 72.4% from "upper R. Petea" (Pt1, Pt2 and Pt3) were not unambiguously allocated, whereas 37% from "lower R. Petea" sample sites (Pt4, Pt5 and Pt6) could not be clearly allocated (Table 4). Based on this test, the amount of "non introgressed" genotypes is highest in L. Matano Outlet and "lower R. Petea" sample sites whereas it is comparatively low in the "upper R. Petea" sample sites

Morphometric analyses

The CVA based on eight PC axes revealed three axes explaining significant variation between the predefined groups. All tested groups form clusters along the first two CV axes, which display 77% of the total variance between the populations relative to the

Table 4 Assignment to populations with MLD of $z = 1$

Sample location	L. Matano Outlet	R. Petea		L. Towuti streams
		Pt1/Pt2/Pt3	Pt4/Pt5/Pt6	
L. Matano Outlet	65 (49.2%)	8 (13.8%)	1 (1.6%)	0
Pt1/Pt2/Pt3	4 (3%)	4 (7%)	3 (4.7%)	0
Pt4/Pt5/Pt6	2 (1.5%)	4 (7%)	37 (57%)	0
L. Towuti streams	(0%)	0	0	8 (80%)
None	61 (46%)	42 (72.4%)	24 (37%)	2 (20%)

R. Petea populations were merged in to two groups "upper Petea" (Pt1/2/3) and "lower Petea" (Pt4/5/6)

variance within the populations (Axis 1: Lambda = 0.20, $\chi^2 = 156.74$, df = 12, $P < 0.01$, explains 77% of total variance; Axis 2: Lambda = 0.63, $\chi^2 = 44.34$, df = 6, $P < 0.01$, explains 11% of total variance; Axis 3: Lambda = 0.85, $\chi^2 = 15.19$, df = 2, $P < 0.01$, explains 11% of total variance; Fig. 4). Overlaps were present between all of the tested groups. The main factors responsible for the morphometric differentiation of the groups are present in the head region, the orientation of the snout and in the length of the caudal peduncle (see deformation grids in Fig. 4). To test for a potential influence of imbalanced sampling on the observed differentiation, CVA scores and centroid sizes were correlated. As this revealed no significant correlation (Spearman signed rank correlation: CV1 vs. Size: $r_S = -0.107$,

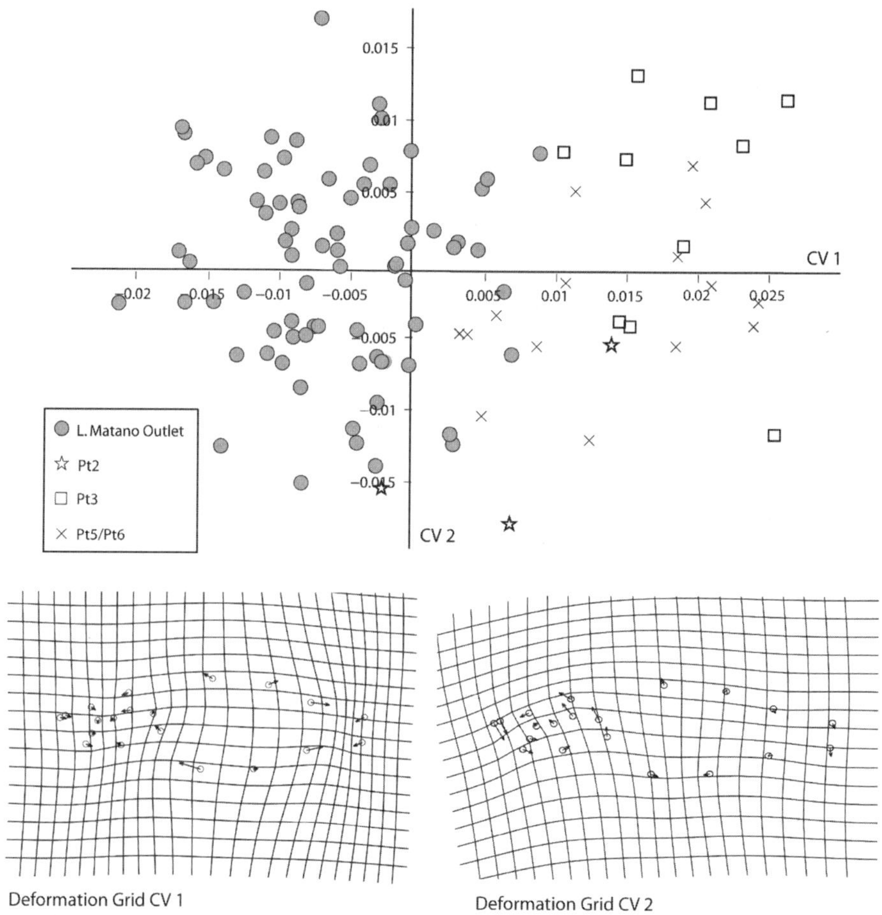

Fig. 4 Results of the Canonical variates analyses. Along the first two CV axes the tested populations are weakly separated, showing overlaps. Deformation grids with relative displacement vectors for each landmark display shape changes captured by CV axes 1 and 2. The main shape changes between lacustrine and riverine populations are present in the position of the dorsal and anal fins, the caudal peduncle length and the head region. All tested populations did not differ significantly in standard length (One way Anova, $P = 0.121$), so an allometric effect on the observed result can be most likely excluded

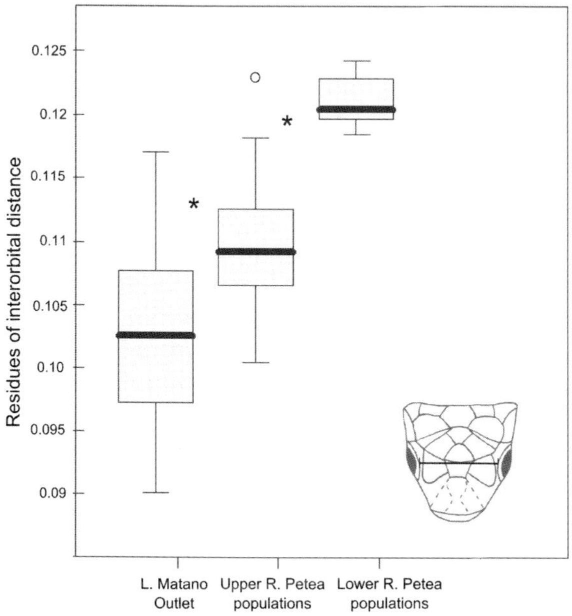

Fig. 5 Differentiation in interorbital width between lacustrine and riverine sharpfin specimens from L. Matano ($N = 74$), upper ($N = 20$) and lower ($N = 9$) R. Petea sampling sites. Significant differences were present between all tested populations (Tukey HSD, all $P < 0.001$, marked by an asterisk). Interorbital width increases in riverine Petea populations with increasing distance to L. Matano

$P = 0.09$, CV2 vs. Size: $r_S = -0.014$, $P = 0.831$), an impact of body size on the observed morphometric differentiation can be excluded.

Analysis of "Interorbital Width" revealed significant differences between "upper R. Petea", "lower R. Petea" and L. Matano populations (one-way ANOVA; $P < 0.001$, Tukey HSD Test, $P < 0.001$) associated with an increasing interorbital width with higher distance to L. Matano (Fig. 5).

Discussion

Hybridization and introgression are currently discussed to potentially increase rather than necessarily retard the generation of biodiversity, especially with regard to adaptive radiations (Seehausen, 2004; Mallet, 2007). Significant signal of introgression by stream-dwelling sailfin silversides is coupled with conspicuously high phenotypic diversity in the radiation of "sharpfin" sailfin silversides of L. Matano, suggesting that introgression might be involved in the generation of diversity in this radiation (Herder et al., 2006a). In this context we hypothesized that substantial gene flow is present between riverine and lacustrine "sharpfins" at the interface between L. Matano and R. Petea.

Our results suggest ongoing gene flow between L. Matano's and R. Petea's "sharpfins", because (1) pairwise population comparisons show comparatively low levels of genetic differentiation (F_{ST}), indicating moderate to high levels of migration (Hartl & Clark, 1997) between all sampled populations (Table 2); (2) R. Petea populations exhibit increasing morphometric (Fig. 5) and genetic similarity (Table 2, Fig. 2) to lacustrine "sharpfins" of the L. Matano Outlet area with decreasing distance to L. Matano (Fig. 2); and (3) the proportion of unassignable riverine Petea individuals increases closer to L. Matano (Table 4).

Presence of unassigned individuals when applying the strict rejection level ($z = 1$) can be explained in two ways. Genomic similarity of these individuals to more than one potential source population could prevent clear assignment, indicative of ongoing gene flow. Alternatively, origin from an unsampled source population would also result in non-assignment under stringent (MLD $z = 1$) conditions and increased the likelihood of assignment by chance under MLD $z = 0$ (Manel et al., 2005). This would imply that those unassigned specimens are of unknown origin, i.e. from an unsampled source population. We consider this as rather unlikely, as extensive explorations in L. Matano, the whole Malili Lakes system and along R. Petea (own surveys in 2002, 2004, 2006; survey by Kottelat in the early 1990th (Kottelat, 1990a, b, 1991)) revealed no hints for important additional neighbouring populations. Moreover, phylogenetic analyses of *Telmatherina*-relatives (Herder et al., 2006a) as well as preliminary morphometric and AFLP-based assignment tests (JS, unpublished data) including all known lacustrine and riverine "sharpfin" specimens of L. Matano (see Herder et al., 2006b) render it comparatively unlikely that other known "sharpfin" populations had an impact on the observed assignment pattern. In conclusion, the observed differences in assignment, based on geographic origin, are most likely related to a decreasing amount of gene flow or a decreasing relatedness between the populations, especially seen against the background of the morphometric results.

Clinal variation and isolation by distance

Our results suggest clinal variation in genetic traits (Fig. 2) and isolation by distance along R. Petea. Remarkably, this pattern is obvious over the very short distance of only about 9 km, and despite an expected high mobility of these agile, non-territorial fish (Gray & McKinnon, 2006). With decreasing distance from the lake, R. Petea populations become genetically more similar to lacustrine "sharpfins" of L. Matano Outlet (Fig. 3, Table 2). This is partially reflected in morphometric characters, too (Fig. 5). Whereas no clinal variation is obvious in body shape (Fig. 4), head width variation also reflects a clinal pattern, with more bulky headed specimens further away from L. Matano. Generally, we observed that riverine specimens develop wider heads than lacustrine L. Matano Outlet specimens (Fig. 5), indicating that general adaptations to lake and stream habitats might be present. This corresponds to previous studies, demonstrating adaptive divergence between lake- and stream-dwelling fish: Hendry et al. (2002) found stream sticklebacks (*Gasterosteus* ssp.) to have deeper bodies and fewer gill rakers than lake sticklebacks; McGuigan et al. (2003) reported adaptive differences for rainbow fish (Melanotaeniidae) which are closely related to telmatherinids (Sparks & Smith, 2004), with lacustrine rainbow fish having more posteriorly positioned first dorsal and pelvic fins than stream-inhabiting ones. However, although we do assume general adaptations to riverine and lacustrine habitats in "sharpfins", it appears unlikely that fine-scale local adaptations to each single R. Petea sample site have evolved, given very high levels of gene flow and the extremely small geographic distances with no obvious habitat changes along the river.

Secondary contact or ongoing divergence?

Two alternative scenarios appear plausible to explain the observed pattern of morphometric and genetic clinal variation of "sharpfins" within R. Petea. Secondary gene flow due to incomplete reproductive isolation between hybridizing but originally separated populations may be one. Following this idea, balance between dispersal and selection against hybrids as explained in the tension-zone model (Barton & Hewitt, 1985) might cause a clinal pattern like the one observed here. Alternatively, ongoing evolution of increasingly limited gene flow between parapatrically diverging populations of L. Matano and R. Petea "sharpfins" would lead to decreasing similarity.

Under the idea of secondary contact, genomic exchange between riverine and lacustrine sailfin silverside populations could potentially contribute to genomic enrichment of the highly diverse lacustrine "sharpfins" of L. Matano with riverine alleles, as suggested for stream-dwelling *T. bonti* by previous analyses (Herder et al., 2006a). Also, the alternative scenario, parapatrically diverging populations of L. Matano and R. Petea, could allow for introgression of alleles adapted to stream conditions into the lacustrine flock. Increasing dissimilarity with increasing geographic distance, as indicated here, would be a result of ongoing divergence. In a recent model, Gavrilets & Vose (2005) point out that adaptive radiation can take place extremely quickly in the presence of ongoing gene flow. Hence, population divergence should potentially proceed without initial evolution of reproductive isolation, and result in a high amount of similarities between diverging populations with few adaptive differences between them. This corresponds to the patterns observed in case of L. Matano and R. Petea "sharpfins", with different eco-phenotypes in lacustrine and riverine habitats.

However, a full reconstruction of the L. Matano introgression/parapatric speciation scenario is beyond the scope of the present study. A larger-scale approach, including population genetic sampling of all riverine populations of the L. Matano-catchments as well as all "sharpfin"-populations of L. Matano, is required to test for the potential impact of introgression by R. Petea's "sharpfins" into the Matano-flock.

A barrier within River Petea?

R. Petea and its potential role as barrier within the Malili Lakes system have recently gained substantial interest. The isolation of L. Matano could play a significant role in its use as a model system to address questions regarding the evolution of species flocks (Roy et al., 2004, 2007a, b; Herder et al., 2006a, b). It has previously been assumed that L. Matano is "separated from downstream lakes by a 72 m waterfall ..." (Roy et al., 2004, p. 1269), citing Brooks (1950) and Kottelat (1990b). In contrast to these assumptions, our fieldwork as well as population genetic analysis unambiguously showed that there is no major waterfall and no major population genetic barrier for stream-dwelling "sharpfins" in R. Petea. During sampling it became evident that R. Petea is very steep and mostly extremely fast-flowing, but does not contain waterfalls which can be regarded as complete upstream barriers for stream-dwelling fishes. Indeed, we observed R. Petea's "sharpfins" directly in strong rapids. Given this observation, low levels of genetic differentiation of sailfin silversides along R. Petea appear highly plausible and are congruent with the lack of a major waterfall. This result is only seemingly in contrast with population genetic studies of *Caridina* freshwater shrimps that supported the "barrier hypothesis" by finding strong distinction of semi-lacustrine *Caridina*-populations from above and below the river course of R. Petea (Roy et al., 2006). In contrast to this study, the analysis presented here is the first one investigating material sampled from the river itself and not comparing lacustrine populations of different lakes separated by strongly riverine habitats. Our results highlight the importance of incorporating both lacustrine and riverine populations when comparing lacustrine populations within the Malili Lakes drainage system.

Implications for conservation

Present results indicate that R. Petea is not an impassable barrier for endemic freshwater fish and correspond to previous results suggesting natural hybridization between stream- and lake-inhabiting populations (Herder et al., 2006a). This has direct implications for conservation issues: besides the lakes, also the rivers connecting them, namely Rivers Petea and Tominanga (connecting L. Mahalona and L. Towuti), should be taken into consideration when attempting to sustainably preserve the worldwide unique ecosystem of "Wallace's Dreamponds" (Herder et al., 2006a). The Malili Lakes area is object to massively increasing surface nickel mining and smelting, with considerable loss or modification of natural habitats and high demands for electric energy, supplied mainly by hydroelectric damming. Given the arguments listed above and the presence of an only recently discovered diversity of fish inhabiting most running waters of the Malili Lakes drainage (Herder et al., 2006b), we strongly recommend to include streams and rivers, especially those connecting the lakes, into management decisions.

Conclusions

The present study suggests ongoing gene flow between "sharpfins" of L. Matano and R. Petea, with decreasing genotypic and phenotypic similarity of stream- to lake fish along this remarkably short river. Although our data do not have the power to differentiate between different scenarios explaining the observed gene flow patterns, the results are one important step beyond previous findings. They indicate strongly that genetic exchange between the lacustrine and riverine "sharpfins" takes place, potentially allowing introgression into L. Matano from lower lakes and streams. However, the significance of the presented results for the evolution of complex diversity patterns in L. Matano remains to be tested. We stress that there is no evidence for a major physical barrier within R. Petea and that overall gene flow of stream "sharpfins" along the river appears high. Implications of the fact that L. Matano's adaptive sailfin silverside radiation is not physically isolated from the downstream telmatherinid diversity should be taken into consideration when analysing evolution of L. Matano's spectacular diversity.

Acknowledgements We thank LIPI for the permit to conduct research in Indonesia. We are deeply grateful to INCO for their outstanding logistic support at the lakes. For invaluable assistance in the field we especially thank J. Pfaender, J. Herder and J. Frommen. Fieldwork at R. Petea greatly benefited from a pre-survey jointly conducted with T. von Rintelen. We thank J. Frommen, M. Haase, C. Klütsch, A. Nolte, R. Schelly and C. Schmidt for critical comments and constructive suggestions. This study was funded by research grants from the Deutsche Forschungsgemeinschaft to UKS and by a graduate fellowship donated by the Rheinische Friedrich-Wilhelms Universität Bonn to FH.

References

Abendanon, E. C., 1915. Midden-Celebes-Expeditie. Geologische en geographische doorkruisingen van Midden-Celebes (1909–1910). E. J Brill, Leiden.

Ahmad, W., 1977. Geology along the Matano Fault Zone East Sulawesi, Indonesia. Regional Conference on the Geology and Mineral resources of South East Asia, Jakarta.

Arnold, M. L., 1997. Natural Hybridization and Evolution. Oxford University Press, New York.

Barton, N. H. & G. M. Hewitt, 1985. Analysis of hybrid zones. Annual Review of Ecology and Systematics 16: 113–148.

Bonnet, E. & Y. Van de Peer, 2002. zt: a software tool for simple and partial Mantel tests. Journal of Statistical software 7: 1–12.

Bookstein, F. L., 1989. Principal warps—Thin-plate splines and the decomposition of deformations. IEEE Transactions on Pattern Analysis and Machine Intelligence 11: 567–585.

Brooks, J. L., 1950. Speciation in ancient lakes. Quarterly Review of Biology 25: 30–60.

Campbell, D., P. Duchesne & L. Bernatchez, 2003. AFLP utility for population assignment studies: analytical investigation and empirical comparison with microsatellites. Molecular Ecology 12: 1979–1991.

Coyne, J. A. & H. A. Orr, 2004. Speciation. Sinauer Associates, Sunderland.

Duchesne, P. & L. Bernatchez, 2002. AFLPOP: a computer program for simulated and real population allocation, based on AFLP data. Molecular Ecology Notes 2: 380–383.

Excoffier, L., P. E. Smouse & J. M. Quattro, 1992. Analysis of molecular variance inferred from metric distances among DNA haplotypes: application to human mitochondrial DNA restriction data. Genetics 131: 479–491.

Feder, J. L., 1998. The apple maggot fly, *Rhagoletis pomonella*: flies in the face of conventional wisdom about speciation? In Howard, D. J. & S. H. Berlocher (eds), Species and Speciation. University Press, New York and Oxford.

Fernandez-Leborans, G., K. Zitzler & R. Gabilondo, 2006. Epibiont protozoan communities on *Caridina lanceolata* (Crustacea, Decapoda) from the Malili lakes of Sulawesi (Indonesia). Zoologischer Anzeiger 245: 167–191.

Gavrilets, S. & A. Vose, 2005. Dynamic patterns of adaptive radiation. Proceedings of the National Academy of Sciences of the United States of America 102: 18040–18045.

Gray, S. M. & J. S. McKinnon, 2006. A comparative description of mating behaviour in the endemic telmatherinid fishes of Sulawesi's Malili Lakes. Environmental Biology of Fishes 75: 471–482.

Haffner, G. D., P. E. Hehanussa & D. Hartoto, 2001. The biology and physical processes of large lakes of Indonesia: Lakes Matano and Towuti. In Munawar, M. & R. E. Hecky (eds), The Great Lakes of the World. Food-web, Health and Integrity. Backhuys Publishers, Leiden: 183–192.

Hartl, D. L. & A. G. Clark, 1997. Principles of population genetics. Sinauer Associates, Inc., Sunderland, Massachusetts.

Hendry, A. P., E. B. Taylor & J. D. McPhail, 2002. Adaptive divergence and the balance between selection and gene flow: Lake and stream stickleback in the misty system. Evolution 56: 1199–1216.

Herder, F., A. Nolte, J. Pfaender, J. Schwarzer, R. K. Hadiaty & U. K. Schliewen, 2006a. Adaptive radiation and hybridisation in Wallace's Dreamponds: evidence from sailfin silversides in the Malili Lakes of Sulawesi. Proceedings of the Royal Society of London, Series B, Biological Sciences 273: 2209–2217.

Herder, F., J. Pfaender & U. K. Schliewen, in press. Adaptive sympatric speciation of polychromatic "roundfin" sailfin silverside fish in Lake Matano (Sulawesi). Evolution. doi: 10.1111/j.1558-5646.2008.00447.x

Herder, F., J. Schwarzer, J. Pfaender, R. K. Hadiaty & U. K. Schliewen, 2006b. Preliminary checklist of sailfin silversides (Pisces:Telmatherinidae) in the Malili Lakes of Sulawesi (Indonesia), with a synopsis of systematics and threats. Verhandlungen der Gesellschaft für Ichthyologie 5: 139–163.

Kottelat, M., 1990a. Sailfin silversides (Pisces: Telmatherinidae) of Lakes Towuti, Mahalona and Wawontoa (Sulawesi, Indonesia) with descriptions of two new genera and two new species. Ichthyological Exploration of Freshwaters 1: 35–54.

Kottelat, M., 1990b. The ricefishes (Oryziidae) of the Malili Lakes, Sulawesi, Indonesia, with a description of a new species. Ichthyological Exploration of Freshwaters 1: 151–166.

Kottelat, M., 1991. Sailfin silversides (Pisces: Telmatherinidae) of Lake Matano, Sulawesi, Indonesia, with descriptions of six new species. Ichthyological Explorations of Freshwaters 1: 321–344.

Lazaruk, K., P. S. Walsh, F. Oaks, D. Gilbert, B. B. Rosenblum, S. Menchen, D. Scheibler, H. M. Wenz, C. Holt & J. Wallin, 1998. Genotyping of forensic short tandem repeat (STR) systems based on sizing precision in a capillary electrophoresis instrument. Electrophoresis 19: 86–93.

Mallet, J., 2007. Hybrid speciation 446: 279–283.

Manel, S., O. E. Gaggiotti & R. S. Waples, 2005. Assignment methods: matching biological questions techniques with appropriate. Trends in Ecology and Evolution 20: 136–142.

McGuigan, K., C. E. Franklin, C. Moritz & M. W. Blows, 2003. Adaptation of rainbow fish to lake and stream habitats. Evolution 57: 104–118.

Nolte, A. W. & H. D. Sheets, 2005. Shape based assignment tests suggest transgressive phenotypes in natural sculpin hybrids (Teleostei, Scorpaeniformes, Cottidae). Frontiers in Zoology 2: 1–12.

Paetkau, D., W. Calvert, I. Stirling & C. Strobeck, 1995. Microsatellite analysis of population structure in Canadian polar bears. Molecular Ecology 4: 347–354.

Potvin, C. & L. Bernatchez, 2001. Lacustrine spatial distribution of landlocked Atlantic salmon populations assessed across generations by multilocus individual assignment and mixed-stock analyses. Molecular Ecology 10: 2375–2388.

Rohlf, F. J., 1990. Rotational fit (Procrustes) methods. Proceedings of the Michigan morphometrics workshop: 227–236.

Rohlf, F. J., 2004a. TpsDig. Department of Ecology and Evolution, State University of New York, Stony Brook, N.Y.

Rohlf, F. J., 2004b. TpsUtil. Department of Ecology and Evolution, State University of New York, Stony Brook, N.Y.

Rohlf, F. J. & L. F. Marcus, 1993. A revolution in morphometrics. Trends in Ecology and Evolution 8: 129–132.

Rohlf, F. J. & D. Slice, 1990. Extensions of the Procrustes method for the optimal superimposition of landmarks. Systematic Zoology 39: 40–59.

Rousset, F., 1997. Genetic differentiation and estimation of gene flow from F-statistics under isolation by distance. Genetics 145: 1219–1228.

Roy, D., M. F. Docker, G. D. Haffner & D. D. Heath, 2007a. Body shape vs. colour associated initial divergence in the Telmatherina radiation in Lake Matano, Sulawesi, Indonesia. Journal of Evolutionary Biology 20: 1126–1137.

Roy, D., M. F. Docker, P. E. Hehanussa, D. D. Heath & G. D. Haffner, 2004. Genetic and morphological data supporting the hypothesis of adaptive radiation in the endemic fish of Lake Matano. Journal of Evolutionary Biology 17: 1268–1276.

Roy, D., D. W. Kelly, C. H. J. M. Fransen, D. D. Heath & G. D. Haffner, 2006. Evidence of small-scale vicariance in *Caridina lanceolata* (Decapoda: Atyidae) from the Malili Lakes, Sulawesi. Evolutionary Ecology Research 8: 1087–1099.

Roy, D., G. Paterson, P. B. Hamilton, D. D. Heath & G. D. Haffner, 2007b. Resource-based adaptive divergence in the freshwater fish Telmatherina from Lake Matano, Indonesia. Molecular Ecology 16: 35–48.

Salzburger, W., S. Bariç & C. Sturmbauer, 2002. Speciation via introgressive hybridization in East African cichlids? Molecular Ecology 11: 619–625.

Schelly, R., W. Salzburger, S. Koblmüller, N. Duftner & C. Sturmbauer, 2006. Phylogenetic relationships of the lamprologine cichlid genus *Lepidiolamprologus* (Teleostei: Perciformes) based on mitochondrial and nuclear sequences, suggesting introgressive hybridization. Molecular Phylogenetics and Evolution 38: 426–438.

Schliewen, U. K. & B. Klee, 2004. Reticulate sympatric speciation in Cameroonian crater lake cichlids. Frontiers in Zoology 1: 6–12.

Schneider, S., D. Roessli & L. Excoffier, 2000. Arlequin ver. 2.000: A Software for Population Genetic Data Analysis. University of Geneva, Switzerland.

Seehausen, O., 2004. Hybridization and adaptive radiation. Trends in Ecology and Evolution 19: 198–207.

Shaw, K. L. & P. D. Danley, 2003. Behavioral genomics and the study of speciation at a porous species boundary. Zoology 106: 261–273.

Shaw, P. W., G. F. Turner, M. R. Idid, R. L. Robinson & G. R. Carvalho, 2000. Genetic population structure indicates sympatric speciation of Lake Malawi pelagic cichlids. Proceedings of the Royal Society of London, Series B, Biological Sciences 267: 2273–2280.

Sheets, D. H., 2002. Integrated Morphometrics Package (IMP). Department of Physics, Canisius College, Buffalo, NY.

Sparks, J. S. & W. L. Smith, 2004. Phylogeny and biogeography of the Malagasy and Australasian rainbowfishes (Teleostei : Melanotaenioidei): Gondwanan vicariance and evolution in freshwater. Molecular Phylogenetics and Evolution 33: 719–734.

Templeton, A. R., 1981. Mechanisms of speciation—a population genetic approach. Annual Review of Ecology and Systematics 12: 23–48.

Turelli, M., N. H. Barton & J. A. Coyne, 2001. Theory and speciation. Trends in Ecology and Evolution 16: 330–343.

von Rintelen, T. & M. Glaubrecht, 2005. Anatomy of an adaptive radiation: a unique reproductive strategy in the endemic freshwater gastropod *Tylomelania* (Cerithioidea: Pachychilidae) on Sulawesi, Indonesia and its biogeographical implications. Biological Journal of the Linnean Society 85: 513–542.

von Rintelen, T. & M. Glaubrecht, 2006. Rapid evolution of sessility in an endemic species flock of the freshwater bivalve *Corbicula* from ancient lakes on Sulawesi, Indonesia. Biology Letters 2: 73–77.

von Rintelen, T., A. B. Wilson, A. Meyer & M. Glaubrecht, 2004. Escalation and trophic specialization drive adaptive radiation of freshwater gastropods in ancient lakes on Sulawesi, Indonesia. Proceedings of the Royal Society of London, Series B-Biological Sciences 271: 2541–2549.

Vos, P., R. Hogers, M. Bleeker, M. Reijans, T. Vandelee, M. Hornes, A. Frijters, J. Pot, J. Peleman, M. Kuiper & M. Zabeau, 1995. Aflp—a new technique for DNA-fingerprinting. Nucleic Acids Research 23: 4407–4414.

Whitten, T., G. S. Henderson & M. Mustafa, 2002. The ecology of Sulawesi. In Indonesia, P. J. B. (ed.), The Ecology of Indonesia Series. Jakarta.

Wright, S., 1943. Isolation by distance. Genetics 28: 114–138.

Wright, S., 1946. Isolation by distance under diverse systems of mating. Genetics 31: 39–59.

Wu, C. I., 2001. The genic view of the process of speciation. Journal of Evolutionary Biology 14: 851–865.

Zelditch, M. L., D. L. Swiderski, D. H. Sheets & W. L. Fink, 2004. Geometric morphometrics for biologists. A primer. Elsevier, Academic Press, New York.

Zitzler, K. & Y. X. Cai, 2006. *Caridina spongicola*, new species, a freshwater shrimp (Crustacea: Decapoda: Atyidae) from the ancient Malili lake system of Sulawesi, Indonesia. Raffles Bulletin of Zoology 54: 271–276.

Hydrobiologia (2008) 615:215–224
DOI 10.1007/s10750-008-9562-2

Processes regulating the community composition and relative abundance of taxa in the diatom communities of the Malili Lakes, Sulawesi Island, Indonesia

Andrew J. Bramburger · Paul B. Hamilton · Peter E. Hehanussa · G. Douglas Haffner

Published online: 13 September 2008

Abstract Although high levels of endemism have been reported on Sulawesi Island for almost 150 years, the aquatic systems of the island have received little attention until recently. The diatoms of the ancient Malili Lakes, located on central Sulawesi, exhibit levels of endemism unequaled elsewhere in the world. Interestingly, the diatom community of the lakes is taxonomically impoverished, and the entire order *Centrales* is conspicuous by its absence. In this article we review the mechanisms contributing to the development and maintenance of community composition and relative abundance of taxa within the system, at within-lake and system-wide scales. Our findings demonstrate that stochastic processes related to biogeography and colonization have had little influence on the diatom flora of the lakes, and deterministic processes related to competition, selection, speciation, and adaptive radiation, functioning on very small spatial scales, have contributed greatly to the diversity, community structure, and endemism of the system.

Keywords Ancient lakes · Diatoms · Malili Lakes · Community ecology

Guest editors: T. Wilke, R. Väinölä & F. Riedel
Patterns and Processes of Speciation in Ancient Lakes: Proceedings of the Fourth Symposium on Speciation in Ancient Lakes, Berlin, Germany, September 4–8, 2006

A. J. Bramburger (✉) · G. D. Haffner
Great Lakes Institute for Environmental Research, University of Windsor, Windsor, ON, Canada N9B3P4
e-mail: ajbram@hotmail.com

P. B. Hamilton
Research Division, Canadian Museum of Nature, Ottawa, ON, Canada K1P6P4

P. E. Hehanussa
Indonesian Institute of Sciences, Kompleks LIPI, Cibinong, West Java, Indonesia

Introduction

Since Wallace (1860), many authors have commented on the striking levels of endemism on Sulawesi Island within many taxonomic groups (Victor & Fernando, 1982; Den Bosch, 1985; White & Bruce, 1986; Reudi, 1995). Sulawesi Island is now reported to have the highest percentage of endemic terrestrial biota in the world, and is considered a global "hotspot" for biodiversity and endemism. Until recently, little was known about the aquatic biota of the island, even though Sulawesi's lakes represent the oldest island-based aquatic ecosystems in the world (Brooks, 1950; Haffner et al., 2001). The Malili Lakes are a system of 5 ancient, tectonically-formed lakes located in central Sulawesi. These lakes are unique among ancient lakes system in the world in that they alone form a hydrological continuum and share a common drainage through the Larona River.

In the late 1930 s, a German expedition to Wallacea, led by Prof. Dr. R. Woltereck collected the aquatic biota of the lakes and the diatom flora was subsequently described by Hustedt (1942). Just as Wallace was impressed by the terrestrial biota of the island, Hustedt was struck by the high proportion of endemic taxa within Sulawesi's lakes, and postulated that the endemic species count would continue to rise as taxonomic work continued. Recent floristic and taxonomic work by Bramburger et al. (2004, 2006) supports this assertion and further ecological work demonstrates that the benthic diatoms of the Malili Lakes represent an ideal system in which we can evaluate the relative importance of mechanisms contributing to the processes of speciation, colonization, and extirpation.

In this article, we review and evaluate the processes regulating community composition and relative abundance of taxa within the diatom communities of the Malili Lakes. We also examine the relative importance of stochastic mechanisms related to dispersal and biogeography, and deterministic mechanisms related to speciation in the development and maintenance of the highly endemic benthic diatom communities of the lakes. In addition to a comprehensive review of these systems, we present new data that contributes to a better understanding of community structure and dynamics within the system.

Stochastic processes—biogeography and colonization

Classically, most diatom taxa were thought to disperse over wide ranges (Darwin, 1841), and exhibit cosmopolitan patterns of distribution (Cleve, 1894, 1895). Under these paradigms, diatom floras would be expected to be homogeneous, at least at a regional scale, consistent with the assertions of Simberloff (1978), Simberloff & Van Holle (1999) and Rahel (2002) that freshwater biota will likely become homogenized as a result of stochastic colonization processes. Recently, however, several authors have shown that diatom distributions are much more restricted than previously thought, and several systems exhibit extraordinary levels of endemism (Moser et al., 1998; Vyvermann, 1991; Bramburger et al., 2004).

The Malili Lakes are unique among ancient lakes systems in that only they form a hydrological continuum (Brooks, 1950). The three main lakes (Matano, Mahalona, and Towuti) are connected by rivers, while the two smaller lakes (Lontoa and Masapi) receive no advective inputs from the main lakes. This situation provides an ideal system from which the influences of proximity and advective propagule inputs on the distribution of taxa within the system can be evaluated. Bramburger et al. (2004) showed that despite lacking a hydrological connection to Lake Matano, Lakes Lontoa and Masapi share a similar proportion of their taxa with Matano to Lakes Mahalona and Towuti (Table 1). The communities of all 5 lakes are characterized by an abundance of endemic taxa unparalleled in other systems. Additionally, similarity indices (SIMI) indicated that the diatom floras of lakes sharing a hydrological connection were no more similar to one another than those of unconnected lakes. In fact, the Malili Lakes bore a degree of similarity to one another that would be typical of lakes located on separate islands. These findings indicated that the diatom floras of the Malili Lakes evolved independently, and advective propagule inputs from Lake Matano have had little influence on the development of the diatom communities of Lakes Mahalona and Towuti. Furthermore, stochastic colonization events and geographic proximity cannot account for the degree of dissimilarity observed among the Malili Lakes diatom floras, and mechanisms operating at the within-lake, or smaller scales are responsible for the development and maintenance of these highly endemic communities.

In a subsequent study limited to Lake Matano, Bramburger et al. (in press a) demonstrated that the lack of a spatial pattern in similarity between diatom floras persists at the within-lake scale. Within Matano, assemblage similarity decreased slightly with distance between sample sites, and littoral

Table 1 Taxonomic richness, percentage of endemic taxa and percentage of taxa shared with Lake Matano for Lakes Matano, Mahalona, Towuti, and Masapi

Lake	Matano	Mahalona	Towuti	Masapi
Taxon richness	140	131	141	36
% Endemic taxa	22	9	27	11
% Taxa shared with L. Matano	NA	54	53	54

Adapted from Bramburger et al. (2004)

assemblages were more similar to one another than they were to planktonic assemblages. Although littoral assemblage similarity is generally related to site proximity, as would be predicted by the distance concept of the Theory of Island Biogeography (MacArthur & Wilson, 1967) and other stochastic models (Simberloff, 1974, 1978; Rahel, 2002), adjacent sample sites frequently support highly dissimilar benthic diatom assemblages, suggesting that benthic diatom assemblages within the lake are strongly influenced by mechanisms operating on a very small spatial scale.

The mechanisms that gave rise to this small-scale patterning, however, are currently poorly understood. A recent small-scale colonization study conducted within the littoral zone of Lake Matano (Bramburger et al., in press a) examined the colonization and succession dynamics of benthic diatoms onto glass slides over a period of 30 days. In the first third of the colonization period, modeled maxima for relative abundance distributions (RAD) of assemblages on the slides became more positive with time, indicating that during early colonization, assemblages on glass slides were approaching the characteristics of the overall Lake Matano community, and colonization was proceeding in accordance with the predictions of stochastic models (Preston, 1948; Patrick, 1967, 1975; Pachepsky et al., 2001). In the second third of the colonization period, however, RAD maxima reached a plateau and standard deviations of RADs decreased, reflective of a shift toward a community dominated by rare taxa, and suggesting that as early succession progressed, stochastic mechanisms related to dispersal and colonization were replaced by deterministic, competition-driven mechanisms in the regulation of species' relative abundance within glass-slide assemblages. Often, adjacent slides sampled at the same time exhibited vastly different assemblages, demonstrating that small-scale variations in environmental conditions can alter community trajectories over relatively short-time intervals. These findings show that although community composition and relative abundance are influenced on the short term by stochastic colonization processes, the overall standing diatom community of Lake Matano (and likely the other Malili Lakes) is regulated largely by deterministic mechanisms operating on a very small spatial scale.

Deterministic processes: selection and speciation

Among lakes—composition

Competition may be described as the mechanism by which the natural environment exerts its influence on a community of biological organisms. Therefore, it follows that for easily dispersible organisms, such as diatoms, highly endemic communities should be reflective of unique environments or sets of selective pressures. With its unique chemical, physical, and limnological characteristics, the high level of endemism observed within this system is not surprising. In fact, a simple comparison of taxonomic richness among several genera within Lakes Matano, Mahalona, Towuti, and Masapi shows that although genera are evenly distributed from lake-to-lake, taxonomic richness varies from genus-to-genus (Fig. 1). This result suggests that environmental pressures are variable at a lake-to-lake scale, and genera exhibit differential levels of success in different lakes. While an examination of species within the genus *Surirella* (Bramburger et al., 2006) provides some valuable insights into the influences of basin morphology and physical limnology on the development of endemic species clusters, the importance of factors, such as nutrients, chemical limnology, and substrata require further evaluation.

With 46 reported taxa, the genus *Surirella* is the most taxonomically rich of all the Malili Lakes diatom genera (Bramburger et al., 2006). Although endemic clusters have been reported within *Surirella* from various locations (Fig. 2; Foged, 1981; Vyvermann, 1991; Cocquyt, 1998; Metzeltin & Lange-Bertalot, 1998), the Malili Lakes remain the world's "hotspot" for *Surirella* diversity and the variety of hooks, spines, papillae, tubular processes, and siliceous tendrils exhibited by the Malili Lakes taxa are unparalleled elsewhere. Presumably, the resource and energetic costs associated with the development of such heavy ornamentations must be substantial and the competitive advantage conferred by these structures must outweigh the costs. The flexible siliceous tendrils characteristic of *Surirella sublinearis* and its variety *suggesta*, (Bramburger & Hamilton) and *S. tenacis* (Bramburger & Hamilton) have been demonstrated to function in attachment of the frustule to littoral zone substrate particles (Fig. 3), and other valvar modifications are anticipated to have similar functions.

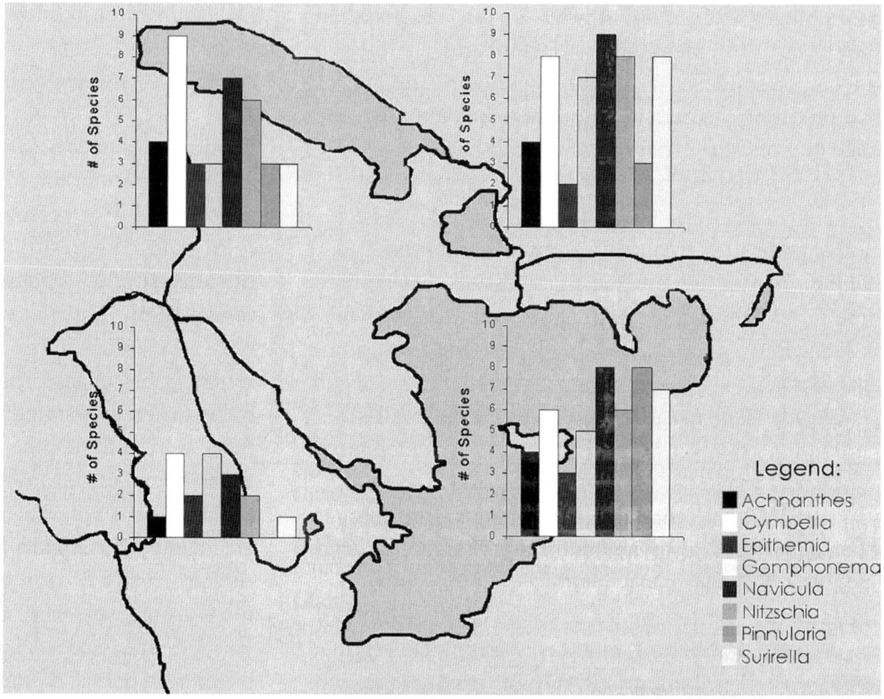

Fig. 1 Taxonomic richness of the 8 most species-rich genera in the Malili Lakes from Lakes Matano (**A**), Mahalona (**B**), Towuti (**C**), and Masapi (**D**), Lake Lontoa (**E**) had not been surveyed when this image was produced. Compiled from data reported by Bramburger et al. (2004)

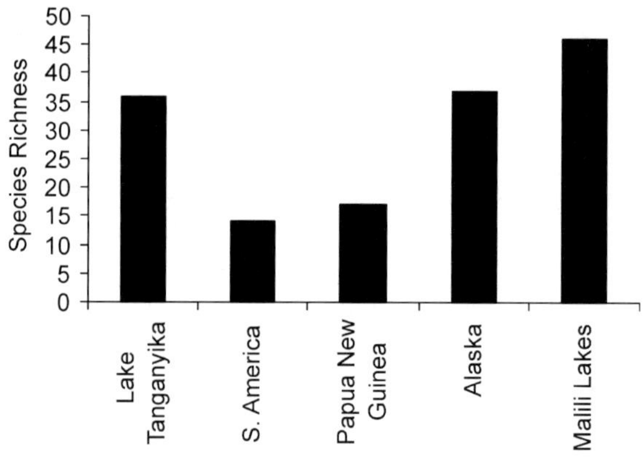

Fig. 2 Species richness within the genus *Surirella* from five global "hotspots" of *Surirella* diversity. Compiled from Cocquyt (1998), Foged (1981), Metzeltin & Lange-Bertalot (1998) and Vyvermann (1991)

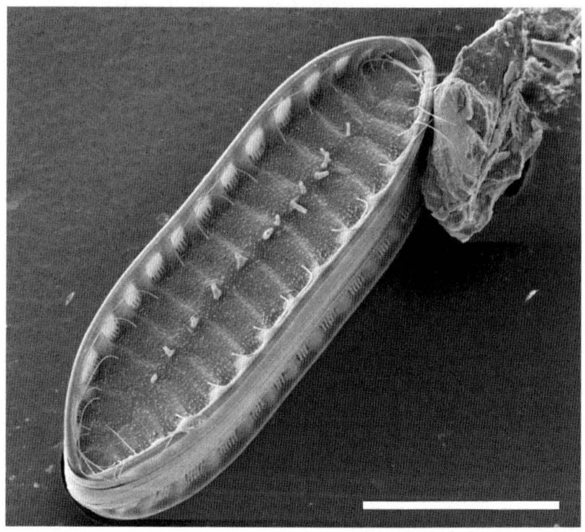

Fig. 3 SEM Micrograph of *Surirella tenacis* illustrating diatom attachment to substrate particle through use of flexible siliceous tendrils. Scale bar = 20 μm (Bramburger et al., 2006)

The importance of the valvar ornamentations of the *Surirella* taxa becomes apparent when the basin morphology and physical limnology of the three large lakes in the system are considered. Though they vary widely in surface area and depth (Table 2), Lakes Matano, Mahalona, and Towuti are similar in that their shoreline profiles are relatively steep, and that the zone of vertical mixing (as indicated by vertical temperature profiles) extends below the euphotic depth (to the bottom in Mahalona and Towuti). A benthic diatom failing to maintain its position within the narrow littoral habitat of the lake risks tumbling downslope, or being mixed downward into the dark below with little chance of re-suspension to a suitable habitat.

Table 2 Lake surface area, maximum depth, elevation, and total phosphorus for the Malili Lakes

Lake	Surface area (km^2)	Max. depth (m)	Elevation (masl)	Total phosphorus (surface, mg/l)
Matano	164	590	396	0.01
Mahalona	25	72	319	0.01
Towuti	560	200	293	0.01
Masapi	4.4	3	434	–
Lontoa	6.3	~7	586	0.01

Total phosphorus has not been measured for Lake Masapi. Instrument detection limit for Phosphorus = 0.0006 mg/l

Ancient lakes have long been recognized as important reserves of aquatic biodiversity and centers of endemism with highly unique biological communities. While we have already dealt with both community composition and relative abundance of taxa at relatively small spatial scales, mechanisms contributing to the endemism and biodiversity of a system exert their influence solely on community composition at the system-wide scale. Cronk (1997) proposed that centers of endemism were either centers of paleo-endemism, created through processes of extinction elsewhere in a taxon's native range, or centers of neo-endemism, created through processes of speciation. While it is impossible to determine whether the Malili Lakes fit the criteria for a center of paleo-endemism (Cronk, 1997), simple floristic and ecomorphological studies have illustrated the importance of extirpation or unsuccessful colonization, and speciation and resource partitioning in the development of a highly unique flora (Bramburger et al., 2004, 2006).

In many taxonomic groups, there is an inverse relationship between absolute latitude and taxonomic richness (Stevens, 1989; Culver & Buzas, 2000). In other words, there are more species in the tropics than there are at higher latitudes. The diatoms show a rough congruence with this trend. Interestingly, with 267 taxa reported, the Malili Lakes diatom flora exhibits a taxonomic diversity typical of lake systems from much higher latitudes. Although the benthic diatom community of the Malili Lakes is relatively diverse, Bramburger et al. (in press a) demonstrated that the planktonic diatom community of Lake Matano may be considered depauperate, with only 31 of the lake's 191 observed taxa reported from the plankton. Most of these occurred at low abundances, or were not alive. Most notably, the entire order *Centrales* (the centric diatoms) are virtually absent from the Malili Lakes flora. Hustedt (1942) and Bramburger et al. (2004) did not encounter any centric diatoms in their surveys of the lakes. Bramburger et al. (in press a) reported no live centric diatoms in the plankton or benthos of the lake, and speculated that elevated metal concentrations and a paucity of nutrients in Lake Matano's water column have inhibited the ability of cosmopolitan and centric forms to successfully colonize the lake. Impoverished planktonic diatom communities have been reported from other ancient lakes, such as Malawi, Tanganyika, Baikal, and Hövsgöl, but their planktonic diatom communities are often dominated by centric forms (Coulter, 1963; Eccles, 1974; Cocquyt, 1998; Edlund et al., 2003, 2006; Flower et al., 2004; Edlund, 2006).

Within the benthic diatom community, the majority of genera can also be considered depauperate, and typically represent <5% of the reported global taxonomic diversity of the genus. In fact, only the most speciose genus within the system (*Surirella*) exhibits >5% (~7%) of its global richness within the Malili Lakes (Fig. 2). Elton (1946) postulated that a depression of species richness at the generic level was reflective of competition for limited resources between ecomorphologically similar species within the genus, consistent with Lack's (1947) concept of competitive exclusion of morphologically redundant taxa. On a coarse scale, the *Surirella* taxa from the Malili Lakes represent a morphologically cohesive genus, and the co-existence of so many taxa in a competitive system is superficially paradoxical. Recent examinations of the Malili Lakes *Surirella* taxa by SEM have revealed novel fine-scale valvar ornamentations and modifications, including various, spines, papillae, and siliceous tendrils (Bramburger et al., 2006), and demonstrated that these structures serve to limit ecomorphological redundancy within the group, thereby limiting interspecific competition and promoting co-existence.

Species flocks, such as the *Surirella* taxa from the Malili Lakes are a recurring phenomenon in ancient lakes system, and several examples of diatom species flocks from ancient lakes have been reported recently. Besides the Malili Lakes (46 taxa, Bramburger et al., 2006), the genus *Surirella* has produced endemic species flocks in the African Rift Lakes (44 taxa,

Cocquyt 1998), South America (14 taxa, Metzeltin & Lange-Bertalot, 1998), Papua New Guinea (17 taxa, Vyvermann, 1991), and Alaska (37 taxa, Foged, 1981), despite its low global taxonomic diversity. Edlund et al. (2006) reported a species flock in the *Navicula reinhardtii* complex from the ancient Lakes Hövsgöl and Prespa in the Baikal Rift system. Examples of ancient lakes species flocks from other taxonomic groups are even more prevalent. Lake Baikal and the surrounding drainage basin support clusters of endemic fishes (Kontula et al., 2003) and turbellarians (Sluys et al., 1998; Fryer, 1996), while the African Rift Lakes support a huge diversity of Cichlid fishes (Sturmbauer, 1998; Seehausen, 2000), which represent the best-documented case of endemic clustering. Within the Malili Lakes, the *Surirellas* are not the only endemic species cluster. Other examples of clustering are observed in the endemic gastropods (Ceruthoidea) (von Rintelen & Glaubrecht, 2003), and Telmatherinid fishes (Roy et al., 2004) of the lakes.

Endemic species clusters within ancient lakes are often cited as examples of adaptive radiation. Schluter (2000) proposed four criteria that the taxa must adhere to in order to be considered as radiating; focal taxa should (i) share common ancestry, (ii) exhibit phenotypic variation in correspondence to their varying habitats, (iii) be differentiated by functional characteristics, and (iv) exhibit rapid rates of speciation. Often authors have been quick to classify endemic species clusters as adaptive radiations without first considering these objective criteria. Due to their explosive rates of evolution and differences in functional trophic morphology, the Cichlid fishes of the African Rift Lakes remain the best example of adaptive radiation (Sturmbauer, 1998; Seehausen, 2000), while in the Malili Lakes, Roy et al. (2004) showed that the Telmatherinid fishes within the genus *Telmatherina* represented a similar radiation on a smaller scale.

For phytoplankton, Schluter's criteria are not as easy to fulfill. In the endemic *Navicula reinhardtii* complex from Lakes Hövsgöl and Prespa (Edlund et al., 2006), for example, variations in valve morphology are subtle, and the functionality of "radiating" structures is not known. Although it is impossible to discern the time of first colonization of the Malili Lakes by *Surirella*, if we assume that most diatom genera colonized the Malili Lakes at approximately the same time, then the higher taxonomic diversity within the genus *Surirella* would reflect a more rapid rate of speciation than that of other genera (Fig. 2) and the utility of the radiating characteristics within the genus have been demonstrated by Bramburger et al. (2006). Although without genetic data it is impossible to assess the ancestry of the *Surirella* taxa from the lakes, the morphological cohesiveness of the genus within the system suggests a common ancestral form. Unfortunately, as the benthic diatom taxa of the Malili Lakes all occupy similar habitat types across a true gradient of environmental variables, we have not yet achieved the level of spatial resolution necessary to evaluate whether the varying *Surirella* morphologies correspond to their respective habitats. Although the *Surirella* taxa from the Malili Lakes seem to exhibit the hallmarks of an adaptive radiation, and satisfy the majority of Schluter's criteria we are not yet able to definitively fulfill all of the prescribed requirements.

Within lakes—relative abundance

In lakes with such an intensive mixing regime, upwelling of deep water must also occur. In a recent study, Bramburger et al. (unpublished data) examined the influence of a simulated upwelling event and subsequent oxygenation of deep water on the littoral epilithic diatom community of Lake Matano. When the two treatment groups in the experiment were subjected to surface water and oxygenated deep water, respectively, they were driven down to two different community trajectories. Interestingly, after perturbation and maintenance of the assemblage under new chemical conditions, there was no detectable change in community composition, but merely a shift in the relative abundance of taxa within the existing endemic community reflective of a shift in the competitive landscape of the assemblage. These findings are in contrast to the assertions of several invasion biologists (Lodge, 1993; Ricciardi, 2001; Rahel, 2002) that increased perturbation facilitates colonization by non-indigenous species, and also demonstrate that biological communities have no fixed end point and suggest that teleological paradigms, such as the biological climax community (Clements, 1916) are not applicable within temporally heterogeneous environments. Unfortunately, due to the "black box" nature of the perturbation imposed in this study, and very low concentrations of nutrients in the Matano water column, it was impossible to determine whether the diatom assemblage was responding to changes in nutrient or metal concentrations in the oxygenated treatment water. Studies conducted with other taxonomic groups from the Malili Lakes, however, have suggested that complex interactions between nutrient limitation and metal toxicity may influence plankton community dynamics within Lake Matano (Sabo et al., 2008).

Although the influences of water chemistry on the diatom community have yet to be thoroughly explored, simple examinations of the standing diatom assemblages of the lake are often sufficient to discern the effects of substrate type and grain size on the composition and relative abundance of diatom assemblages within a system at a relatively coarse scale. We sampled diatoms from 5 substrate types within 10 sites around the littoral zone of Lake Matano. Sample sites were characterized by their predominant bedrock type (ultrabasic, limestone, sandstone, and conglomerate), and their primary substrate grain caliber (silt, sand, cobble, boulder, and bedrock). Individual substrate samples were combined into an amalgamated sample representative of each sample site, and diatom valves were enumerated from amalgamated samples. Similarity indices (SIMI) were calculated for pairings of each site with the other 9, and multiple one way ANOVAs with post-hoc Bonferroni tests were used to determine whether sites with similar substrate characteristics were more similar to one another than to other sites. Additionally, principal components analysis (PCA) performed using relative abundance data from the benthic diatom assemblage at each site was used to determine if sites were clustered together based on bedrock type or substrate grain caliber.

In this analysis, one way ANOVAS were unable to detect any significant differences in mean SIMI among various pairings of bedrock type, or dominant grain size ($n = 45$, $P > 0.05$; Fig. 4A, B). Similarly, PCA was unable to compile any components with sufficient eigenvalues to successfully discriminate among assemblages. These results indicate that coarse scale bedrock type and dominant grain caliber at a sample site are poor predictors of diatom community characteristics within a site, and the diatoms are likely responding to selective pressures at a finer scale.

Discussion

Extraordinary levels of terrestrial endemism on Sulawesi Island have been reported since approximately 150 years ago (Wallace, 1860). Only recently, however, have authors begun to document similar patterns within the island's aquatic habitats (Hustedt, 1942; Kottelat, 1990a, b, c; Haffner et al., 2001; von Rintelen & Glaubrecht, 2003; Roy et al., 2004) The diatom flora of the island's Malili Lakes, although taxonomically impoverished, supports a level of endemism estimated to be higher than that of any other system (Bramburger et al., 2004, 2006). Our understanding of the mechanisms contributing to this striking endemism is still in its infancy.

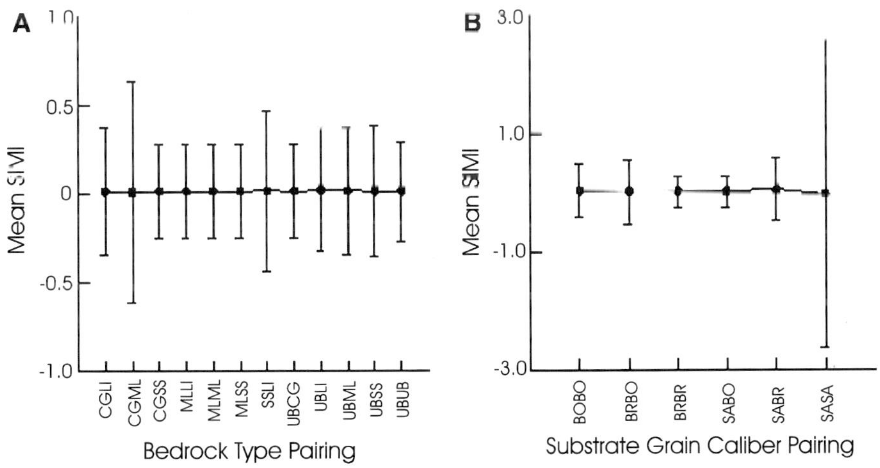

Fig. 4 ANOVA results for mean SIMI scores for site pairings within Lake Matano. (**A**) Grouped by bedrock type comparison. Bedrock type codes are as follows: CG—Conglomerate, LI—Limestone, ML—Metamorphosed Limestone, SS—Sandstone, UB—Ultrabasic Rock. (**B**) Grouped by dominant substrate caliber. Substrate caliber codes are as follows: BO—Boulder, BR—Bedrock, SA—Sand

Bramburger et al. (2004, in press b) have demonstrated that although stochastic colonization events influence initial community composition and relative abundance of taxa at a small scale, deterministic mechanisms related to competition and resource partitioning exert a more potent influence on community dynamics on a long-term basis. Despite intensive colonization vectors between lakes, advective propagule input has had little effect on the development of the diatom floras of the lakes, in contrast to the predictions of currently accepted stochastic models of community development and homogenization (Simberloff, 1978; Simberloff & Van Holle, 1999; Rahel, 2002).

The selective pressures that exert their influence on the Malili Lakes diatom flora do so at vanishingly small spatial scales. Bramburger et al. (in press a) showed that small scale disturbance events were sufficient to drive early successional communities down widely divergent community trajectories on substrata located within centimeters of one another. These findings, coupled with a shift in species relative abundance brought about by a simulated perturbation event (Bramburger et al., unpublished data.), demonstrate that small-scale variations in the abiotic environment and stochastic disturbance events can have broad impacts at a community scale, and suggest that successional paradigms of community dynamics do not apply in temporally unstable environments.

Within the community, some selective pressures are more favorable to specific groups of species. In the Malili Lakes, taxonomic diversity within genera is not evenly distributed across lakes, and the genus *Surirella* exhibits almost 7% of its global diversity within the five lakes of the system. The diversity of siliceous ornamentations exhibited by the Malili Lakes *Surirella* taxa is unparalleled elsewhere in the world, and although this genus appears to exhibit rapid speciation, functionality of radiating structures, and a monophyletic ancestry in this system, we are unable to quantify the rate at which these taxa evolved. Although the Malili Lakes *Surirellas* are a good candidate as an example of adaptive radiation in an algal group, we are limited in our ability to classify it as such due to the stringent nature of Schluter's (2000) criteria.

Although the mechanisms that regulate the taxonomic composition and relative abundance of taxa within biological communities remain poorly understood, the benthic diatom communities of the ancient Malili Lakes provide an ideal system to evaluate the relative importance of stochastic and deterministic factors. Our results to date demonstrate that stochastic mechanisms related to dispersal and colonization has not played a significant role in the development of the highly endemic flora of the lakes, even at very small scales. Deterministic processes related to competition, extirpation, and speciation, operating on the within-lake and site-to-site scales have made an important contribution to the development and maintenance of highly endemic communities in these unique ancient systems.

Acknowledgements This project was financially supported by an NSERC grant to G. D. Haffner, an RAC grant to P. B. Hamilton, and an Ontario Graduate Scholarship to A. J. Bramburger. Further financial and logistical supports were furnished by Inco Ltd. Canada, and PT Inco, TBK, Indonesia.

References

Bramburger, A. J., G. D. Haffner & P. B. Hamilton, 2004. Examining the distributional patterns of the diatom flora of the Malili Lakes, Sulawesi, Indonesia. In Poulin, M. (ed.), Proceedings of the 17th International Diatom Symposium. Biopress Limited, Bristol, U.K.: 11–25.

Bramburger, A. J., P. B. Hamilton, P. E. Hehanussa, F. Hinz & G. D. Haffner, 2006. An examination of species within the genus *Surirella* from the Malili Lakes, Sulawesi Island, Indonesia, with descriptions of 11 new taxa. Diatom Research 21: 1–56.

Bramburger, A. J., P. B. Hamilton, P. E. Hehanussa & G. D. Haffner, in press a. Spatial patterns of planktonic and benthic diatom distribution and assemblage similarity in Lake Matano (Sulawesi Island, Indonesia). Proceedings of the 19th International Diatom Symposium, 28 pp.

Bramburger, A. J., P. B. Hamilton, P. E. Hehanussa & G. D. Haffner, in press b. Colonization of artificial substrates by the littoral diatom community of an ancient tropical lake. Journal of Phycology.

Brooks, L., 1950. Speciation in ancient lakes (concluded). The Quarterly Review of Biology 25: 131–176.

Clements, F. E., 1916. Plant succession: an analysis of the development of vegetation. Carnegie Institution, Washington, D.C., Publication No. 242.

Cleve, P. T., 1894. Synopsis of the naviculoid diatoms part I. Kongliga Svenska Vetenskaps Akademiens Handlingar 26: 1–194.

Cleve, P. T., 1895. Synopsis of the naviculoid diatoms part II. Kongliga Svenska Vetenskaps Akademiens Handlingar 27: 1–219.

Cocquyt, C., 1998. Diatoms from the northern basin of Lake Tanganyika. Bibliotheca Diatomologica 39: 1–275.

Coulter, G. W., 1963. Hydrological changes in relation to biological production in southern Lake Tanganyika. Limnology and Oceanography 8: 463–477.

Cronk, Q. C. B., 1997. Islands: stability, diversity, conservation. Biodiversity and Conservation 6: 477–493.

Culver, S. J. & M. A. Buzas, 2000. Global latitudinal species gradient in deep-sea benthic foraminifera. Deep Sea Research Part 1: Oceanographic Research Papers 47: 259–275.

Darwin, C. R., 1841. The Voyage of the Beagle. J Murray, London.

Den Bosch, H. A. J., 1985. Snakes of Sulawesi: checklist, key, and additional biogeographical remarks. Zoologische Verhandlingen 217: 1–50.

Eccles, D. H., 1974. An outline of the physical limnology of Lake Malawi (Lake Nyasa). Limnology and Oceanography 19: 730–742.

Edlund, M. B., 2006. Persistent low diatom phytoplankton diversity within the otherwise highly diverse Lake Baikal ecosystem. Nova Hedwigia 130: 339–356.

Edlund, M. B., R. M. Williams & N. Soninkhishig, 2003. The planktonic diatom diversity of ancient Lake Hövsgöl, Mongolia. Phycologia 42: 232–260.

Edlund, M. B., Z. Levkov, N. Soninkhishig, S. Krstic & T. Nakov, 2006. Diatom species flocks in large ancient lakes: the *Navicula reinhardtii* complex from Lakes Hövsgöl (Mongolia) and Prespa (Macedonia). In Witkowski, A. (ed.), Proceedings of the 18th International Diatom Symposium 2004. Biopress Limited, Bristol, U.K.: 61–74.

Elton, C. S., 1946. Competition and the structure of ecological communities. Journal of Animal Ecology 15: 54–68.

Flower, R. J., G. Pomazkina, E. Rodionova & D. M. Williams, 2004. Local and mesoscale diversity patterns of benthic diatoms in Lake Baikal. In Poulin, M. (ed.), Proceedings of the 17th International Diatom Symposium, Ottawa 25th–31st August 2002. Biopress Limited, Bristol, U.K.: 69–92.

Foged, N., 1981. Diatoms in Alaska. Bibliotheca Phycologica 53: 1–318.

Fryer, G., 1996. Evolution in ancient lakes. Hydrobiologia 321: 245–254.

Haffner, G. D., P. E. Hehanussa & D. Hartoto, 2001. The biology and physical processes of large lakes of Indonesia. In Munawar, M. & R. E. Hecky (eds), The Great Lakes of the World. Backhuys Publishers, Leiden: 183–192.

Hustedt, F., 1942. Süßwasser-Diatomeen des indomalayischen Archipels und der Hawaii-Inseln. Internationale Revue der Gesamten Hydrobiologie und Hydrographie 42: 1–252.

Kontula, T., S. V. Kirilchik & R. Vainola, 2003. Endemic diversification of the monophyletic cottoid fish species flock in Lake Baikal explored with mtDNA sequencing. Molecular Phylogenetics and Evolution 1: 143–155.

Kottelat, M., 1990a. Synopsis of the endangered Buntingi (Osteichthyes: Adrianichthydae and Oryziidae) of Lake Poso, Central Sulawesi, Indonesia, with a new reproductive guild and descriptions of three new species. Ichthyological Exploration of Freshwaters 1: 49–67.

Kottelat, M., 1990b. The ricefishes (Oryziidae) of the Malili Lakes, Sulawesi, Indonesia, with description of a new species. Ichthyological Exploration of Freshwaters 1: 321–344.

Kottelat, M., 1990c. Sailfin silversides (Pisces: Telmatherinidae) of Lakes Towuti, Mahalona, and Wawontoa (Sulawesi, Indonesia) with descriptions of two new species. Ichthyological Exploration of Freshwaters 1: 35–54.

Lack, D., 1947. Darwin's Finches. Cambridge University Press, Cambridge, UK.

Lodge, D. M., 1993. Species invasions and deletions: community effects and responses to climate and habitat change. In Kareiva, O. M., J. G. Kingsolver & R. B. Huey (eds), Biotic Interactions and Global Change. Sinauer Associates, Sunderland, MA: 367–387.

MacArthur, R. H. & E. O. Wilson, 1967. The Theory of Island Biogeography. Princeton University Press, Princeton.

Metzeltin, D. & H. Lange-Bertalot, 1998. Tropical diatoms of South America I. About 700 predominantly rarely known or new taxa representative of the neotropical flora. Iconographia Diatomologica 5: 1–695.

Moser, G., H. Lange-Bertalot & D. Metzeltin, 1998. Insel der Endemiten. Geobotanisches Phänomen Neukaledonien. Diatomeenflora einer Tropeninsel. Bibliotheca Diatomologica 38: 1–464.

Pachepsky, E., J. W. Crawford, J. L. Bown & G. Squire, 2001. Towards a general theory of biodiversity. Nature 410: 923–926.

Patrick, R., 1967. The effect of invasion rate, species pool, and size of area on the structure of the diatom community. Proceedings of the National Academy of Sciences of the United States of America 58: 1335–1342.

Patrick, R., 1975. The formation and maintenance of benthic diatom communities. Proceedings of the American Philosophical Society 120: 475–484.

Preston, F. W., 1948. The commonness, and rarity, of species. Ecology 29: 254–283.

Rahel, F. J., 2002. Homogenization of freshwater faunas. Annual Review of Ecology and Systematics 33: 291–315.

Reudi, M., 1995. Taxonomic revision of shrews of the genus Crocidura from the Sunda Shelf and Sulawesi with descriptions of two new species (Mammalia: Soricidae). Zoological Journal of the Linnean Society 115: 211–265.

Ricciardi, A., 2001. Facilitative interactions among aquatic invaders: is an "invasional meltdown" occurring in the Great Lakes? Canadian Journal of Fisheries and Aquatic Sciences 58: 2513–2525.

Roy, D., M. F. Docker, D. D. Heath & G. D. Haffner, 2004. Genetic and morphological data supporting the hypothesis of adaptive radiation in the endemic fish of Lake Matano. Journal of Evolutionary Biology 17: 1268–1276.

Sabo, E., D. Roy, P. B. Hamilton, P. E. Hehanussa, R. McNeely & G. D. Haffner, 2008. The plankton community of Lake Matano: factors regulating plankton composition and relative abundance in an ancient tropical lake of Indonesia. Hydrobiologia (this volume).

Schluter, D., 2000. The Ecology of Adaptive Radiation. Oxford University Press, Oxford.

Seehausen, O., 2000. Explosive speciation rates and unusual species richness in haplochromine cichlid fishes: effects of sexual selection. Advances in Ecological Research 31: 237–274.

Simberloff, D. S., 1974. Equilibrium theory of island biogeography and ecology. Annual Review of Ecology and Systematics 5: 161–182.

Simberloff, D. S., 1978. Using island biogeographic distributions to determine if colonization is stochastic. American Naturalist 112: 713–726.

Simberloff, D. S. & B. Van Holle, 1999. Positive interactions of nonindigenous species: invasional meltdown? Biological Invasions 1: 21–32.

Sluys, R., O. A. Timoshkin & M. Kawakatsu, 1998. A new species of giant planarian from Lake Baikal, with some remarks on character states in the Dendrocoelidae (Platyhelminthes, Tricladida, Paludicola). Hydrobiologia 383: 69–75.

Stevens, G. C., 1989. The latitudinal gradient in geographical range: how so many species coexist in the tropics. American Naturalist 133: 240–256.

Sturmbauer, C., 1998. Explosive speciation in cichlid fishes of the African Great Lakes: a dynamic model of adaptive radiation. Journal of Fish Biology 53: 18–36.

Victor, R. & C. H. Fernando, 1982. Distribution of freshwater Ostracoda (Crustacea) in Southeast Asia. Journal of Biogeography 9: 281–288.

von Rintelen, T. & M. Glaubrecht, 2003. New discoveries in old lakes: three new species of *Tylomelania* Sarasin & Sarasin, 1897 (Gastropoda: Cerithioidea: Pachychilidae) from the Malili lake system on Sulawesi, Indonesia. Journal of Molluscan Studies 69: 3–17.

Vyvermann, W., 1991. Diatoms from Papua New Guinea. Bibliotheca Diatomologica 22: 1–225.

Wallace, A. R., 1860. On the zoological geography of the Malay Archipelago. Journal of the Proceedings of the Linnean Society (Zoology) 4: 172–184.

White, C. M. & M. Bruce, 1986. The birds of Wallacea. Checklist No. 7. British Ornithologists Union, London.

Hydrobiologia (2008) 615:225–235
DOI 10.1007/s10750-008-9560-4

SPECIATION IN ANCIENT LAKES

The plankton community of Lake Matano: factors regulating plankton composition and relative abundance in an ancient, tropical lake of Indonesia

Elisabeth Sabo · Denis Roy · Paul B. Hamilton · Peter E. Hehanussa · Roger McNeely · G. Douglas Haffner

Published online: 14 September 2008

Abstract Recent evidence reveals that food webs within the Malili Lakes, Sulawesi, Indonesia, support community assemblages that are made up primarily of endemic species. It has been suggested that many of the species radiations, as well as the paucity of cosmopolitan species in the lakes, are related to resource limitation. In order to substantiate the possibility that resource limitation is playing such an important role, a study of the phytoplankton and zooplankton communities of Lake Matano was implemented between 2000 and 2004. We determined species diversity, relative abundances, size ranges, and total biomass for the phytoplankton and zooplankton, including the distribution of ovigerous individuals throughout the epilimnion of Lake Matano in three field seasons. The phytoplankton community exhibited very low biomass (<15 μg l^{-1}) and species richness was depressed. The zooplankton assemblage was also limited in biomass (2.5 mg l^{-1}) and consisted only of three taxa including the endemic calanoid *Eodiaptomus wolterecki* var. *matanensis*, the endemic cyclopoid, *Tropocyclops matanensis* and the rotifer *Horaella brehmi*. Zooplankton were very small (<600 μm body length), and spatial habitat partitioning was observed, with *Tropocylops* being confined to below 80 m, while rotifer and calanoid species were consistently observed above 80 m. Less than 0.1% of the calanoid copepods in each year were egg-bearing, suggesting very low population turnover rates. It was concluded that chemical factors as opposed to physical or biological processes were regulating the observed very low standing crops of phytoplankton which in turn supports a very minimal zooplankton community restricted in both species composition and abundance. As chemical factors are a function of the catchment basin of Lake Matano, it is predicted that resource limitation has long played an important role in shaping the unique endemic assemblages currently observed in the food web of the lake.

Guest editors: T. Wilke, R. Väinölä & F. Riedel
Patterns and Processes of Speciation in Ancient Lakes: Proceedings of the Fourth Symposium on Speciation in Ancient Lakes, Berlin, Germany, September 4–8, 2006

E. Sabo · D. Roy · G. D. Haffner (✉)
Great Lakes Institute for Environmental Research, University of Windsor, 401 Sunset Ave., Windsor, ON, Canada N9B 3P4
e-mail: haffner@uwindsor.ca

E. Sabo
Environment Canada, 351 St. Joseph Blvd, Gatineau, QC, Canada K1A 0H3

P. B. Hamilton · R. McNeely
Research Division, Canadian Museum of Nature, P.O. Box 3443, Station D, Ottawa, ON, Canada K1P 6P4

P. E. Hehanussa
APCE c/o Limnology-LIPI Jalan Prof. Dr. Dody Tisna Amidjaja, Cibinong Science Centre, LIPI, Cibinong 16911, Indonesia

Keywords Lake Matano · Phytoplankton · Zooplankton · Community structure · Tropical limnology

Introduction

The relationship between biodiversity and productivity has been extensively studied in order to understand the interactions and interdependence of processes regulating the composition and abundance of species in aquatic ecosystems. Many studies have focused on factors determining the temporal and spatial variability of plankton (Talling, 1957; Hutchinson, 1967) with an emphasis on temperate lakes. A deficit of information still exists, however, for tropical freshwater systems where community assemblages are often rich in endemic species. As a result, there is insufficient knowledge of the relationship between biodiversity and productivity in tropical lakes, and this is especially true for ancient, tropical lakes (>1 million years old).

As described by Lodge (1993) community composition is a function of dispersal, habitat quality and biological interactions. The relative simplicity of pelagic plankton communities in deep tropical lakes has been related to the homogeneity of the pelagic zone (Dumont & Segers, 1996) that assumes planktonic species can disperse to most lakes, but in-lake processes, chemical and biological, act as filters that determine the final assemblage. There is considerable evidence, however, that predation pressure can also have a direct impact on the structuring of plankton communities, and predation is particularly important in lakes at low latitudes (Ogutu-Ohwayo & Hecky, 1991; Pinto-Coelho et al., 2005). It has been argued that while temperate zooplankton can decouple their life cycles from predators, this strategy is compromised in the tropics as a result of the lack of seasonality in breeding seasons (Brooks & Dodson, 1965; Hall et al., 1976; Fernando, 2002; Hobaek et al., 2002). For example, year-round predation by sardine populations has virtually eliminated Cladocerans from Lake Tanganyika (Ogutu-Ohwayo & Hecky, 1991).

Resolving the relative importance of habitat quality on regulating the composition and relative abundance of species in biological assemblages is difficult as physical, chemical and biological factors are often interactive and interdependent. Lake Matano, south-central Sulawesi, Indonesia, presents a novel system in which to determine the relative importance of factors governing the relationship between biodiversity and production in plankton communities. The depth (590 m) (Haffner et al., 2001) and steeply sloping sides of this classical graben lake (Brooks, 1950) minimizes the influence of the littoral zone. Lake Matano is an ancient lake, estimated at approximately 1–4 million years of age, and only recently have detailed limnological investigations on the lake commenced. These studies have revealed a highly endemic lake community, recorded notably in the diatoms (Hustedt, 1942; Bramburger, 2004), gastropods (Sarasin & Sarasin, 1897; von Rintelen & Glaubrecht, 2003; von Rintelen et al., 2004) and fish (Kottelat, 1990a, b, 1991; Roy et al., 2004, 2006a, b, 2007).

To date, however, other than the preliminary surveys and reviews by Lehmusluoto et al. (1999) and Haffner et al. (2001), no detailed accounts have focused on the phytoplankton and zooplankton assemblages in this system. The importance of a better understanding of the factors regulating the biological structure of Lake Matano is amplified by recent increases in anthropogenic activity in and around the lake. Increasing water demands on the lake, as a source of freshwater as well as being the receiving basin for wastewater and mining runoff, pose potential threats to the highly endemic communities through changes in nutrient dynamics, species introductions and toxicological stress.

In this study, we determined the species composition and biomass of the standing phytoplankton community as well as the composition, relative abundances, size ranges, and total biomass of the zooplankton community throughout the upper 110 m of Lake Matano in August 2000, January 2002 and August 2004. The August season samples represented 'dry season' characteristics of the lake whereas the January samples were collected during the wet season. We used these measurements along with physical and chemical data to provide the first detailed characterization of the plankton community of the lake, and to assess the relative importance of factors regulating primary and secondary production. Specifically, the goals of this study were: (1) To quantify the composition, relative abundance and distribution of phytoplankton and zooplankton in the top 100 m of Lake Matano and (2) to examine the relative importance of physical, chemical and biological factors that regulate composition and relative abundance of species in the plankton assemblage of this ancient lake ecosystem.

Methods

Study area

With a maximum area of 16,408 ha, and a maximum depth of ~590 m, Lake Matano is one of the deepest lakes in the world. Lake Matano spans a still active fault line and tectonic movement approximates 2 cm y^{-1} (Haffner et al., 2001). Precipitation during the year is intense with almost 3 m of rainfall, and this results in an estimated inflow of 630 × 10^6 m^3 y^{-1}. Lake volume and rainfall contribute to a lake retention time of approximately 100 years (Fig. 1).

Water column structure and chemistry

Water chemistry samples were collected each year at the Master Station near the center of the lake (2.280°S, 121.190°E). Secchi depth was determined and a Brancker XR-420-CDT (Conductivity, Depth, Temperature) probe was used to measure in situ temperature. Oxygen, conductivity and pH were measured using a minsonde Hydrolab® with a 100 m cable. Water samples were collected throughout the water column to a final depth of 590 m, using a 2 l acrylic Kemmerer sampler. Water from these depths was immediately overflowed into sampling containers using Tygon tubing to minimize contact with the air for later analyses of total phosphorus (TP) and nitrate/nitrite. Total phosphorus was measured on unfiltered water samples, with an automated ascorbic acid reduction method technique (Standard Methods, 1995), using a Bran and Luebbe continuous flow autoanalyzer. Nitrate/nitrite was measured using an automated Hydrazine sulphate reduction method (Standard Methods, 1995), with a Bran and Luebbe continuous flow autoanalyzer. Total nitrogen (TN) was determined as the total of nitrate, nitrite, ammonia and organic nitrogen as according to Standard Methods (1995).

Phytoplankton composition and biomass

The sampling site for collection of phytoplankton in Lake Matano was confined to the Master Station. Kemmerer bottle samples were collected at successive 10 m depth intervals, down to a depth of 110 m. Five hundred millilitre of water was collected at each depth and immediately preserved with Lugol's solution in sedimentation jars. Sedimentation of the undisturbed plankton samples commenced for 2 weeks under cool-dark conditions. Subsequently, the supernatant was gently siphoned off and the

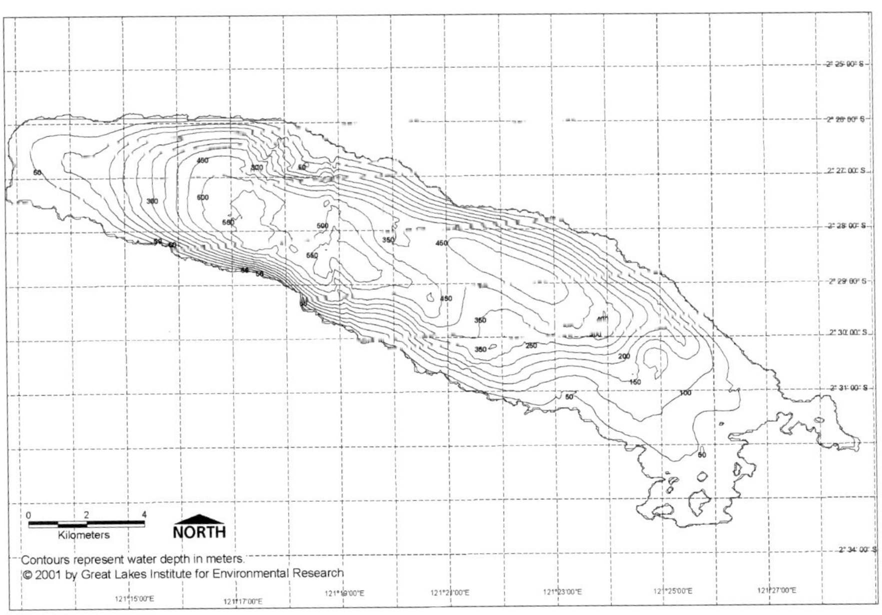

Fig. 1 Bathymetry of Lake Matano, South Sulawesi, revealing the eastern and western basins

concentrated plankton samples were stored for later analysis. Random sub-samples of the supernatant were also taken and examined to verify the efficacy of the sedimentation process.

One millilitre aliquots from the concentrated phytoplankton samples were settled in sedimentation chambers for 24 h prior to analysis (Utermöhl, 1958). Phytoplankton were enumerated under 200 and 600× magnifications using a Leica® inverted microscope and identified to genus according to Bramburger (2004), Hustedt (1942) and Wehr & Sheath (2003). Three subsamples were analyzed for each sample, which exceeded 100 counts of individuals of the most dominant species. The volumetric dimensions of up to 30 individuals from each species were measured using Openlab® 3.15 image analysis software and the mean dimensions were used to calculate the biovolume for each species, in accordance with the methodology and formulation presented in Sun & Liu (2003). Based on these volumetric relationships, total phytoplankton biomass was estimated with depth.

Zooplankton community analysis

The zooplankton sampling regime for the years 2000 and 2002 consisted of four approximately equidistant sampling locations across the lake, from 121.230°E to 121.170°E. In 2004, zooplankton sampling was implemented at only one of these sampling sites (121.190). Sampling periods spanned the wet and dry seasons, commencing in the mid-August (2000 and 2004), and in mid-January 2002. Sampling times ranged from early morning to afternoon throughout the sampling periods. The same sampling procedure was applied to all collections. A vertical closing net with a 64 μm mesh was used to successively filter 10 m intervals of the water column, from 0 to 110 m. These vertical transects were designed to quantify the distribution of zooplankton in the water column across the lake and throughout the depth of the epilimnion. Filtered material was rinsed into plastic scintillation vials and immediately preserved in 95% ethanol. Vials were refrigerated at 4°C for storage purposes.

In the laboratory, zooplankton samples from successive depths were subsampled for analysis of community composition and abundance. Sample vials were mixed thoroughly but gently and subsamples of 1 ml were removed using a wide-bore pipette. Subsamples stood for 24 h in sedimentation chambers before microscopic observation. After detailed surveys revealed that the spatial community composition was consistent across the lake at all depths, sampling was restricted to the Master Station to produce vertical profiles of community composition and abundance. Counting and identification were performed using a Leica® inverted microscope at 50–600× magnifications. In order to produce the vertical profile, counts of each species were replicated three times from each depth interval such that 100 individuals of the dominant species were enumerated within each sample. For composition and abundance data, post-naupliar copepods were identified to species level according to Ranga Reddy (1994), while naupliar stages of copepods were counted as a single group. In order to verify identifications, trophi of selected adult rotifer were dissected out and identified to species, according to Ruttner-Kolisko (1974). A lake-wide size profile of copepod body length measurements was obtained using Openlab® 3.15 image analysis software. Detailed zooplankton body morphs were measures for each species ($N > 100$). Lengths were measured from the anterior end point of the head, between antennae, to the pre-setae edge of the caudal rami. Standing biomasses of rotifers and copepods were calculated with depth interval, using standard formulae for dry weight estimates according to Dumont et al. (1975). In addition, the abundance and depth locations of ovigerous females were documented. ANOVA was used to determine whether significant partitioning of all species between depths occurred, and ANCOVA to specifically compare calonoid and rotifer depth distributions.

Results

Physical/chemical

During the 'dry' season of June to September, Lake Matano was well mixed to 100 m (Fig. 2A). A distinct thermal layer between 40 and 50 m was observed during the 'rainy' period, as regional evaporation rates decreased and the surface waters were warmed by approximately 1°C. Conductivity profiles in the upper 100 m, however, did not show any seasonal pattern (Fig. 2B). Oxygen profiles revealed a weak photosynthetic profile extending

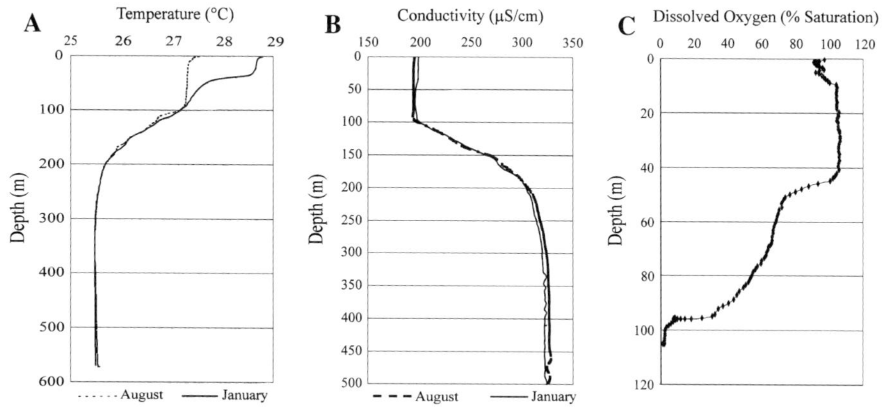

Fig. 2 Physical water parameters in Lake Matano: (**A**) temperature, (**B**) conductivity and (**C**) dissolved oxygen. Between 100 and 280 m Lake Matano is permanently stratified restricting chemical transfer between the upper and lower waters

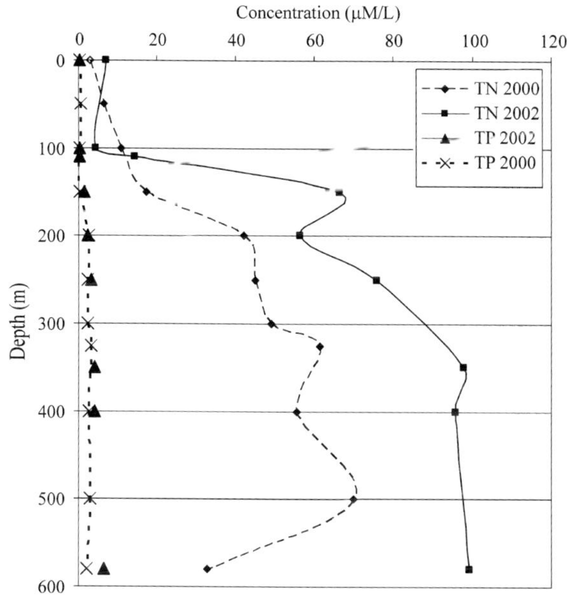

Fig. 3 Total phosphorus and total nitrogen in Lake Matano. Total phosphorus concentrations were very low in the epilimnion, and in conjunction with the relatively high Fe concentrations, bioavailable P is predicted to be minimal

down to approximately 40 m, and after that depth oxygen declined to below 1% saturation (Fig. 2C).

Lake Matano is oligotrophic with respect to TN and TP in the upper 100 m of the water column (Fig. 3). Nitrogen and phosphorus concentrations were very low (TP < 0.2 µmol l^{-1}; TN < 5 µmol l^{-1}), and often below detection. Specifically, total phosphorus measurements increased to 4 µmol l^{-1} in the hypolimnion, and total N increased to 100 µmol l^{-1}. There is little evidence of seasonal variation in TP concentrations below 100 m.

Phytoplankton

The biomass of the phytoplankton community of Lake Matano was dominated by a few diatoms (*Brachysira longirostris*, *Achnanthidium minutisimum* sensu lato, *Nitzschia* spp., *Surirella woltereckі*), and a variety of chlorophyte taxa (including filamentous forms and desmids) as well as one or two *Peridinium* taxa (*Peridinium palustre* var. *raciborskii*; *Peridinium* cf. *inconspicuum* var. *contracta*). Non-nitrogen fixing cyanobacteria, such as *Merismopedia* spp., *Snowella* cf. *lacustris*, *Aphanothece* spp. dominated numerically (>20,000 cells l^{-1}). The peak mean biomass of 1.3×10^{-2} mg l^{-1} (se = 7.0×10^{-3}) occurred at a depth of 10 m (Fig. 4). There is some evidence of seasonality in the phytoplankton community, with cyanophytes tending to dominate during the rainy season when there was some thermal structure, and chlorophytes, diatoms and dinoflagellates more prominent during the summer 'mixed' period (data not presented). Although not abundant (<400 cells l^{-1}), *Peridinium palustre* var. *raciborskii* was a dominant component of the biomass at the surface with a second smaller biomass peak at 70 m depth. Below this depth, most species declined to rarity (total biomass <0.2×10^{-2} mg l^{-1}). In samples from 110 m, cell biomass and counts were negligible and all cells detected were moribund and in an advanced state of decay. Entire classes of phytoplankton, such as the Cryptophyceae and Chrysophyceae were absent from the assemblage. Furthermore, although the lake supported a relatively rich assemblage of endemic pennate diatoms, the centric diatoms were exceedingly rare. These 'missing' taxa included species that are well adapted to

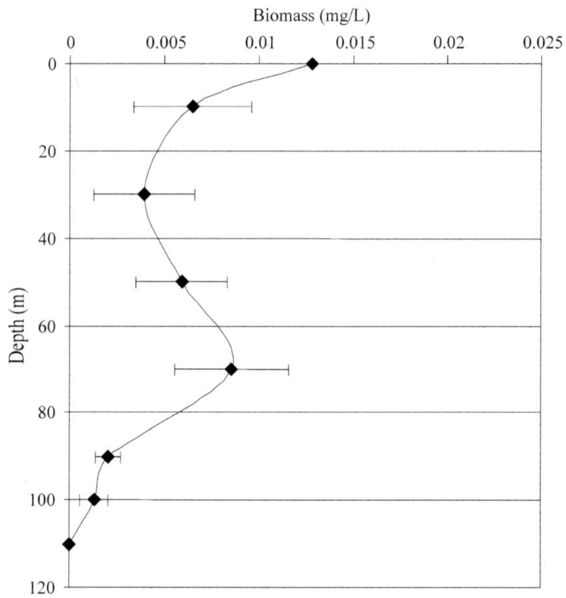

Fig. 4 Phytoplankton biomass sampled by Kemmerer in 10 m increments of depth, from surface to 100 m in Lake Matano during August, 2004. Mean peak biomass occurred at a depth of 10 m with a mixture of cyanophytes and chlorophytes

ultra-oligotrophic conditions, and were found in nearby watersheds.

Zooplankton

There were few taxa of zooplankton observed, and populations maintained relatively low abundance with little difference in distribution between the wet and dry seasons. The zooplankton community did comprise the endemic calanoid, *Eodiaptomus woltere-recki* var. *matanensis*, an endemic cyclopoid, *Tropocyclops matanensis* (Defaye, 2007) and the tropical rotifer *Horaella brehmi*. Vertical distributions of these taxa revealed distinct depth stratification (Figs. 5–7). *Eodiaptomus wolterecki* var. *matanensis* numerically dominated the pelagic zone during all three sampling occasions. In the dry seasons of 2002 and 2004, *E. wolterecki* var. *matanensis* and the rotifer *H. brehmi* were distributed throughout the 0–65 m zone at both stations 2 and 4 with the highest numbers and biomass in the upper 0–30 m (*E. wolterecki* var. *matanensis* biomass: 0.9–2.5 mg l^{-1}).

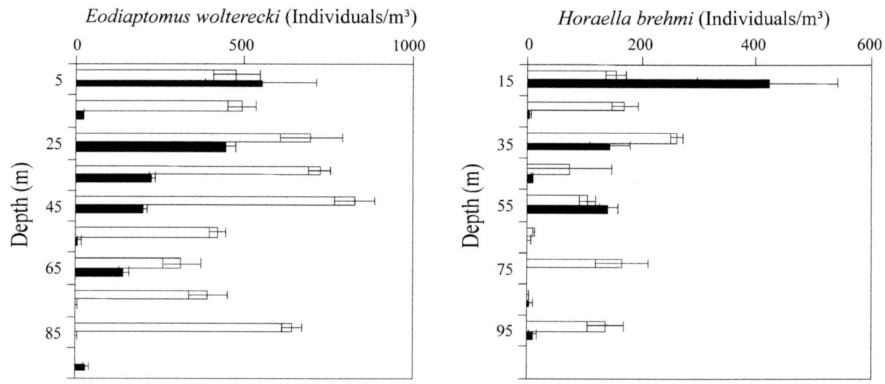

Fig. 5 Abundances of zooplankton species (±se), with depth, at sampling sites 2 and 4, August 2000. Upper bar: site 2, Lower bar: site 4. Note the lack of *Tropocyclops matanensis*

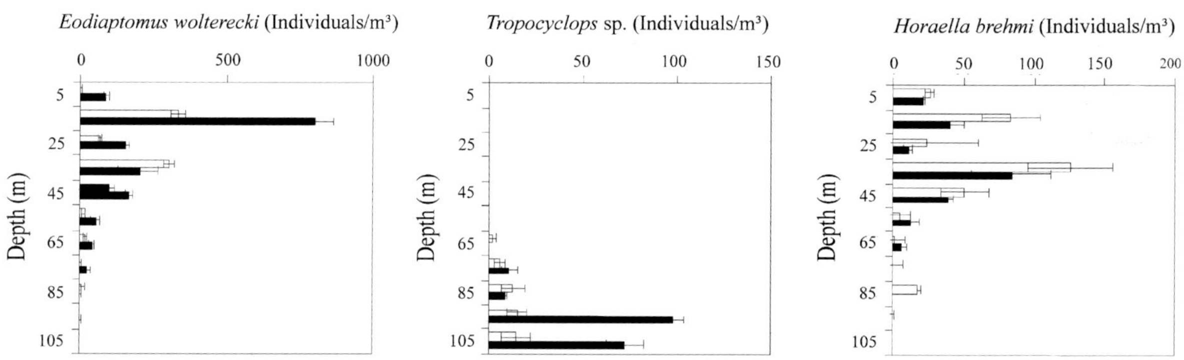

Fig. 6 Abundances of zooplankton species (±se), with depth, at sampling sites 2 and 4, January 2002. Upper bar: site 2, Lower bar: site 4

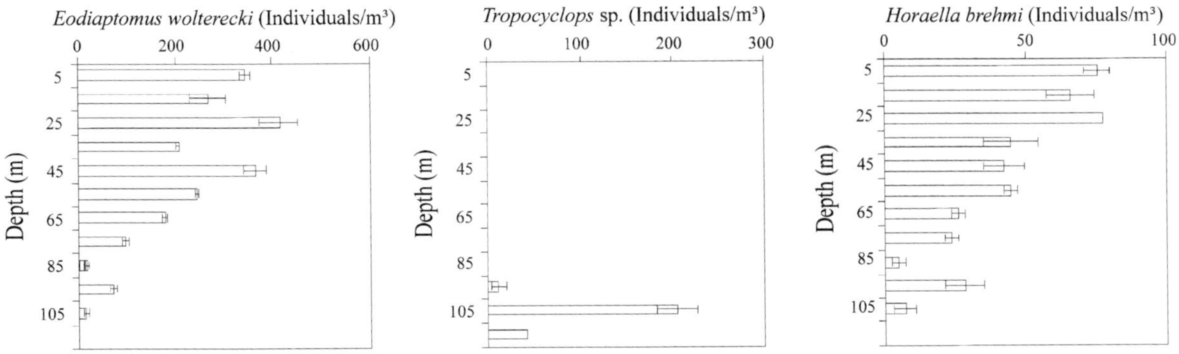

Fig. 7 Abundances of zooplankton community (±se), with depth, at site 2, August 2004

Both species showed a significant decline in numbers below 35 m at station 4 and 45 m at station 2. In the wet season (January) of 2002, *E. wolterecki* var. *matanensis* was more prominent in the 10–20 m net haul, while *H. brehmi* was significantly more abundant in the 30–40 m net haul (Fig. 6). Although not observed in 2000, *Tropocyclops* sp. was present only in the lower waters (>70 m) in 2002 and 2004 (Figs. 5, 7). An ANOVA demonstrated that there were significant differences in relative abundance with depth for the three species ($P < 0.05$). In addition, an ANCOVA indicated that for calanoids and rotifers there were significant differences with depth ($P < 0.01$), and highly significant differences in abundance between species ($P < 0.001$), while cyclopoids were clearly confined only to lower depths.

Calanoid body lengths ranged from approximately 200 μm to 900 μm with significantly larger body size predominating during the dry season. In the years 2000, 2002 and 2004, respectively, the mean calanoid body lengths were 5.8×10^2 μm (se = 28), 4.9×10^2 μm (se = 32) and 5.8×10^2 μm (se = 11). Cylopoid body length also showed a significant difference between seasons, with a larger mean length of 3.9×10^2 μm (se = 38) in 2002 compared to 3.2×10^2 μm (se = 17) in the dry season of 2004.

Ovigerous copepods were depauperate in all seasons. For the year 2000, the mean abundance of ovigerous calanoids was 1.4 individuals m^{-3} (se = 0.41), with ovigerous individuals rarely observed in samples from the upper portion of the water column, and more commonly observed between 90 and 100 m. In the wet season of 2002, ovigerous copepods (both calanoids and cyclopoids) were restricted to samples from 80 to 110 m depths of the lake. The mean abundance of ovigerous calanoids in the water column was 0.65 individuals m^{-3} (se = 0.33) and the mean relative abundance of ovigerous cyclopoids throughout the water column was 1.2 individuals m^{-3} (se = 0.43). In 2004, however, ovigerous calanoids were not confined to the 80–110 m depths but were found in low numbers throughout the water column, with a mean abundance of 1.4 individuals m^{-3} (se = 0.41). The mean cylopoid abundance in 2004 was 0.20 individuals m^{-3} (se = 0.16).

Discussion

The water chemistry of Lake Matano is very unique as a result of the iron rich laterite catchment basin. The top 100 m of the lake behave much like a typical monomictic lake with a thermal structure developing during the wet season when surface cooling by evaporation is minimal. Conductivity profiles, however, did not show any seasonal patterns. The deep Secchi depths of 23 m revealed a system with very little suspended biotic or abiotic material. Oxygen profiles exhibited a photosynthetic curve that extended down to 40 m, consistent with the predicted depth of the euphotic zone. After 40 m the oxygen concentrations declined, and it is postulated that this decline was in response to high microbial respiration rates associated with the metabolism of sedimenting organic matter.

Below 100 m there was a persistent thermal/chemical gradient that suggests Lake Matano could be defined as meromictic (Crowe et al., 2008), with very little chemical exchange occurring between the upper and lower waters. Such a structure would contribute significantly to the continued loss of nutrients from the epilimnion via the sedimentation of planktonic material. Furthermore, settled algal cells would have very low potential for returning to the epilimnion to seed future growths.

Phosphorus and nitrogen concentrations in the upper 100 m were measured at levels where nutrient limitation is possible. The molar TN:TP ratio of 25:1 indicates that TP is the more limiting nutrient. The relatively high concentration of Fe, as well as Cr(VI) (180 nmol l^{-1}; Crowe et al., 2008) observed in the surface waters of Lake Matano can regulate the bioavailability of nutrients and potentially exert toxicological stress suppressing primary production. The ratio of Fe:TP is 16:1 and is fairly constant with depth suggesting that Fe cycling is a critical component of the nutrient dynamics of Lake Matano (Holtan et al., 1988). In the oxic epilimnion, it is predicted that much TP will either be lost as precipitate or will not be biologically available when bound with Fe in colloidal form. The water chemistry of the lake is very much dominated by the lateritic rich soils of the catchment basin. Physical factors, such as light availability and mixing are not limiting factors, although a lack nutrient exchange through the deep thermal layer might be playing an important role, where sedimentation of biological and geological material can continuously deplete the nutrient pool of the upper surface waters.

Phytoplankton density and biomass estimates from Lake Matano, represent some of the smallest standing crops of phytoplankton ever reported for either tropical or temperate lakes (Reynolds et al., 2000). Apart from persistent populations of small cyanoprokaryota (picoplankton) and the endemic diatom (*Brachysira longirostris*), the pelagic community is sparse, with a scattered collection of desmids, single cell chlorophytes, colonial chlorophytes, and a few species of dinoflagellates. Surprisingly, apart from diatoms, the flora is not endemic as compared with the other trophic levels that comprise the food web of Lake Matano. Many of the taxa observed in the open water are prominent species within the littoral zone of Lake Matano and might not represent "true" plankters. The scattered collection of taxa within pelagic waters, at all depths, combined with the evident grouping of taxa into very small (nanoplankton) and large (netplankton) size fractions suggests that primary producers in the phytoplankton (bottom-up dynamics) are not an adequate food source for sustaining large populations of primary consumers. Indeed, *Peridinium palustre* var. *raciborskii* often has the largest biomass and is not that well suited as a food source and may even be competing for niche space with primary consumers (Bucka & Zurek, 1992; Sanderson & Frost, 1996). One of the most significant observations is the absence of selected phytoplankton classes (Chrysophyceae, Cryptophyceae) that have global distributions (Kugrens & Clay, 2003; Nicholls & Wujek, 2003). It was also surprising to note the absence of centric diatoms, both in the pelagic zone and in the littoral zones around the lake (Bramburger et al., in press). Centric diatoms are found in other lakes on the island of Sulawesi, thus the absence of these cosmopolitan planktonic groups supports the conclusion that Lake Matano's water chemistry is playing an important role in regulating both the composition and biomass of the phytoplankton community.

In Lake Matano, the peak phytoplankton biomass is 0.013 mg l^{-1}, indicating an impoverished production base that will have ramifications for the zooplankton community. This low biomass estimate supports the observations of Lehmusluoto et al. (1999), where a biomass of 2×10^{-3} mg l^{-1} was reported for Lake Matano. According to Reynold et al.'s. (2000) classification of lakes, Lake Matano ought to exhibit production in keeping with its tropical nature. For instance, Lake Tanganyika, of similar depth and latitude sustains a biomass of 0.9 mg l^{-1} (Reynolds et al., 2000). Rating Matano's primary production using Reynold's scheme for large lakes of the world puts it in the range of the most unproductive, comparable to arctic or alpine type lakes. The very low phytoplankton standing crop in Lake Matano substantiates the importance of resource-base control in this system. Specifically, the critical range of phytoplankton required to support obligate filter feeders, such as *Daphnia* is 0.1–0.4 mg C l^{-1} (Reynold et al., 2000), an order of magnitude above the phytoplankton biomass in Lake Matano. Moreover, the phytoplankton assemblage of Lake Matano largely consists of unpalatable species, such as *Peridinium* and *Tribonema* cf. sp., and therefore Lake Matano does not have the capacity to sustain a significant zooplankton community.

Previous studies of the pelagic zooplankton in Lake Matano are limited. Brehm's account of the endemic calanoid, *Eodiaptomus woltereckí* var. *matanensis*, described during the Woltereck Expedition to Wallacea of 1932 (Brehm, 1933), constitutes the earliest record of the composition of the pelagic plankton. Brooks (1950) reviewed the distribution of endemics in Lake Matano and downstream lakes and included the endemic calanoid as the only known pelagic microzooplankter. Fernando (1987) visited the lake and mentioned the absence of Cladocera, and postulated this might be a result of Cr toxicity. Simple zooplankton assemblages, such as that observed in Lake Matano, are commonly observed features of tropical freshwater systems (Dumont & Segers, 1996). The low species richness of these lakes has often been attributed to top-down control exerted by diverse and abundant fish communities. This top-down control can limit species richness in the secondary production base, even under conditions favourable to primary production (Lewis, 1996). It has also been demonstrated that the tendency towards smaller zooplankton, such as those found in Lake Matano, is often a result of fish predation. Top-down predation effects, however, do not appear to be adequate to regulate the low species richness and abundance of the zooplankton in Lake Matano. The lake supports limited top-level predators (Whitten et al., 1987; Haffner et al., 2001; Roy et al., 2004) and the forage fish community, though endemic, is simple and not abundant, consisting mainly of very small fish (<10 cm) that are confined to the near-shore zone (<40 m depth) (Haffner et al., 2001; Roy et al., 2004). Indeed, within the last 40–50 years, the low fish production potential of the lake has been a concern to local communities. The size spectra, low biomass and poor reproductive potential (as demonstrated by the low number of ovigerous females) of the zooplankton community of Lake Matano reflect that of a resource limited system, where bottom up controls strongly regulate food web dynamics.

Egg production and recruitment capability of a population has been established as an endpoint signifying the concentration and quality of food available to adults and nauplii (Poulet et al., 1995). Carlotti et al. (1997) concluded that energy is allotted to reproduction only after the female has attained a critical weight and nutritional fulfillment. Therefore, greater resource quality and quantity will yield faster maturation times and higher fecundity. We observed distinct changes in body length with *Eodiaptomus wolterecki* var. *matanensis* becoming smaller and *Tropocyclops matanensis* larger during the wet season, thus indicating the presence of annual changes in feeding and possibly life cycle dynamics. In the present study, exceptionally low fecundity was observed throughout the study period, with ovigerous individuals comprising less than 1% of the calanoid population. Due to the dependence of clutch size in copepods on their habitat quality (Fernando, 2002), this yearly low population fecundity supports the conclusion that strong resource limitations are imposed on the zooplankton community of Lake Matano. Total copepodite and nauplii contribute less than 20% of the total copepod counts in all years, indicative of very low biomass turnover rates. The proportion of total copepodite and nauplii is quite high relative to ovigerous copepods, and this observation is consistent with habitats that present a food limitation to adults but not to nauplii, so that egg production is stilted while naupliar growth continues at a higher rate (Poulet et al., 1995). This resource limitation probably plays a major role in the vertical structuring observed in the zooplankton community of Lake Matano.

Although the pelagic zone of tropical lakes is often too homogeneous to sustain high levels of zooplankton diversity (Martens & Schön, 1999), this mechanism is not sufficient to explain the observed zooplankton community in Lake Matano. Given the poor food base in Lake Matano, it is postulated that resource partitioning is a key process in regulating the composition and relative abundance of species in Lake Matano. Significant spatial resource partitioning is supported by the significant differences in depth distribution of the three species in the water column at three different time periods.

Although Lake Matano has a very high proportion of endemic species, it exhibits relatively low species richness. Within both the fish and gastropod communities there is evidence that ecological speciation is being driven by resource partitioning (von Rintelen et al., 2004; Roy et al., 2007). It is probable that the water chemistry of Lake Matano not only minimizes the number of successful colonization events as observed within the plankton communities, but also contributes to severe resource limitation that results in strong competition for resources by the those species that can tolerate the unique water chemistry. In essence, speciation processes in Lake Matano are predicted to be driven primarily by resource limitation. This link between the geochemistry of the basin and evolutionary processes makes Lake Matano an exciting laboratory for future study.

Conclusions

This study provides the first comprehensive description of the pelagic phytoplankton and zooplankton community composition in Lake Matano, and proposes that bottom-up processes are most influential in regulating both phytoplankton and zooplankton community composition and abundance. It is concluded that water chemistry inhibits primary production primarily through P limitation, although toxicological stress associated with Cr(VI) cannot be ruled out. The minimal standing crops of phytoplankton, as well as food quality associated with the phytoplankton composition, regulates the species richness and abundance in the zooplankton community. The data presented here support the conclusion that many of the species radiations observed in Lake Matano have been initiated by resource limitation and partitioning.

References

Bramburger, A. J., 2004. Floristics and taxonomy of the endemic diatom flora of the ancient Malili Lakes, Sulawesi Island, Indonesia. Thesis, Faculty of Graduate Studies and Research, University of Windsor, Windsor.

Bramburger, A. J., P. B. Hamilton, P. E. Hehanussa & G. D. Haffner, 2008. Spatial patterns of planktonic and benthic diatom distribution and assemblage in Lake Matano (Sulawesi Island, Indonesia). Proceedings of the 19th International Diatom Symposium (in press).

Brehm, H., 1933. Mitteilungen von der Wallacea-Expedition Woltereck. Mitteilung IV. Einige neue Diaptomiden. Zoologischer Anzeiger 103: 295–304.

Brooks, J. L., 1950. Speciation in ancient lakes (concluded). The Quarterly Review of Biology 25: 131–176.

Brooks, J. L. & S. I. Dodson, 1965. Predation, body size and composition of plankton. Science 150: 28–35.

Bucka, H. & R. Zurek, 1992. Trophic relations between phyto- and zooplankton in a field experiment in the aspect of the formation and decline of water blooms. Acta Hydrobiologica 34: 139–155.

Carlotti, F. & H. J. Hirche, 1997. Growth and egg production of female *Calanus finmarchicus*: An individual-based physiological model and experimental validation. Marine Ecology Progress Series 149: 91–104.

Crowe, S. A., A. H. O'Neill, S. Katsev, P. Hehanussa, G. D. Haffner, B. Sundby, A. Mucci & D. Fowle, 2008. The biogeochemistry of tropical lakes: A case study from Lake Matano, Indonesia. Limnology and Oceanography 53: 319–331.

Defaye, D., 2007. A new Tropocyclops (Copepoda, Cyclopidae) from Lake Matano, Indonesia. Zootaxa 54: 17–29.

Dumont, H. J. & H. Segers, 1996. Estimating lacustrine zooplankton species richness and complementarity. Hydrobiologia 341: 125–132.

Dumont, H. J., I. van de Velde & S. Dumont, 1975. The dry weight estimate of biomass in a selection of Cladocera, Copepoda and Rotifera from the plankton, periphyton, and benthos of continental waters. Oecologica 19: 75–97.

Fernando, C. H., 1987. Tropical freshwater zooplankton with special reference to South East Asia (Oriental Region). Reports of the National Geographic Society 87: 287–311.

Fernando, C. H., 2002. A Guide to Tropical Freshwater Zooplankton. Identification, Ecology and Impact on Fisheries. Backhuys Publishers, Leiden, Netherlands: 291 pp.

Haffner, G. D., P. E. Hehanussa & D. Hartoto, 2001. The biology and physical processes of large lakes of Indonesia. In Munawar, M. & R. E. Hecky (eds), The Great Lakes of the World: Food-web, Health, and Integrity. Backhuys Publishers, Leiden, Netherlands: 183–194.

Hall, D. J., S. T. Threlkeld, C. W. Burns & P. H. Crowley, 1976. Size-efficiency hypothesis and size structure of zooplankton communities. Annual Review of Ecology and Systematics 7: 177–208.

Hobaek, A., M. Manca & T. Andersen, 2002. Factors influencing species richness in lacustrine zooplankton. Acta Oecologica 23: 155–163.

Holtan, H., L. Kamp-Nielsen & A. O. Stuanes, 1988. Phosphorus in soil, water and sediment: An overview. Hydrobiologia 170: 19–34.

Hustedt, F., 1942. Süßwasser-Diatomeen des indomalayischen Archipels und der Hawaii-Inseln. Internationale Revue der Gesamten Hydrobiologie und Hydrographie 42: 1–252.

Hutchinson, G. E., 1967. A Treatise on Limnology, Vol. II. Introduction to Lake Biology and the Limnoplankton. John Wiley & Sons, New York.

Kottelat, M., 1990a. Synopsis of the endangered Buntingi (Osteichthyes: Adrianichthydae and Oryziidae) of Lake Poso, Central Sulawesi, Indonesia, with a new reproductive guild and descriptions of three new species. Ichthyological Exploration of Freshwaters 1: 49–67.

Kottelat, M., 1990b. The ricefishes (Oryziidae) of the Malili Lakes, Sulawesi, Indonesia, with description of a new species. Ichthyological Exploration of Freshwaters 1: 321–344.

Kottelat, M., 1991. Sailfin silversides (Pisces: Telmatherinidae) of Lake Matano, Sulawesi, Indonesia, with descriptions of 6 new species. Ichthyological Exploration of Freshwaters 1: 321–344.

Kugrens, P. & B. L. Clay, 2003. Cryptomonads. In Wehr, J. D. & R. G. Sheath (eds), Freshwater Algae of North America Ecology and Classification. Academic Press, Amsterdam: 715–755.

Lehmusluoto, P., B. Machbub, N. Teranna, F. Achmad, L. Boer, S. Brahmana, B. Steiadji, B. Pridadie, K. H. Timotius & F. Goeltenboth, 1999. Limnology in Indonesia.

From the legacy of the past to the prospects for the future. In Wetzel, R. G. & B. Gopal (eds), Limnology in Developing Countries, Vol. 2. International Scientific Publications, New Delhi, India: 119–234.

Lewis, W., 1996. Tropical lakes: How latitude makes a difference. In Schiemer, F. & K. T. Boland (eds), Perspectives in Tropical Limnology. SPB Academic Publishers, Amsterdam, Netherlands: 43–64.

Lodge, D. M., 1993. Species invasions and deletions: Community effects and responses to climate and habitat change. In Kareiva, P. M., J. G. Kingsolver & R. B. Huey (eds), Biotic Interactions and Global Change. Sinauer Associates, Sunderland, MA: 367–387.

Martens, K. & I. Schön, 1999. Crustacean biodiversity in ancient lakes: A review. Crustaceana 72: 899–910 (Part 8).

Nicholls, K. H. & D. E. Wujek, 2003. Crysophycean algae. In Wehr, J. D. & R. G. Sheath (eds), Freshwater Algae of North America Ecology and Classification. Academic Press, Amsterdam: 471–509.

Ogutu-Ohwayo, R. & R. E. Hecky, 1991. Fish introductions to Africa and some of their implications. Canadian Journal of Fisheries and Aquatic Sciences 48: 8–12.

Pinto-Coelho, R., B. Pinel-Alloul, G. Méthot & K. E. Havens, 2005. Crustacean zooplankton in lakes and reservoirs of temperate and tropical regions: variation with tropical status. Canadian Journal of Fisheries and Aquatic Sciences 62: 348–361.

Poulet, S. A., A. Ianora, M. Laabir & W. C. M. Klein-Breteler, 1995. Towards the measurement of secondary production and recruitment in copepods. Journal of Marine Sciences 52: 359–368.

Ranga Reddy, Y., 1994. Copepoda: Calanoid: Diaptomidae. Guide to the Identification of the Microinvertebrates of the Continental Waters of the World, Vol. 5. Academic Publishing, The Hague, Netherlands: 221.

Reynolds, C. S., S. N. Reynolds, I. F. Munawar & M. Munawar, 2000. The regulation of phytoplankton population dynamics in the world's largest lakes. Aquatic Ecosystem Health Management 3: 1–21.

Roy, D., M. F. Docker, G. D. Haffner & D. D. Heath, 2007. Body shape vs. colour associated initial divergence in the Telmatherina radiation in Lake Matano, Sulawesi, Indonesia. Journal of Evolutionary Biology 20: 1126–1137.

Roy, D., M. F. Docker, P. Hehanussa, D. D. Heath & G. D. Haffner, 2004. Genetic and morphological data supporting the hypothesis of adaptive radiation in the endemic fish of Lake Matano. Journal of Evolutionary Biology 17: 1268–1276.

Roy, D., D. W. Kelly, C. H. J. M. Fransen, D. D. Heath & G. D. Haffner, 2006b. Evidence of small scale vicariance in *Caridinia lanceolata* (Decapoda: Atyidae) from the Malili Lakes, Sulawesi. Evolutionary Ecology Research 8: 1087–1099.

Roy, D., G. Paterson, P. B. Hamilton, D. D. Heath & G. D. Haffner, 2007. Resource-based adaptative divergence in the freshwater fish Telmatherina from Lake Matano, Indonesia. Molecular Ecology 16: 35–48.

Ruttner-Kolisko, A., 1974. Plankton Rotifers: Biology and Taxonomy. Die Binnengewässer Vol. 26/1 (Suppl.): 1–146.

Sanderson, B. L. & T. M. Frost, 1996. Regulation of dinoflagellate populations: Relative importance of grazing, resource limitation, and recruitment from sediments. Canadian Journal of Fisheries and Aquatic Sciences 53: 1409–1417.

Sarasin, P. & F. Sarasin, 1897. Reisebericht aus Celebes. IV. Reise durch central-Celebes vom Golf von Boni nach den Golf von Tomieni. Gesellschaft für Erdkunde zu Berlin 30: 312–352.

Standard Methods, 1995. For the examination of water and wastewater. American Public Health Association: 1268.

Sun, J. & D. Liu, 2003. Geometric models for calculating cell biovolume and surface area for phytoplankton. Journal of Plankton Research 25: 1331–1346.

Talling, J. F., 1957. Some observations on the stratification of Lake Victoria. Limnology and Oceanography 2: 231–221.

Utermöhl, H., 1958. Zur Vervollkommnung der quantitativen phytoplankton Methodik. Mitteilungen der Internationale Vereinigung für Limnologie, Mitteilungen 9: 1–38.

Von Rintelen, T. & M. Glaubrecht, 2003. New discoveries in old lakes: Three new species of *Tylomelania* Sarasin & Sarasin, 1879 (Gastropoda: Cerithoidea: Pachychilidae) from the Malili Lake system on Sulawesi, Indonesia. Journal of Molluscan Studies 69: 3–17.

Von Rintelen, T., A. B. Wilson, A. Meyer & M. Glaubrecht, 2004. Escalation and trophic specialization drive adaptive radiation of freshwater gastropods in ancient lakes on Sulawesi, Indonesia. Proceedings of the Royal Society of London, Series B, Biological Sciences 271: 2541–2549.

Wehr, J. D. & R. G. Sheath, 2003. Freshwater Algae of North America, Ecology and Classification. Academic Press, Amsterdam: 918 pp.

Whitten, T., M. Mustafa & G. S. Henderson, 1987. The Ecology of Sulawesi. Gadjah Mada University Press, Singapore.